Internal Combustion Engines
Applied Thermosciences

Internal Combustion Engines

Applied Thermosciences

Second Edition

Colin R. Ferguson

Mechanical Engineering Department
Colorado State University

Allan T. Kirkpatrick

Mechanical Engineering Department
Colorado State University

John Wiley & Sons, Inc.
New York / Chichester / Weinheim / Brisbane / Singapore / Toronto

EDITOR   *Joe Hayton*
MARKETING MANAGER   *Katherine Hepburn*
PRODUCTION DIRECTOR   *Pam Kennedy*
PRODUCTION EDITOR   *Leslie Surovick*
SENIOR DESIGNER   *Kevin Murphy*
COVER DESIGNER   *Lynn Rogan*
PHOTO EDITOR   *Lisa Gee*
ILLUSTRATION COORDINATOR   *Sandra Rigby*
ILLUSTRATION STUDIO   *Wellington Studios*

This book is printed on acid-free paper. ∞

The paper in this book was manufactured by a mill whose forest management programs include sustained yield harvesting of its timberlands. Sustained yield harvesting principles ensure that the numbers of trees cut each year does not exceed the amount of new growth.

Library of Congress Cataloging in Publication Data:
Ferguson, Colin R.
    Internal combustion engines: applied thermodynamics / Colin R. Ferguson, Allan T. Kirkpatrick.—2nd ed.
        p. cm.
    Includes bibliographical references and index.
    ISBN 0-471-35617-4 (cloth : alk. paper)
    1. Internal combustion engines. 2. Thermodynamics. I. Kirkpatrick, Allan Thomson. II. Title.

TJ756 .F47 2000
621.43—dc21

00-040432

10 9 8 7 6 5 4 3 2 1

Preface

Introduction

This textbook presents a modern approach to the study of internal combustion engines. Internal combustion engines have been, and will remain for the foreseeable future, a vital and active area of engineering education and research. The purpose of this book is to apply the principles of thermodynamics, fluid mechanics, and heat transfer to the analysis of internal combustion engines. This book is intended first to demonstrate to the student the application of engineering sciences, especially the thermal sciences, and second, it is a book about internal combustion engines. Considerable effort is expended making the requisite thermodynamics accessible to students. This is because most students have little, if any, experience applying the first law to unsteady processes in open systems or in differential form to closed systems and have experience with only the simplest of reacting gas mixtures.

The text is designed for a one-semester course in internal combustion engines at the advanced undergraduate level. At Colorado State University, this second edition is used in a senior level class in internal combustion engines. The class meets for a lecture two times per week and a recitation/laboratory once a week, for a term of fifteen weeks. The course follows the subject matter in the text, and finishes with a student project.

New in the Second Edition

Building upon the foundation of the first edition, the book has been completely revised, with each chapter reorganized and updated. The content changes include up-to-date discussion of new engine technologies and presentation of new material on thermodynamic modeling, intake and exhaust flow, combustion analysis, alternative fuels, emissions, instrumentation and control systems. The text also features modern web-based computational methods. Java based computational applets are used for solution of problems in engine combustion and thermodynamics, heat transfer, and friction. The computational applets are useful for open-ended design oriented problems and projects, and are accessed using a web browser, such as Netscape Communicator or Microsoft Internet Explorer, at www.wiley.com/college/ferguson or www.engr.colostate.edu/ ~allan/engines.html.

The text is also useful as a reference text by practicing engineers. Each chapter has detailed reference lists to guide entry into the research literature. The second edition of the text includes additional problems, and a fully worked out solution manual.

The chapter sequence is the same as in the first edition, with some minor changes in the chapter headings to reflect new material. A chapter by chapter break down of specific changes follows:

Chapter One Some historical details, including important inventors of internal combustion engines, have been added. Operational parameters such as bore, stroke, mean effective pressure, power, volumetric efficiency, etc., have been reorganized into one section. A discussion of engine dynamics and balancing has been added. A guide to engine configurations and components has been added. The engine examples have been reorganized into one section, and have been revised to include an automotive spark ignition engine, a heavy duty truck diesel engine, and a large stationary natural gas engine. The alternative power plants section has been updated with a discussion of electric vehicles, fuel cells, gas turbines, and steam engines.

Chapter Two   The Miller cycle has been added. Java applets are used to numerically compute the heat release, pressure, temperature, optimum spark timing, etc., for a finite heat release model. The analysis of four stroke cycles has been revised, and Java applets for the computation of compression and exhaust strokes, volumetric efficiency, residual fraction, etc. for an ideal gas four stroke cycle have been added.

Chapter Three   The sections on ideal gas equations of state, and liquids and liquid-vapor-gas mixtures have been rearranged into concurrent sections. The combustion calculations in the low temperature stoichiometry section have been changed to a per mole of fuel basis. Equilibrium combustion mole fractions and mixture properties for a variety of fuels at given temperatures, pressures, and equivalence ratios are computed using Java applets. Java applets are also used to compute isentropic changes of state, constant pressure and constant volume combustion states for fuel-air mixtures; and to compute adiabatic flame temperatures for a variety of fuels as function of pressure and residual fraction. Examples illustrating fuel-air equilibrium combustion calculations have been added.

Chapter Four   The comparison of first and second law efficiency has been revised. A fuel-air ideal Otto cycle is computed using Java applets. Java applets are also used to compute state properties, volumetric efficiency, residual fraction, etc., for a fuel-air four stroke Otto cycle for a variety of fuels. A section comparing fuel-air cycle PV diagrams with actual spark ignition and compression ignition cycle PV diagrams has been added.

Chapter Five   This chapter has a new title to reflect the emphasis on engine testing and control. The dynamometer and fluid flow measurement sections have been expanded. Material on cylinder pressure measurements, combustion analysis, vehicle emissions testing, engine sensors and acutators, and engine digital control systems has been added.

Chapter Six   The chapter has been revised to present analysis and modeling of friction first, followed by engineering correlations for various engine components. Updated friction models and correlations are introduced. Use of Java applets to compute the friction mean effective pressure for various components is described. Graphs have been added to compare the relative magnitudes of the various frictional processes.

Chapter Seven   The topics of valve flow, discharge coefficients, valve timing, and their effect on volumetric efficiency have been revised and expanded. The new material on intake and exhaust flow includes discussions of intake and exhaust tuning and computational fluid dynamics codes. Material on PIV measurement systems, turbulent length scales and turbulence models has been added. The discussion of spark ignition and compression fuel injection systems has been expanded.

Chapter Eight   The chapter has been reorganized to put engine cooling systems and energy balance first, followed by measurements, heat transfer modeling, and then engineering correlations. Java applets are used to compute effect of heat transfer on finite heat release models, and to compare the predictions of two widely used heat transfer correlations. Recent material on radiation heat transfer is added.

Chapter Nine   The chapter has been revised with spark ignition engine combustion experiments covered first, followed by knock, combustion experiments in compression ignition engines, combustion analysis, and ending with emissions. Additional material in the chapter includes combustion models for compression ignition engines, and emission control for vehicles, primarily the catalytic converter.

Chapter Ten   The chapter has been reorganized to first introduce hydrocarbon chemistry, then refining. The section on alternative fuels has been significantly expanded to include propane, natural gas, hydrogen, methanol, and ethanol. Reformulated gasoline is also discussed.

Chapter Eleven   The chapter has undergone partial reorganization. A section on vehicle performance testing has been added.

Acknowledgments

The approach and style of this text reflects our experiences as students at the Massachusetts Institute of Technology. In particular, we have learned a great deal from MIT Professors John B. Heywood, C. F. Taylor, and Jean F. Louis.

In the preparation of the second edition, several people have made major contributions by reviewing drafts and making suggestions. Thanks are due to the following individuals: Professor Chris Atkinson, University of West Virginia; Dr. John Dec, Sandia Research Laboratories; Professor Jon Van Gerpen, Iowa State University; Professor S. R. Gollahalli, University of Oklahoma; Professor Justin Poland, University of Maine; Professor Ahmet Selamet, Ohio State University; and Professor Etim Ubong, Kettering University. Professor Tim Tong, former CSU Mechanical Engineering Department Head, and Professor Bryan Willson, Director of the CSU Engines and Energy Conversion Laboratory, provided a collegial working environment for work on this text.

Many thanks go to the editorial and production staff at John Wiley & Sons, Inc. for their work on the second edition. Mr. Joe Hayton and Mr. Steve Peterson deserve special acknowledgement for their leadership and sponsorship of this project.

Finally, Allan Kirkpatrick would like to thank his family: Susan, Anne, and Rob, for their support.

Fort Collins, Colorado
Allan T. Kirkpatrick (allan@engr.colostate.edu)
Colin Ferguson

Contents

Appendices   353

Index   367

About the Authors

Dr. Colin R. Ferguson received his M.S. and Ph.D. (1975) degrees in Mechanical Engineering from the Massachusetts Institute of Technology. He taught thermal science courses at Purdue University for twelve years, performing research and publishing in the internal combustion engines area, and is currently living in California. He is an Adjunct Professor of Mechanical Engineering at Colorado State University.

Dr. Allan T. Kirkpatrick, P.E., received his B.S. (1972) and Ph.D. (1981) degrees in Mechanical Engineering from the Massachusetts Institute of Technology, and has been at Colorado State University since 1980. Dr. Kirkpatrick teaches and performs research in the engines and buildings areas. He has received teaching and research awards from Colorado State University, the American Society for Engineering Education, and Sigma Xi. Dr. Kirkpatrick has worked both in industry and in national laboratories on energy related research. He has published two books, over 75 conference and journal articles, and has one patent. Dr. Kirkpatrick is currently a Professor of Mechanical Engineering at Colorado State University.

Chapter 1

Introduction to Internal Combustion Engines

1.1 INTRODUCTION

In this chapter we discuss the performance characteristics of internal combustion engines. Major engine cycles and types are discussed. Establishing engine nomenclature and explaining how various types of engines work are emphasized. An internal combustion engine is defined as an engine in which the chemical energy of the fuel is released inside the engine and used directly for mechanical work, as opposed to an external combustion engine in which a separate combustor is used to burn the fuel.

The internal combustion engine was conceived and developed in the late 1800s. It has had a significant impact on society, and is considered one of the most significant inventions of the last century. The internal combustion engine has been the foundation for the successful development of many commercial technologies. For example, consider how the internal combustion engine has transformed the transportation industry, allowing the invention and improvement of automobiles, trucks, airplanes, and trains.

Internal combustion engines can deliver power in the range from 0.01 kW to 20×10^3 kW, depending on their displacement. They compete in the market place with electric motors, gas turbines, and steam engines. The major applications are in the vehicular (automobile and truck), railroad, marine, aircraft, home use, and stationary areas. The vast majority of internal combustion engines are produced for vehicular applications, requiring a power output on the order of 10^2 kW.

Internal combustion engines have become the dominant prime mover technology in several areas. For example, in 1900 most automobiles were steam or electrically powered, but by 1920 most automobiles were powered by gasoline engines. As of the year 2000, in the United States alone there are about 200 million motor vehicles powered by internal combustion engines. In 1900, steam engines were used to power ships and railroad locomotives; today two- and four-stroke diesel engines are used. Prior to 1950, aircraft relied almost exclusively on piston engines. Today gas turbines are the power plant used in large planes, and piston engines continue to dominate the market in small planes. The adoption and continued use of the internal combustion engine in different application areas has resulted from its relatively low cost, favorable power to weight ratio, high efficiency, and relatively simple and robust operating characteristics. Alternatives to the internal combustion engine are discussed in Section 1.6.

The first internal combustion engines used the reciprocating piston-cylinder principle shown in Figure 1-1, in which a piston oscillates back and forth in a cylinder and transmits power to a drive shaft through a connecting rod and crankshaft mechanism. Valves are used to control the flow of gas into and out of the engine. The components of a reciprocating internal combustion engine, block, piston, valves, crankshaft, and connecting rod, have remained basically unchanged since the late 1800s.

The main differences between a modern day engine and one built 100 years ago are the thermal efficiency and the emissions level. For many years, internal combustion engine research was aimed at improving thermal efficiency and reducing noise and vibration.

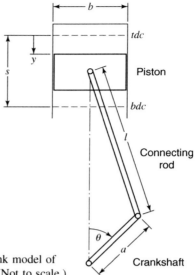

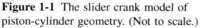

Figure 1-1 The slider crank model of piston-cylinder geometry. (Not to scale.)

As a consequence, the thermal efficiency has increased from about 10% to values as high as 50%. Since 1970, with the recognition of the importance of air quality, there has also been a great deal of work devoted to reducing emissions from engines. Currently, emission control requirements are one of the major factors in the design and operation of internal combustion engines.

1.2 ENGINE CYCLES

There are two major cycles used in internal combustion engines: Otto and Diesel. The Otto cycle is named after Nikolaus Otto (1832–1891) who developed a four-stroke engine in 1876. It is also called a spark ignition (SI) engine, since a spark is needed to ignite the fuel-air mixture. Otto is considered the inventor of the modern internal combustion engine, and the founder of the internal combustion engine industry. The Diesel cycle is named after Rudolph Diesel (1858–1913) who in 1897 developed an engine designed for the direct injection of liquid fuel into the combustion chamber. The Diesel cycle engine is also called a compression ignition (CI) engine, since the fuel will auto-ignite when injected into the combustion chamber. The Otto and Diesel cycles operate on either a four- or two-stroke cycle.

Otto Cycle

As shown in Figure 1-2, the four-stroke Otto cycle has the following sequence of operations:

1. An intake stroke that draws a combustible mixture of fuel and air past the throttle and the intake valve into the cylinder.

2. A compression stroke with the valves closed which raises the temperature of the mixture. A spark ignites the mixture toward the end of the compression stroke.

3. An expansion or power stroke resulting from combustion of the fuel-air mixture.

4. An exhaust stroke that pushes out the burned gases past the exhaust valve.

   Air enters the engine through the intake manifold, a bundle of passages to distribute the air mixture to individual cylinders. The fuel, typically gasoline, is mixed using a fuel injector or carburetor with the inlet air in the intake manifold, intake port, or directly

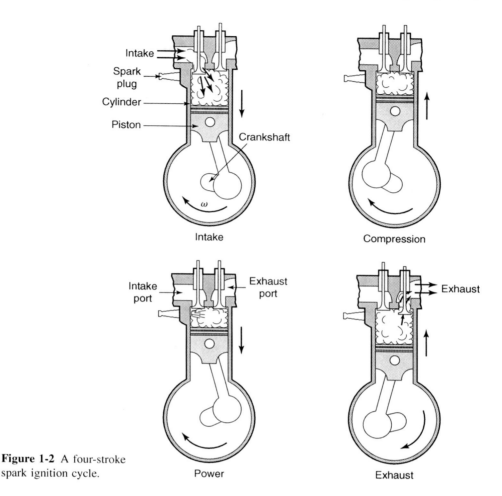

Figure 1-2 A four-stroke spark ignition cycle.

injected into the cylinder, resulting in the cylinder filling with a homogeneous mixture. When the mixture is ignited by a spark, a turbulent flame develops and propagates through the mixture, raising the cylinder temperature and pressure. The flame is extinguished when it reaches the cylinder walls. If the initial pressure is too high, the compressed gases ahead of the flame will auto-ignite, causing a problem called knock. Knock limits the maximum compression ratio of Otto cycle engines. The burned gases exit the engine past the exhaust valves through the exhaust manifold. The exhaust manifold channels the exhaust from individual cylinders into a central exhaust pipe.

In the Otto cycle, a throttle is used to control the amount of air inducted. As the throttle is closed, the amount of air entering the cylinder is reduced, causing a proportional reduction in the cylinder pressure. Since the fuel flow is metered in proportion to the air flow, the throttle in an Otto cycle, in essence, controls the power.

Diesel Cycle

The four-stroke Diesel cycle (see Figure 1-3) has the following sequence:

1. An intake stroke that draws inlet air past the intake valve into the cylinder.

2. A compression stroke that raises the air temperature above the auto-ignition temperature of the fuel. Diesel fuel is sprayed into the cylinder near the end of the compression stroke.

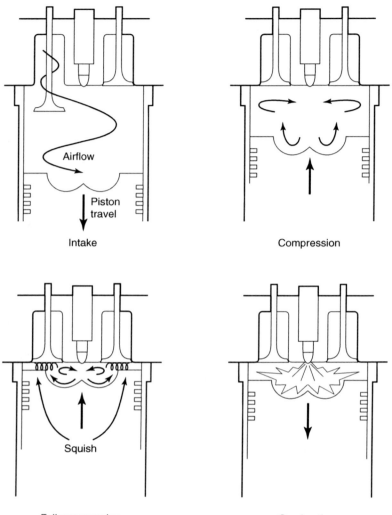

Figure 1-3 Diesel cycle intake, compression and combustion processes (Merrion, 1994). Reprinted with permission © 1994. Society of Automotive Engineers, Inc.

3. Evaporation, mixing, ignition, and combustion of the diesel fuel during the later stages of the compression stroke and the first part of the expansion stroke.

4. An exhaust stroke that pushes out the burned gases past the exhaust valve.

The inlet air in the diesel engine is unthrottled. The power is controlled by the amount of fuel injected into the cylinder. In order to ignite the fuel-air mixture, diesel engines are required to operate at a higher compression ratio, resulting in a higher theoretical efficiency compared to spark ignition (SI) engines. Since the diesel fuel is mixed with cylinder air just before combustion is to commence, knock similar to that in SI engines does not limit the compression ratio of a diesel engine. Diesel engine performance is limited by the formation of smoke, which forms if there is inadequate mixing of the fuel and air. As we shall see, many different diesel combustion chamber designs have been invented to achieve adequate mixing.

Two-Stroke Cycle

The two-stroke cycle was developed by Dugald Clerk in 1878. As the name implies, two-stroke engines need only two strokes of the piston or one revolution to complete a cycle. There is a power stroke every revolution instead of every two revolutions as for four-stroke engines. Two-stroke engines are mechanically simpler than four-stroke engines, and have a higher power to weight ratio. The principle of operation of a crankcase scavenged two-stroke cycle is illustrated in Figure 1-4.

During compression of the crankcase scavenged two-stroke cycle, a sub-atmospheric pressure is created in the crankcase. In the example shown, this opens a reed valve letting air rush into the crankcase. Once the piston reverses direction during combustion and expansion begins, the air in the crankcase closes the reed valve so that the air is compressed. As the piston travels further, it uncovers holes or exhaust ports, and exhaust gases begin to leave, rapidly dropping the cylinder pressure to that of the atmosphere. Then the intake ports are opened and compressed air from the crankcase flows into the cylinder pushing out the remaining exhaust gases. This pushing out of exhaust by the incoming air is called scavenging.

Herein lies one problem with two-stroke engines: the scavenging is not perfect; some of the air will go straight through the cylinder and out the exhaust port, a process called short circuiting. Some of the air will also mix with exhaust gases and the remaining incoming air will push out a portion of this mixture. The magnitude of the problem is strongly dependent on the port designs and the shape of the piston top.

Less than perfect scavenging is of particular concern if the engine is a carbureted gasoline engine, for instead of air being in the crankcase there is a fuel-air mixture. Some of this fuel-air mixture will short circuit and appear in the exhaust, wasting fuel and increasing the hydrocarbon emissions. Carbureted two-stroke engines are used where efficiency is not of primary concern and advantage can be taken of the engine's simplicity; this translates into lower cost and higher power per unit weight. Familiar examples include motorcycles, chain saws, outboard motors, and model airplane engines. With a two-stroke

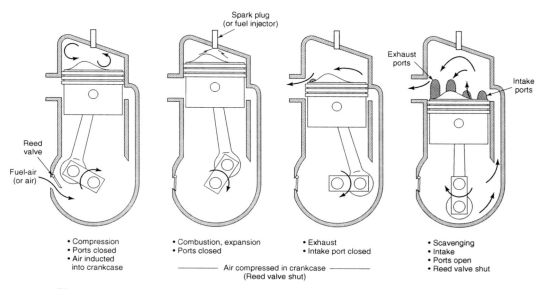

Figure 1-4 A cross-scavenged two-stroke cycle.

diesel or fuel injected gasoline engine, loss of fuel is not a problem since air only is used for scavenging. Two-stroke engines are discussed further in Chapter 7.

1.3 OPERATIONAL PARAMETERS

The performance of the internal combustion engine is characterized with several geometric and thermodynamic parameters. For any one cylinder, the crankshaft, connecting rod, piston, and head assembly can be represented by the mechanism shown in Figure 1-1. Of particular interest are the following geometric parameters: bore, b; connecting rod length, l; crank radius, a; stroke, s; and crank angle, θ. The crank radius is one-half of the stroke. The top dead center (tdc) of an engine refers to the crankshaft being in a position such that $\theta = 0°$. The volume in this position is minimum and is often called the clearance volume, V_c. Bottom dead center (bdc) refers to the crankshaft being at $\theta = 180°$. The volume V_1 is maximum at bottom dead center.

The compression ratio, r, is defined as the ratio of the maximum to minimum volume.

$$r = \frac{V_{bdc}}{V_{tdc}} = \frac{V_1}{V_c} \tag{1.1}$$

The displacement volume, V_d, is the difference between the maximum and minimum volume; for a single cylinder,

$$V_d = V_1 - V_c = \frac{\pi}{4} b^2 s \tag{1.2}$$

For multicylinder engines, the total displacement is the product of the number of cylinders, n_c, and the displacement volume of a cylinder.

The instantaneous volume at any crank angle is

$$V(\theta) = V_c + \frac{\pi}{4} b^2 y \tag{1.3}$$

where y is the instantaneous stroke

$$y = l + a - [(l^2 - a^2 \sin^2 \theta)^{1/2} + a \cos \theta] \tag{1.4}$$

The mean piston speed is an important parameter in engine design since stresses and other factors scale with piston speed rather than with engine speed. Since the piston travels a distance of twice the stroke per revolution it should be clear that

$$\overline{U}_p = 2Ns \tag{1.5}$$

The engine speed, N, refers to the rotational speed of the crankshaft and is expressed in revolutions per minute. The engine frequency, ω, also refers to the rotation rate of the crankshaft but in units of radians per second. The instantaneous piston speed is

$$\frac{U_p(\theta)}{\overline{U}_p} = \frac{\pi}{2} \sin \theta \left[1 + \frac{\cos \theta}{\left(\left(\frac{l}{a}\right)^2 - \sin^2 \theta\right)^{1/2}} \right] \tag{1.6}$$

The brake power, $\dot{W}_b$, is the rate at which work is done; and the engine torque, τ, is a measure of the work done per unit rotation (radians) of the crank. The brake power is the power output of the engine, and measured by a dynamometer. Early dynamometers were simple brake mechanisms. The brake power is less than the boundary rate of work done by the gas, called indicated power, partly because of friction. As we shall see when discussing dynamometers in Chapter 5, the brake power and torque are related by

$$\dot{W}_b = 2\pi\tau N \tag{1.7}$$

The net power is from the complete engine; whereas gross power is from an engine without the cooling fan, muffler, and tail pipe.

The mean effective pressure (mep) is the work done per unit displacement volume. It scales out the effect of engine size. Two useful mean effective pressure parameters are imep and bmep. The indicated mean effective pressure (imep) is the net work per unit displacement volume done by the gas during compression and expansion. The pressure in the cylinder initially increases during the expansion stroke due to the heat addition from the fuel, and then decreases due to the volume increase. The brake mean effective pressure (bmep) is the external shaft work per unit volume done by the engine.

The mean effective pressure is the average pressure that results in the same amount of indicated or brake work produced by the engine. Based on torque, the bmep is

$$\text{bmep} = \frac{4\pi\tau}{V_d} \quad \text{(4 stroke)} \tag{1.8}$$

$$= \frac{2\pi\tau}{V_d} \quad \text{(2 stroke)}$$

and in terms of power the bmep is

$$\text{bmep} = \frac{2\dot{W}_b}{V_d N} \quad \text{(4 stroke)} \tag{1.9}$$

$$= \frac{\dot{W}_b}{V_d N} \quad \text{(2 stroke)}$$

$$\left(\frac{\text{kPa}}{\text{cycle}}\right) = \left(\frac{\text{kW}}{\text{m}^3 \cdot \dfrac{\text{rev}}{\text{min}} \cdot \dfrac{\text{min}}{60 \text{ s}} \cdot \dfrac{\text{cycle}}{\text{rev}}}\right)$$

since the four-stroke engine has two revolutions per cycle and the two-stroke engine has one revolution per cycle.

The brake specific fuel consumption (bsfc) is the fuel flow rate $\dot{m}_f$, divided by the brake power $\dot{W}_b$

$$\text{bsfc} = \frac{\dot{m}_f}{\dot{W}_b} = \frac{\dot{m}_f}{2\pi\tau N} \tag{1.10}$$

The bsfc is a measure of engine efficiency. In fact, bsfc and engine efficiency are inversely related, so that the lower the bsfc the better the engine. Engineers use bsfc rather than thermal efficiency primarily because a more or less universally accepted definition of thermal efficiency does not exist. We will explore the reasons why in Chapter 4. Note for now only that there is an issue with assigning a value to the energy content of the fuel. Let us call that energy the heat of combustion q_c; the brake thermal efficiency is then

$$\eta = \frac{\dot{W}_b}{\dot{m}_f q_c} = \frac{1}{\text{bsfc} \cdot q_c} \tag{1.11}$$

Inspection of Equation 1.11 shows that bsfc is a valid measure of efficiency provided q_c is held constant. Thus two different engines can be compared on a bsfc basis provided that they are operated on the same fuel.

Another performance parameter of importance is the volumetric efficiency, e_v. It is defined as the mass of fuel and air inducted into the cylinder divided by the mass that would occupy the displaced volume at the density ρ_i in the intake manifold. Note that volumetric efficiency is a mass ratio and not a volume ratio. For a four-stroke engine the volumetric efficiency is

$$e_v = \frac{2(\dot{m}_a + \dot{m}_f)}{\rho_i V_d N} \tag{1.12}$$

In Equation 1.12, $\dot{m}_f$ is the flow rate of the fuel inducted. For a direct injection engine $\dot{m}_f = 0$. The factor 2 accounts for the two revolutions per cycle in a four-stroke engine. The intake manifold density is used as a reference condition instead of the standard atmosphere, so that supercharger performance is not included. For two-stroke cycles, a parameter related to volumetric efficiency called the delivery ratio is defined in terms of the air flow only and the ambient air density instead of the intake manifold density.

It is desirable to maximize the volumetric efficiency of an engine since the amount of fuel that can be burned and power produced for a given engine displacement (hence size and weight) is maximized. The volumetric efficiency depends on the intake manifold configuration, valve size, lift, and timing. Although it does not influence in any way the thermal efficiency of the engine, it will influence the efficiency of the system in which it is installed. Clearly, heavier engines in a vehicle will reduce the fuel economy. If one can realize the same power from an engine that is lighter but of thermal efficiency comparable to the heavier one, the fuel economy will go up.

The performance characteristics of a small two-stroke compression ignition model airplane engine (see Figure 1-5), a four-stroke spark ignition automobile engine (see

Figure 1-5 Model airplane engine. (Courtesy Roger J. Schroeder/Classic Model Airplane Engine Construction Kits.)

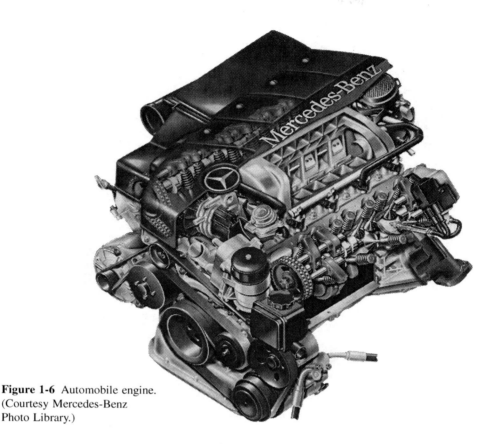

Figure 1-6 Automobile engine.
(Courtesy Mercedes-Benz
Photo Library.)

Figure 1-6), and a large two-stroke stationary compression ignition engine (see Figure 1-7) are compared in Table 1-1. These engines represent the smallest, average, and largest sizes of internal combustion engines. It should be noted that when scaled correctly, all piston engines are remarkably similar. As Table 1-1 illustrates, the mean piston speed is about 5 m/s, and the bmep is about 5 bar for the three engines.

Table 1-1   Comparison of Three Internal Combustion Engines

Characteristics	Model Airplane	Automotive	Marine
Bore (m)	0.0126	0.089	0.737
Stroke (m)	0.0131	0.080	1.016
Displacement per cylinder (m^3)	1.6×10^{-6}	4.98×10^{-3}	0.433
Power per cylinder (kW)	0.1	16.8	529
Engine speed (rpm)	11,400	2500	160
Mass per cylinder (kg)	0.12	34.3	3.56×10^4
Mean piston speed (m/s)	5.0	6.6	5.6
Bmep (bar)	3.2	8.0	4.5
Power/Volume (kW/m^3)	6.3×10^4	3.4×10^4	1.2×10^3
Mass/Volume (kg/m^3)	7.5×10^{-2}	8.2×10^{-2}	6.9×10^{-2}
Power/Mass (kW/kg)	8.4×10^5	4.1×10^5	1.7×10^4

Figure 1-7 Marine engine. (Courtesy Man B&W Diesel.)

There is good reason for this; all engines tend to be made from similar materials. The small differences noted could be attributed to different service criteria for which the engine was designed. The automotive and model airplane mobile engines are more highly stressed and lighter than the diesel stationary engine. Since material stresses in an engine depend to a first order only on the bmep and mean piston speed, it follows that for the same stress limit imposed by the material, all engines should have the same bmep and mean piston speed. Finally, since the engines geometrically resemble one another independent of size, the mass per unit displacement volume is more or less independent of engine size.

The wide open throttle performance of a 2.0 L automotive four-stroke engine is plotted in Figure 1-8. As with most engines, the torque and power both exhibit maxima with engine speed. Viscous friction effects increase quadratically with engine speed, causing the torque curve to decrease at high engine speeds. The maximum torque occurs at lower speed than maximum power, since power is the product of torque and speed. The speed at which peak torque occurs is called the maximum brake torque (MBT). Notice that the torque curve is rippled. This is due to both inlet and exhaust airflow dynamics and mechanical friction, discussed in Chapters 6 and 7.

The volumetric efficiency versus engine speed for various intake manifold configurations of an automotive four-stroke engine is shown in Figure 1-9. The volumetric

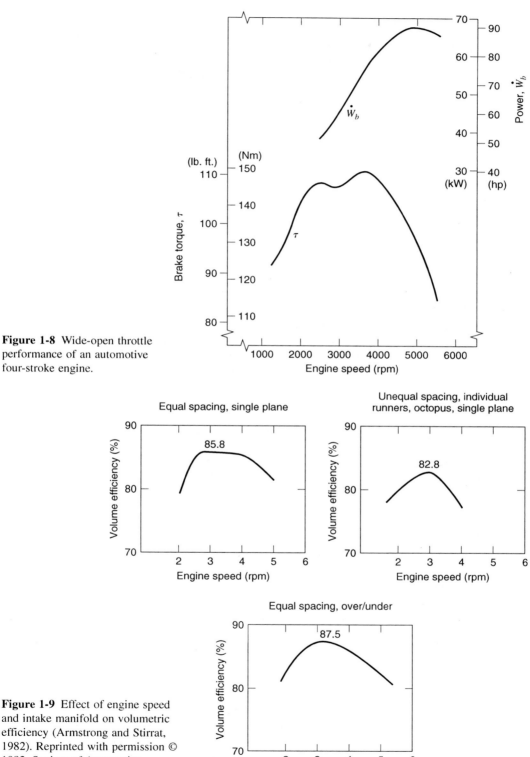

Figure 1-8 Wide-open throttle performance of an automotive four-stroke engine.

Figure 1-9 Effect of engine speed and intake manifold on volumetric efficiency (Armstrong and Stirrat, 1982). Reprinted with permission © 1982. Society of Automotive Engineers, Inc.

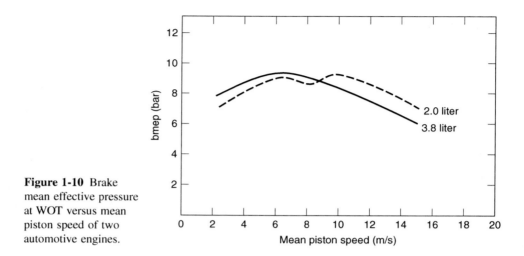

Figure 1-10 Brake mean effective pressure at WOT versus mean piston speed of two automotive engines.

efficiency is also influenced by the valve size, valve lift, and valve timing. The shape and location of the peaks of the volumetric efficiency curve are very sensitive to the manifold configuration. Some configurations produce a flat curve, others produce a very peaked and asymmetric curve.

The bmep of two different displacement automobile engines is compared versus mean piston speed in Figure 1-10. Notice that when performance is scaled to be size independent there is considerable similarity.

1.4 ENGINE CONFIGURATIONS

Internal combustion engines can be built in many different configurations. For a given engine, using a four- or two-stroke Otto or Diesel cycle, the configurations are characterized by the piston-cylinder geometry, the inlet and exhaust valve geometry, the use of super or turbochargers, the type of fuel delivery system, and the type of cooling system.

Piston-Cylinder Geometry

Since the invention of the internal combustion engine, many different piston-cylinder geometries have been designed, as shown in Figure 1-11. The choice of a given arrangement depends on a number of factors and constraints, such as engine balancing and available volume. The *in-line* engine is the most prevalent as it is the simplest to manufacture and maintain. The *V* engine is formed from two in-line banks of cylinders set at an angle to each other, forming the letter V. A horizontally opposed or flat engine is a V engine with 180° offset piston banks. The W engine is formed from three in-line banks of cylinders set at an angle to each other, forming the letter W.

A *radial* engine has all of the cylinders in one plane with equal spacing between cylinder axes. Radial engines are used in air-cooled aircraft applications since each cylinder can be cooled equally. Since the cylinders are in a plane, a master connecting rod is used for one cylinder, and articulated rods are attached to the master rod.

Alternatives to the reciprocating piston-cylinder arrangement have also been developed, such as the rotary Wankel engine. However, the reciprocating piston-cylinder combination remains the dominant form of the internal combustion engine.

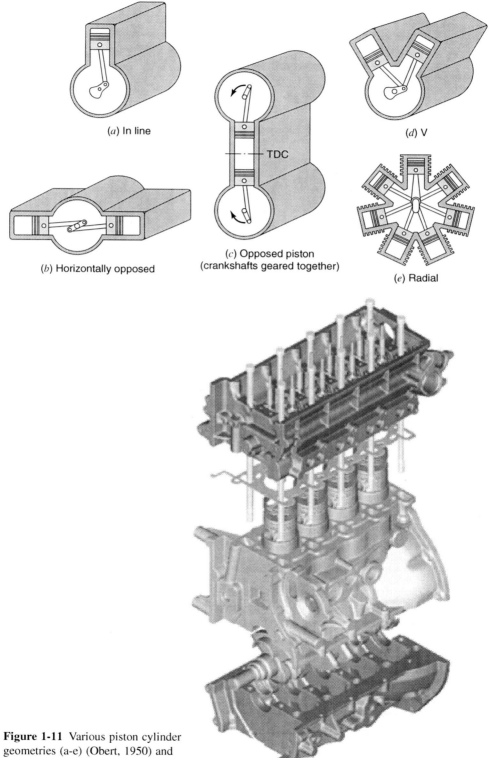

(a) In line

(b) Horizontally opposed

TDC

(c) Opposed piston
(crankshafts geared together)

(d) V

(e) Radial

Figure 1-11 Various piston cylinder geometries (a-e) (Obert, 1950) and (f) (Courtesy of Adapco, Inc.).

The reciprocating motion of the connecting rod and piston creates inertial forces and moments that need to be considered in the choice of an engine configuration. The instantaneous acceleration of the reciprocating masses can be obtained by differentiating Equation 1.4 with respect to time. Rearrangement of Equation 1.4 and replacing θ with ωt gives the instantaneous position y:

$$y = a(1 - \cos \omega t) + l\left[1 - \left(1 - \left(\frac{a}{l}\right)^2 \sin^2 \omega t\right)^{1/2}\right] \tag{1.13}$$

The a/l values for the slider-crank geometries used in modern engines are approximately $1/3$, so the term $(a/l)^2 \sim 1/9$. Using the series expansion approximation $(1 - \varepsilon)^{1/2} \sim 1 - \varepsilon/2$, the displacement y can be approximated as

$$y = a(1 - \cos \omega t) + \frac{a^2}{2l} \sin^2 \omega t \tag{1.14}$$

From the trigonometric identity $\sin^2 \omega t = (1 - \cos 2\omega t)/2$, we have

$$y = \left(a + \frac{a^2}{4l}\right) - a \cos \omega t - \frac{a^2}{4l} \cos 2\omega t \tag{1.15}$$

The velocity and acceleration are therefore

$$\frac{dy}{dt} = a\omega\left(\sin \omega t + \frac{a}{2l} \sin 2\omega t\right) \tag{1.16}$$

$$\frac{d^2y}{dt^2} = a\omega^2\left(\cos \omega t + \frac{a}{l} \cos 2\omega t\right) \tag{1.17}$$

If these terms are multiplied by the mass of the piston and the effective mass of the connecting rod, the vertical (y-direction) momentum and inertia terms result. Note that these terms have two components, one varying with the same speed as the crankshaft, known as the primary term, and the other varying at twice the crankshaft speed, known as the secondary term. In the limit of an infinitely long connecting rod, the motion is simple harmonic.

In multicylinder engines, the cylinder arrangement and firing order are chosen to minimize the primary and secondary forces and moments. Complete cancellation is possible for the following four-stroke engines: in-line 6- and 8-cylinder engines; horizontally opposed 8- and 12-cylinder engines, and 12- and 16-cylinder V engines.

Intake and Exhaust Valve Arrangement

Gases are admitted and expelled from the cylinders by valves that open and close at the proper times, or by ports that are uncovered or covered by the piston. There are many design variations for the intake and exhaust valve type and location.

Poppet valves (see Figure 1-12) are the primary valve type used in internal combustion engines since they have excellent sealing characteristics. Sleeve valves have also been used, but do not seal the combustion chamber as well as poppet valves. The poppet valves can be located either in the engine block or in the cylinder head, depending on manufacturing and cooling considerations. Older automobiles and small four-stroke engines have the valves located in the block, a configuration termed *underhead* or *L-head*. Currently, most engines use valves located in the cylinder head, an *overhead* or *I-head* configuration, as this configuration has good inlet and exhaust flow characteristics.

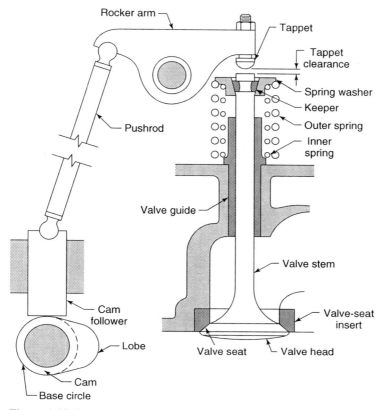

Figure 1-12 Poppet-valve nomenclature (Taylor, 1985).

The valve timing is controlled by a camshaft that rotates at half the engine speed for four-stroke engine. A valve timing profile is shown in Figure 1-13. Lobes on the camshaft along with lifters, pushrods, and rocker arms control the valve motion. Some engines use an overhead camshaft (see Figure 1-14) to eliminate pushrods. The valve timing can be varied to increase volumetric efficiency through the use of advanced camshafts that have moveable lobes, or with electric valves. With a change in the load, the valve opening duration and timing can be adjusted. The gas flow through an engine is discussed in Chapter 7.

Superchargers and Turbochargers

All the engines discussed so far are naturally aspirated, not *supercharged* or *turbocharged*. Supercharging is mechanical compression of the inlet air to a pressure higher than standard atmosphere by a compressor powered by the crankshaft. The compressor raises the density of the incoming charge so that more fuel and air can be delivered to the cylinder to increase the power.

The concept of turbocharging is illustrated in Figure 1-15. Exhaust gas leaving an engine is further expanded through a turbine that drives a compressor. The benefits are twofold: (1) the engine is more efficient because energy that would have otherwise been wasted is recovered from the exhaust gas; and (2) a smaller engine can be constructed to produce a given power because it is more efficient and because the density of the incoming

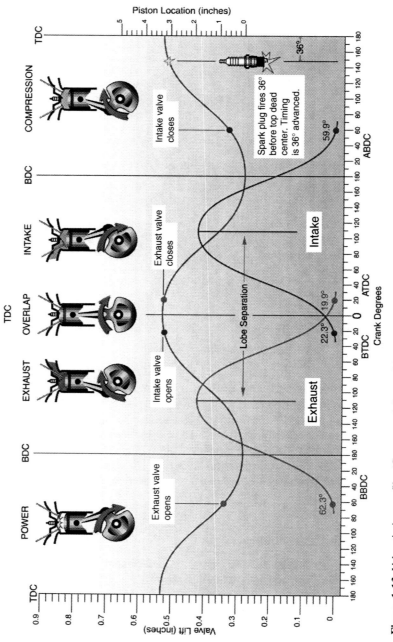

Figure 1-13 Valve timing profile. (Courtesy of Competition Cams, Inc.)

16

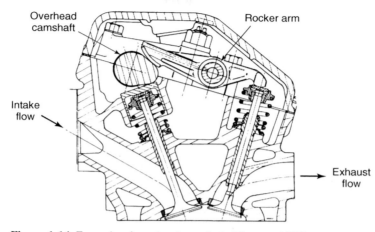

Figure 1-14 Example of overhead camshaft (Newton, 1983).

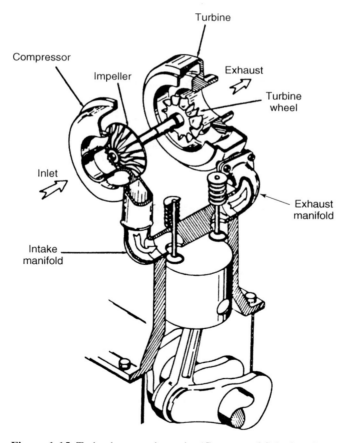

Figure 1-15 Turbocharger schematic. (Courtesy of Schwitzer.)

charge is greater. The power available to drive the compressor when turbocharging is a nonlinear function of engine speed such that at low speeds there is little, if any, boost (density increase), whereas at high speeds the boost is maximum. It is also low at part throttle and high at wide-open throttle. These are desirable characteristics for an automotive engine since throttling or pumping losses are minimized. Most large and medium size diesel engines are turbocharged to increase their efficiency. Super and turbocharging are discussed further in Chapter 7.

Fuel Injectors and Carburetors

Revolutionary changes have taken place with computerized engine controls and fuel delivery systems in recent years and the progress continues. For example, the ignition and fuel injection are computer controlled in engines designed for vehicular applications. Conventional carburetors in automobiles were replaced by throttle body fuel injectors in the 1980s, which in turn were replaced by port fuel injectors in the 1990s. A throttle body injector is a fuel injector located at the intake manifold before the manifold branches to the individual cylinders. Due to its distance from the cylinders, it injects a continuous spray of fuel into the manifold. Port fuel injectors are located in the intake port of each cylinder just upstream of the intake valve, so there is an injector for each cylinder. The port injector does not need to maintain a continuous fuel spray, since the time lag for fuel delivery is much less than that of a throttle body injector. A port fuel injector is shown in Figure 1-16.

Direct injection spark ignition engines are available on some production engines. With direct injection, the fuel is sprayed directly into the cylinder during the late stages of the compression stroke. Compared with port injection, direct injection engines can be operated at a higher compression ratio, and therefore will have a higher theoretical efficiency, since they will not be knock limited. They will also be unthrottled, so they will have a greater volumetric efficiency at part load. The evaporation of the injected fuel in the combustion chamber will have a charge cooling effect, which will also increase its volumetric efficiency.

Cooling Systems

Some type of cooling system is required to remove the approximately 30% of the fuel energy rejected as waste heat. There are two main types of cooling systems: water and air cooling.

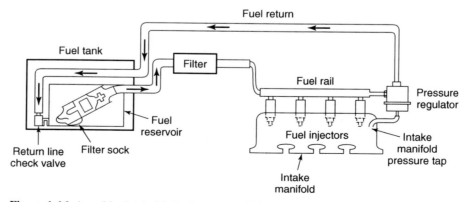

Figure 1-16 A multipoint fuel injection system (Allen and Rinschler, 1984). Reprinted with permission © 1984. Society of Automotive Engineers, Inc.

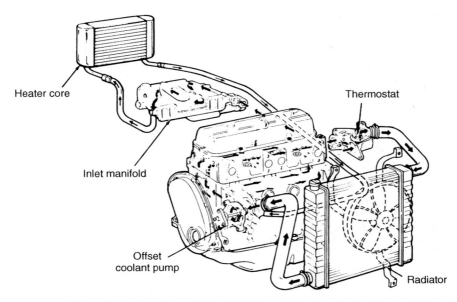

Figure 1-17 Parallel coolant flow, GM L-4 (Walker, 1982). Reprinted with permission ©
1982. Society of Automotive Engineers, Inc.

As shown in Figure 1-17, the water cooling system is usually a single loop where a water
pump sends coolant to the engine block, and then to the head. Warm coolant flows through
the intake manifold to warm it and thereby assist in vaporizing the fuel. The coolant will then
flow to a radiator or heat exchanger, reject the waste heat to the atmosphere, and flow back
to the pump. When the engine is cold, a thermostat prevents coolant from returning to the
radiator, resulting in a more rapid warm-up of the engine. Water-cooled engines are quieter
than air-cooled engines, but have leaking, boiling, and freezing problems.

Engines with relatively low power output, less than 20 kW, primarily use air cooling.
Air cooling systems use fins to lower the air side surface temperature (see Figure 1-5).
An early engine, the Mors, had a finned air-cooled cylinder and water-cooled heads. Engine
heat transfer is the topic of Chapter 8.

1.5   ENGINE EXAMPLES

Automotive Spark Ignition Four-Stroke Engine

A 60° V6 3.2 L automobile engine is shown in Figures 1-18 and 1-19. The engine has
a 89 mm bore and a stroke of 86 mm. The maximum power is 165 kW (225 hp) at 5550 rpm.
The engine has a single overhead camshaft per piston bank with four valves per cylinder. The
pistons are flat with notches for valve clearance. The fuel is mixed with the inlet air by spray-
ing the fuel into the intake port at the Y-junction just above the intake valves. As shown in
Figure 1-20, the overhead camshaft acts on both the intake and exhaust valves via rocker
arms. Note that roller bearings are used on the rocker arms to reduce friction. The clearance
volume is formed by an angled "pent roof" in the cylinder head, with the valves also angled.

The engine has variable valve timing applied to the intake valves with a shift from
low-lift short duration cam lobes to high-lift long duration cam lobes above 3500 rpm. In
the low-lift short duration cam operation the two intake valves have staggered timing which
creates additional swirl to increase flame propagation and combustion stability.

Figure 1-18 Photograph of a 3.2 L V6 automobile engine. (Courtesy of Honda Motor Co.)

Heavy Duty Truck Diesel Engine

Heavy duty truck diesel engines are shown in Figures 1-21 and 1-22. The engine in Figure 1-22 is an inline six cylinder turbocharged engine with a 137-mm bore and 165-mm stroke for a total displacement of 14.6 L. The rated engine power is 373 kW (500 hp). The compression ratio is 16.5 to 1. The engine has electronically controlled, mechanically actuated fuel injectors, and an overhead camshaft. Note that the cylinder head is flat, with the diesel fuel injector mounted in the center of the combustion chamber. The inlet ports impart a swirl to the air in the combustion chamber to improve mixing with the radial fuel spray.

The top of the piston has a "Mexican hat"-shaped crater bowl, so that the initial combustion will take place in the piston bowl. The injection nozzles have three to six holes through which the fuel sprays into the piston bowl. The pressure required to spray the diesel fuel into the combustion chamber is of the order of 1000 bar, for adequate spray penetration into the bowl and subsequent atomization of the diesel fuel. The fuel injection pressure is generated by a plunger driven by the camshaft rocker arm.

Stationary Gas Engine

A stationary natural gas engine is shown in Figures 1-23 and 1-24. Typical applications for stationary engines include co-generation, powering gas compressors, and power generation. The engine shown in Figure 1-24 is an in-line eight-cylinder turbocharged engine, with rated power of 1200 kW, bore of 240 mm, and stroke of 260 mm for a total displacement of 94 L. The compression ratio is 10.9 to 1. This type of engine is designed to operate at a

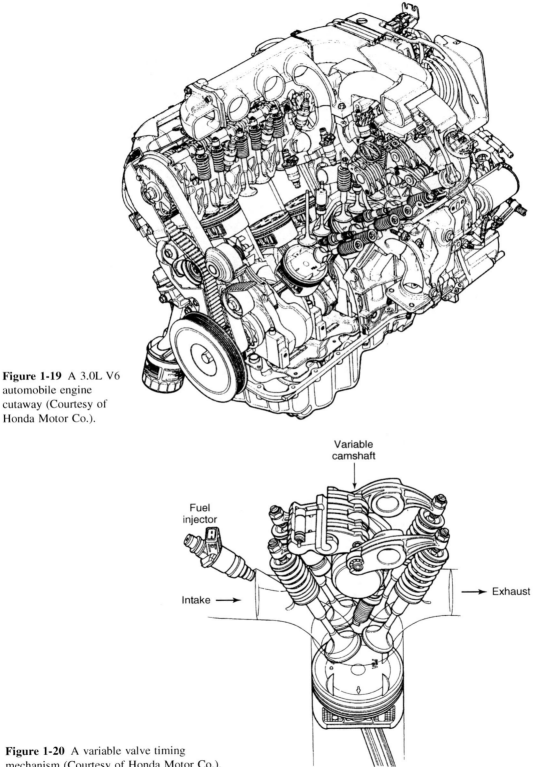

Figure 1-19 A 3.0L V6 automobile engine cutaway (Courtesy of Honda Motor Co.).

Figure 1-20 A variable valve timing mechanism (Courtesy of Honda Motor Co.).

Figure 1-21 A 5.9 L L6 on-highway diesel engine (Courtesy PriceWeber.).

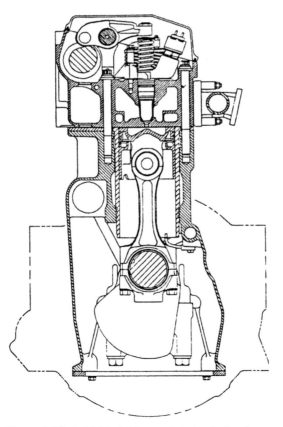

Figure 1-22 A 14.6 L L6 heavy duty truck diesel
engine cross section (Balek and Heitzman, 1994).

Figure 1-23 A stationary natural gas engine (Courtesy of Cooper Energy
Services, Inc.).

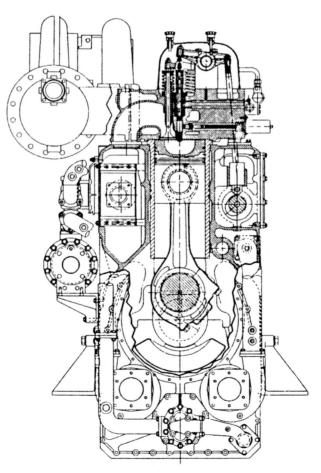

Figure 1-24 A 94 L L8 stationary
natural gas engine cross section
(Courtesy of Cooper Energy
Services, Inc.).

constant speed condition, typically 1200 rpm. Each cylinder has two intake and two exhaust valves. The piston has a combustion bowl with a deep dish concentrated near the center of the piston, so most of the clearance volume is in the piston bowl.

Since natural gas engines are operated lean to reduce nitrogen oxides (NO_x), prechambers are used to initiate a stable combustion process. Pressurized natural gas is injected into a prechamber above the piston, and a spark plug in the prechamber is used to ignite the natural gas. The increase in pressure projects the burning mixture into the main combustion chamber, where the final stages of the combustion take place. Prechambers are also used in high-speed diesel engines to achieve acceptable mixing.

1.6   ALTERNATIVE POWER PLANTS

In this section alternative power plants will be discussed in terms of a particular application where they dominate the field by having some advantage over the internal combustion engine.

First, consider electric motors which compete in the range of powers less than about 500 kW. They are used, for example, in forklifts operated within a factory or warehouse. Internal combustion engines are not applied in this case because they would build up high levels of pollutants such as carbon monoxide or nitric oxide. Electric motors are found in a variety of applications, such as where the noise and vibration of a piston engine or the handling of a fuel are unacceptable. Other examples are easy to think of in both industrial and residential sectors. Electric motors will run in the absence of air, such as in outer space or under water; they are explosion proof; and they can operate at cryogenic temperatures. If one can generalize, one might state with respect to electric motors that internal combustion engines tend to be found in applications where mobility is a requirement or electricity is not available.

Proponents of electric vehicles point out that almost any fuel can be used to generate electricity, therefore we can reduce our dependence upon petroleum by switching to electric vehicles. There would be no exhaust emissions emitted throughout an urban environment. The emissions produced by the new electric generating stations could be localized geographically so as to minimize the effect. The main problem with electric vehicles is the batteries used for energy storage. The electric vehicles that have been built to date have a limited range of only 50 to 100 mi (80 to 160 km), on the order of one-fifth of what can be easily realized with a gasoline engine powered vehicle. It is generally recognized that a breakthrough in battery technology is required if electric vehicles are to become a significant part of the automotive fleet. Batteries have about 1% of the energy per unit mass of a typical vehicular fuel, and a life span of about 2 years.

Hybrid electric vehicles (HEV), which incorporate a small internal combustion engine with an electric motor and storage batteries, have been the subject of recent research, and as of the year 2000, have reached the production stage, primarily due to their low fuel consumption and emission levels. The first HEV was the Woods "Dual Power" automobile, introduced in 1916. A hybrid electric vehicle has more promise than an electric vehicle, since the HEV has an internal combustion engine to provide the energy to meet vehicle range requirements. The battery then provides the additional power needed for acceleration and climbing hills. The fuels used in the HEV engines in current production include gasoline, diesel, and natural gas.

As shown in Figure 1-25, the engine and electric motor are placed in either a series or parallel configuration. In a series configuration only the electric motor with power from the battery or generator is used to drive the wheels. The internal combustion engine is

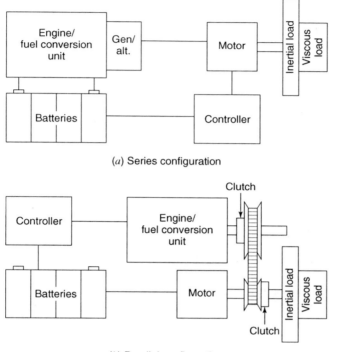

(*a*) Series configuration

(*b*) Parallel configuration

Figure 1-25 Hybrid electric vehicle powertrain configurations.

maintained at its most efficient and lowest emission operating points to run the generator and charge the storage batteries. With the parallel configuration, the engine and electric motor can be used separately or together to power the vehicle. The motors can be used as generators during braking to increase vehicle efficiency.

The fuel cell electric vehicle (FCEV) is currently in the development phase. The chemical reaction in a fuel cell produces lower emissions relative to combustion in an internal combustion engine. Recent developments in proton exchange membrane (PEM) technology have been applied to vehicular fuel cells. Current PEM fuel cells are small enough to fit beneath a vehicle's floor next to storage batteries and deliver 50 kW to an electric motor. The PEM fuel cell requires a hydrogen fuel source to operate. Since there is presently no hydrogen fuel storage infrastructure, on-board reforming of methanol fuel to hydrogen and CO_2 is also required. The reforming efficiency is about 60%, so coupled with a fuel cell efficiency of 70%, and a motor efficiency of 90%, the overall fuel cell engine efficiency will be about 40%, comparable with high efficiency internal combustion engines. Current estimates are that fuel cell vehicles will be in production by 2005.

Gas turbine engines compete with internal combustion engines on the other end of the power spectrum, at powers greater than about 500 kW. The advantages offered depend on the application. Factors to consider are the efficiency and power per unit weight. A gas turbine consists basically of a compressor-burner-turbine combination that provides a supply of hot, high-pressure gas. This may then be expanded through a nozzle (turbojet), through a turbine, to drive a fan, and then through a nozzle (turbofan), through a turbine, to drive a propeller (turboprop), or through a turbine to spin a shaft in a stationary or vehicular application.

One advantage a gas turbine engine offers to the designer is that the hardware responsible for compression, combustion, and expansion are three different devices, whereas in a piston engine all these processes are done within the cylinder. The hardware for each process in a gas turbine engine can then be optimized separately; whereas in a piston engine compromises must be made with any given process, since the hardware is expected to do three tasks. However, it should be pointed out that turbochargers give the designer of conventional internal combustion engines some new degrees of freedom toward optimization.

With temperature limits imposed by materials, the reciprocating engine can have a greater peak cycle temperature than the gas turbine engine. In an internal combustion engine, the gases at any position within the engine vary periodically from hot to cold. Thus the average temperature during the heat transfer to the walls is neither very hot nor cold. On the other hand, the gas temperature at any position in the gas turbine is steady, and the turbine inlet temperature is always very hot, thus tending to heat material at this point to a greater temperature than anywhere in a piston engine.

The thermal efficiency of a gas turbine engine is highly dependent upon the adiabatic efficiency of its components, which in turn is highly dependent upon their size and their operating conditions. Large gas turbines tend to be more efficient than small gas turbines. That airliners are larger than automobiles is one reason gas turbines have displaced piston engines in airliners, but not in automobiles. Likewise gas turbines are beginning to penetrate the marine industry, though not as rapidly, as power per unit weight is not as important with ships as with airplanes.

Another factor favoring the use of gas turbines in airliners (and ships) is that the time the engine spends operating at part or full load is small compared to the time the engine spends cruising, therefore the engine can be optimized for maximum efficiency at cruise. It is a minor concern that at part load or at take-off conditions the engine's efficiency is compromised. Automobiles, on the other hand, are operated over a wide range of load and speed so a good efficiency at all conditions is better than a slightly better efficiency at the most probable operating condition and a poorer efficiency at all the rest.

Steam- or vapor-cycle engines are much less efficient than internal combustion engines, since their peak temperatures are about 800 K, much lower than the peak temperatures ($\sim$2500 K) of an internal combustion engine. They are used today almost totally in stationary applications and where the energy source precludes the use of internal combustion engines. Such energy sources include coal, waste feed stocks, nuclear, solar, and waste heat in the exhaust gas of combustion devices including internal combustion engines.

In some applications, engine emission characteristics might be a controlling factor. In the 1970s, in fact, a great deal of development work was done toward producing an automotive steam engine when it was not known whether the emissions from the internal combustion engine could be reduced enough to meet the standards dictated by concern for public health. However, the development of catalytic converters, as discussed in Chapter 9, made it possible for the internal combustion engine to meet emission standards at that time, and remain a dominant prime mover technology.

1.7   SOURCES OF ADDITIONAL INFORMATION

The Society of Automotive Engineers (SAE) and the American Society of Mechanical Engineers (ASME) hold regular national and regional meetings, as well as publish conference and journal proceedings on topics related to internal combustion engines. Their Internet web addresses are *www.sae.org*, and *www.asme.org*. The text web page has a listing of web sites related to internal combustion engines. The sites include internal combustion engine manufacturers, professional societies, university research laboratories, national laboratories, and application areas such as automobiles, trucks, motorcycles, and airplanes.

1.8 REFERENCES

ALLEN, K. and P. RINSCHLER (1984), "Turbocharging the Chrysler 2.2 Liter Engine," SAE paper 840252.

ARMSTRONG, D. and G. STIRRAT (1982), "Ford's 1982 3.8 L V6 Engine," SAE paper 820112.

BALEK, S. and R. HEITZMAN (1994), "The Caterpillar 3406E Heavy Duty Diesel Engine," ASME ICE-Vol. 22, pp. 177-186.

MERRION, D. (1994), "Diesel Engine Design for the 1990s," SAE SP-1011.

NEWTON, K., W. STEEDS, and T. GARRETT (1983), *The Motor Vehicle*, Butterworths, London, England.

OBERT, E. (1950), *Internal Combustion Engines*, International Textbook Co., Scranton, Pennsylvania.

TAYLOR, C. (1985), *The Internal Combustion Engine in Theory and Practice*, Vol. 2, MIT Press, Cambridge, Massachusetts.

WALKER, J. (1982), "The GM 1.8 Liter L4 Gasoline Engine Designed by Chevrolet," SAE Paper 820111.

1.9 HOMEWORK

1. (a) Derive a formula for the dimensionless cylinder volume $V(\theta)/V(0)$ as a function of the crank angle, θ, compression ratio, r, and dimensionless stroke, $\varepsilon = s/2l$.
(b) Plot the dimensionless cylinder volume for the case of $r = 10$, $s = 100$ mm, and $l = 150$ mm.

2. (a) Derive Equation 1.6.
(b) Plot the dimensionless piston velocity for $s = 100$ mm and $l = 150$ mm.

3. To illustrate the effect of combustion chamber geometry on swirl amplification consider an axisymmetric engine where at bottom center the velocity field of the air inside the cylinder is approximately $v_r = v_z = 0$ and $v_\theta = V_o(2r/b)$. The cylinder has a bore, b, and the piston has a disk-shaped bowl of diameter, d, and depth, h. The motion is said to be solid body since the gas is swirling as though it were a solid. If at top dead center the motion is also solid body and angular momentum is conserved during compression, what is the ratio of the initial to final swirl speed, ω, as a function of the compression ratio and the cylinder geometry? The moment of inertia of solid body rotation of a disk of diameter, d, and depth, h, is $I = \pi \rho h d^4/32$.

4. Assuming that the mean effective pressure, mean piston speed, power per unit piston area, and mass per unit displacement volume are all size independent, how will the power per unit weight of an engine depend upon the number of cylinders if the total displacement is constant? To make the analysis easier, assume that the bore and stroke are equal.

5. Compute the mean piston speed, bmep (bar), torque (Nm), and the power per piston area for the following engines:

Engine	Bore (mm)	Stroke (mm)	Cylinders	Speed (rpm)	Power (kW)
Marine	136	127	12	2600	1118
Dragster	108	95	8	6400	447
Formula One	86	57	8	10500	522

6. The volumetric efficiency of the fuel injected marine engine in Problem 5 is 0.80 and the inlet manifold density is 50% greater than the standard atmospheric density of $\rho_{amb} = 1.17$ kg/m^3. If the engine speed is 2600 rpm, what is the air flow rate (kg/s)?

7. A 3.8 L four-stroke fuel-injected automobile engine has a power output of 88 kW at 4000 rpm and volumetric efficiency of 0.85. The bsfc is 0.35 kg/kW hr. If the fuel has a heat of combustion of 42 MJ/kg, what are the bmep, thermal efficiency, and air to fuel ratio? Assume atmospheric conditions of 298 K and 1 bar.

8. A six-cylinder two-stroke engine produces a torque of 1100 Nm at a speed of 2100 rpm. It has a bore of 123 mm and a stroke of 127 mm. What is its bmep and mean piston speed?

9. A 380 cc single-cylinder two-stroke motorcycle engine is operating at 5500 rpm. The engine has a bore of 82 mm and a stroke of 72 mm. Performance testing gives a bmep of 6.81 bar, bsfc of 0.49 kg/kW hr, and delivery ratio of 0.748. (a) What is the fuel to air ratio? (b) What is the air flow rate (kg/s)?

Chapter 2

Gas Cycles

2.1 INTRODUCTION

A study of gas cycles as models of internal combustion engines is useful for illustrating some of the important parameters influencing engine performance. Gas cycle calculations treat the combustion process as an equivalent heat release. The combustion processes are modeled as constant volume, constant pressure, and finite heat release processes. By modeling the combustion process as a heat release process, the analysis is simplified since the details of the physics and chemistry of combustion are not required.

This chapter also provides a review of thermodynamics. The gas cycle serves to introduce many of the ideal processes that are also used in more complex combustion cycle models, such as the fuel-air cycle, that are introduced in Chapter 3. This chapter first uses closed-system first law analysis for the compression and expansion strokes and then incorporates open-system analysis for the intake and exhaust strokes.

2.2 CONSTANT VOLUME HEAT ADDITION

This cycle is often referred to as the Otto cycle and considers the special case of an internal combustion engine whose combustion is so rapid that the piston does not move during the combustion process, and thus combustion is assumed to take place at constant volume. A spark initiates the combustion, so the Otto cycle engine is also called a spark ignition engine. The Otto cycle is a four-stroke process, with intake, compression, expansion, and exhaust strokes, so that there are two crankshaft revolutions per cycle.

The Otto cycle is named after Nikolaus Otto (1832–1891) who developed a four-stroke engine in 1876. Otto is considered the inventor of the modern internal combustion engine, and founder of the internal combustion engine industry. His production engine produced about 1.5 kW (2 hp) at a speed of 160 rpm.

The working fluid in the Otto cycle is assumed to be an ideal gas. Additionally, let us assume, to keep our mathematics simple, that it has constant specific heats. This assumption results in simple analytical expressions for the efficiency as a function of the compression ratio. The specific heat ratio, $\gamma = c_p/c_v$, of air is plotted in Figure 2-1 as a function of temperature. Typical values of γ chosen for an Otto air cycle calculation range from 1.3 to 1.4, to correspond with measured cylinder temperature data.

The Otto cycle for analysis is shown in Figure 2-2. The four basic processes are:

1 to 2   isentropic compression
2 to 3   constant-volume heat addition
3 to 4   isentropic expansion
4 to 1   constant-volume heat rejection

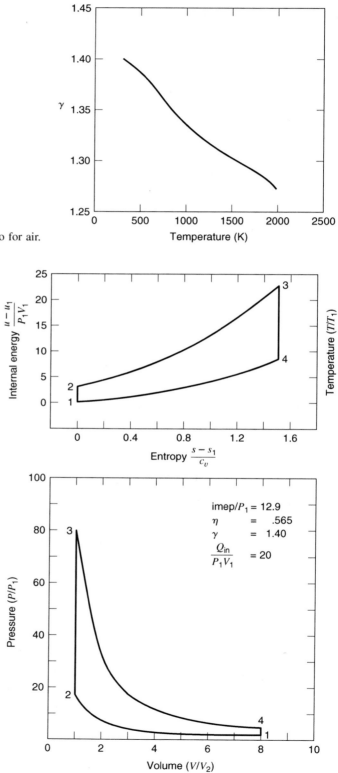

Figure 2-1 Specific heat ratio for air.

Figure 2-2 The Otto cycle.

The reader should be able to show that the following relations are valid:

Heat addition

$$Q_{in} = mc_v(T_3 - T_2) \tag{2.1}$$

Heat rejection

$$Q_{out} = mc_v(T_4 - T_1) \tag{2.2}$$

Compression stroke

$$\frac{P_2}{P_1} = r^\gamma \quad \frac{T_2}{T_1} = r^{\gamma - 1} \tag{2.3}$$

Expansion stroke

$$\frac{P_4}{P_3} = \left(\frac{1}{r}\right)^\gamma \quad \frac{T_4}{T_3} = \left(\frac{1}{r}\right)^{\gamma - 1} \tag{2.4}$$

where

m = mass of gas in the cylinder, $P_1 V_1 / RT_1$
c_v = constant volume specific heat
r = compression ratio
γ = specific heat ratio

The compression ratio of an engine is defined to be

$$r = \frac{V_1}{V_2} \tag{2.5}$$

For a thermodynamic cycle such as this, the thermal efficiency is

$$\eta = \frac{W_{out}}{Q_{in}} = 1 - \frac{Q_{out}}{Q_{in}} \tag{2.6}$$

If we introduce the previously cited relations for Q_{in} and Q_{out}, we get

$$\eta = 1 - \frac{(T_4 - T_1)}{(T_3 - T_2)} = 1 - r^{1 - \gamma} \tag{2.7}$$

The indicated mean effective pressure (imep) is

$$\frac{imep}{P_1} = \frac{Q_{in}}{P_1 V_1}\left(\frac{r}{r - 1}\right)\eta \tag{2.8}$$

The imep is nondimensionalized by the initial pressure P_1, and the heat input Q_{in} is nondimensionalized by $P_1 V_1$. The indicated mean effective pressure is plotted versus compression ratio and the heat input in Figure 2-3. As expected, the imep is proportional to the heat addition. The thermal efficiency of the Otto cycle depends only on the specific heat ratio and on the compression ratio. Figure 2-3 plots the thermal efficiency versus compression ratio for a range of specific heat ratios from 1.2 to 1.4. The efficiencies we have computed; for example, $\eta \sim 0.56$, for $r = 8$, and $\gamma = 1.4$, are about twice as large as measured for actual engines. There are a number of reasons for this. We have not accounted for internal friction and combustion of a fuel within the engine; and we have ignored heat losses.

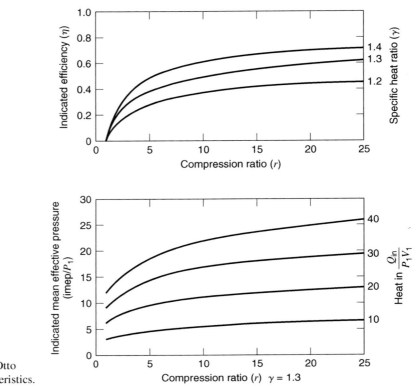

Figure 2-3 Otto cycle characteristics.

Compression ratios found in actual spark ignition engines typically range from 9 to 11. The compression ratio is limited by two practical considerations: material strength and engine knock. The maximum pressure, P_3, of the cycle scales with compression ratio as r^γ. Engine heads and blocks have a design maximum stress, which should not be exceeded, thus limiting the compression ratio. In addition, the maximum temperature T_3 also scales with the compression ratio as r^γ. If T_3 exceeds the auto-ignition temperature of the air-fuel mixture, combustion will occur ahead of the flame, a condition termed "knock". The pressure waves that are produced are damaging to the engine, and they reduce the combustion efficiency. Auto-ignition is discussed further in Chapter 9.

2.3 CONSTANT PRESSURE HEAT ADDITION

This cycle is often referred to as the Diesel cycle and models the special case of an internal combustion engine whose combustion is controlled so that the beginning of the expansion stroke occurs at constant pressure. The Diesel cycle engine is also called a compression ignition engine. The Diesel cycle is named after Rudolph Diesel (1858–1913), who in 1897 developed an engine designed for the direct injection of liquid fuel into the combustion chamber. The compression ratio is higher in a Diesel engine, with typical values in the range of 15 to 20, than that of an Otto engine, so that the cylinder temperature will be high enough for the air-fuel mixture to self-ignite. The duration of the combustion process is controlled by the injection and mixing of the fuel spray, which is discussed further in Chapter 9. Fuel is sprayed directly into the cylinder by a high-pressure fuel injector beginning at about 15° before top dead center, and ending about 5° after top dead center. The injection duration is a function of the engine load. Since the compressed air

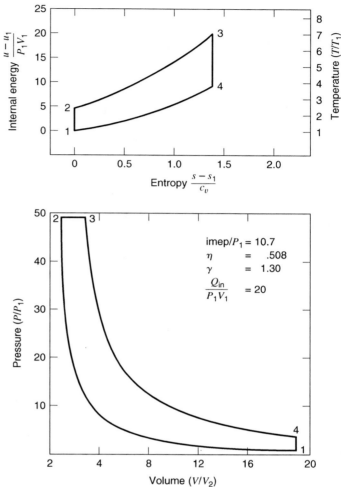

Figure 2-4 The Diesel cycle.

is at a temperature above the auto-ignition temperature, combustion will begin in regions of the fuel spray that have an air-fuel ratio close to stoichiometric. The other processes in the Diesel cycle are the same as in an Otto cycle.

The cycle for analysis is shown in Figure 2-4. The four basic processes are:

1 to 2   isentropic compression
2 to 3   constant pressure heat addition
3 to 4   isentropic expansion
4 to 1   constant volume heat rejection

Again assuming constant specific heats, the student should recognize the following equations:

Heat addition

$$Q_{in} = mc_p(T_3 - T_2) \tag{2.9}$$

Expansion stroke

$$\frac{P_4}{P_3} = \left(\frac{\beta}{r}\right)^{\gamma} \qquad \frac{T_4}{T_3} = \left(\frac{\beta}{r}\right)^{\gamma - 1} \tag{2.10}$$

where we have defined the parameter β as

$$\beta = \frac{V_3}{V_2} = \frac{T_3}{T_2} \tag{2.11}$$

In this case the indicated efficiency is

$$\eta = 1 - \frac{1}{r^{\gamma-1}} \frac{\beta^\gamma - 1}{[\gamma(\beta - 1)]} \tag{2.12}$$

The term in brackets in Equation 2.12 is greater than one, so that for the same compression ratio, r, the efficiency of the Diesel cycle is less than that of the Otto cycle. However, since Diesel cycle engines are not knock limited, they operate at about twice the compression ratio of Otto cycle engines. For the same maximum pressure, the efficiency of the Diesel cycle is greater than that of the Otto cycle. Diesel cycle efficiencies are shown in Figure 2-5 for a specific heat ratio of 1.3. They illustrate that high compression ratios are desirable and that the efficiency decreases as the heat input increases. As β approaches one, the Diesel cycle efficiency approaches the Otto cycle efficiency.

Although Equation 2.12 is correct, the utility suffers somewhat in that β is not a natural choice of independent variable. Rather, in engine operation, we think more in terms of the heat transferred in. The two are related according to Equation 2.13.

$$\beta = 1 + \frac{\gamma - 1}{\gamma} \left(\frac{Q_{\text{in}}}{P_1 V_1} \right) \frac{1}{r^{\gamma-1}} \tag{2.13}$$

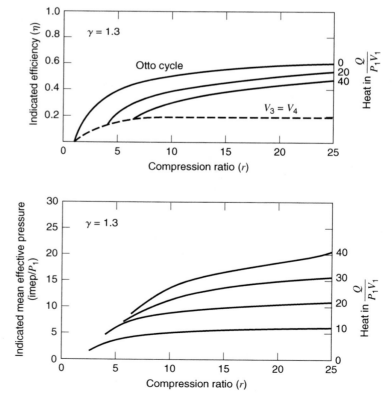

Figure 2-5 Diesel cycle characteristics.

2.4   DUAL CYCLE

Modern compression ignition engines resemble neither the constant-volume nor the constant-pressure cycle, but rather an intermediate cycle in which some of the heat is added at constant volume and then the remaining heat is added at constant pressure. The Dual cycle is a gas cycle model that can be used to more accurately model combustion processes that are slower than constant volume, but more rapid than constant pressure. The Dual cycle also can provide algebraic equations for performance parameters such as the thermal efficiency. The distribution of heat added in the two processes is something the designer can specify approximately by choice of fuel, the fuel injection system, and the engine geometry, usually to limit the peak pressure in the cycle. Consequently this cycle is also referred to as the limited-pressure cycle.

The cycle notation is illustrated in Figure 2-6. In this case we have the following difference:

Heat addition

$$Q_{in} = mc_v(T_{2.5} - T_2) + mc_p(T_3 - T_{2.5}) \tag{2.14}$$

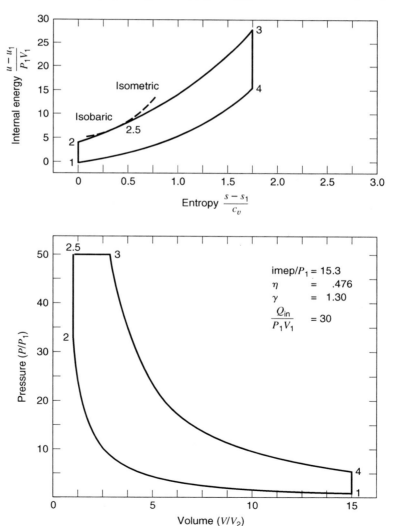

Figure 2-6 The Dual cycle.

The expansion stroke is still described by Equation 2.11 provided we write $\beta = V_3/V_{2.5}$. In terms of $\alpha = P_3/P_2$, a pressure rise parameter, it can be shown that

$$\eta = 1 - \left(\frac{1}{r}\right)^{\gamma-1}\left[\frac{\alpha\beta^\gamma - 1}{(\alpha - 1) + \gamma\alpha(\beta - 1)}\right] \tag{2.15}$$

The constant-volume and constant-pressure cycles can be considered as special cases of the Dual cycle in which $\beta = 1$ and $\alpha = 1$, respectively. The use of the Dual cycle model requires information about either the fractions of constant volume and constant pressure heat addition, or the maximum pressure, P_3. A common assumption is to equally split the heat addition. Results for the case of $P_3/P_1 = 50$ and $\gamma = 1.3$ are shown in Figure 2-7, showing efficiencies and imep that are between the Otto and Diesel limits. For the same compression ratio, the Otto cycle has the largest net work, followed by the Dual, and the Diesel. Transformation of β and α to more natural variables yields

$$\beta = 1 + \frac{\gamma - 1}{\alpha\gamma}\left[\frac{Q_{in}}{P_1 V_1}\frac{1}{r^{\gamma-1}} - \frac{\alpha - 1}{\gamma - 1}\right] \tag{2.16}$$

$$\alpha = \frac{1}{r^\gamma}\frac{P_3}{P_1} \tag{2.17}$$

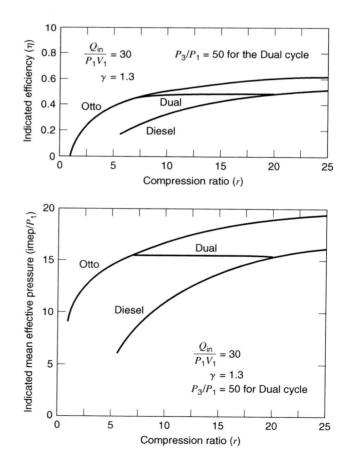

Figure 2-7 The Dual cycle compared with Diesel and Otto cycles.

2.5   MILLER CYCLE

The efficiency of an internal combustion engine will be increased if the expansion ratio is greater than the compression ratio. There have been many mechanisms designed and patented to produce different compression and expansion ratios. The Miller cycle (R. H. Miller, 1890–1967) uses early or late inlet valve closing to decrease the compression ratio.

The Miller cycle is shown in Figure 2-8. The air intake is unthrottled and is controlled by the intake valve. As the piston moves downward on the intake stroke, the cylinder pressure follows the constant pressure line from point 6 to point 1. For early inlet valve closing, the inlet valve is closed at point 1 and the cylinder pressure decreases during the expansion to point 7. As the piston moves upward on the compression stroke, the cylinder pressure retraces the path from point 7 through point 1 to point 2. The net work done along the two paths 1-7 and 7-1 cancel, so that the effective compression ratio $r_c = V_1/V_2$ is therefore less than the expansion ratio $r_e = V_4/V_3$.

For late inlet valve closing, a portion of the intake air is pushed back into the intake manifold before the intake valve closes at point 1. A related cycle, the Atkinson cycle, is one in which the expansion stroke continues until the cylinder pressure at point 4 decreases to atmospheric pressure.

The parameter, λ, is defined as the ratio of the expansion ratio to the compression ratio:

$$\lambda = r_e/r_c \tag{2.18}$$

The heat rejection has two components

$$Q_{\text{out}} = mc_v(T_4 - T_5) + mc_p(T_5 - T_1) \tag{2.19}$$

In this case the thermal efficiency is

$$\eta = 1 - (\lambda r_c)^{1-\gamma} - \frac{\lambda^{1-\gamma} - \lambda(1-\gamma) - \gamma}{(\gamma - 1)} \frac{P_1 V_1}{Q_{\text{in}}} \tag{2.20}$$

Equation 2.20 reduces to the Otto cycle thermal efficiency as $\lambda \to 1$. The imep is

$$\frac{\text{imep}}{P_1} = \frac{Q_{\text{in}}}{P_1 V_1} \left(\frac{r_c}{\lambda r_c - 1} \right) \eta \tag{2.21}$$

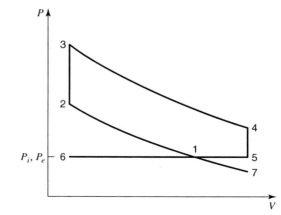

Figure 2-8 The Miller cycle.

The thermal efficiency of the Miller cycle is not only a function of the compression ratio and specific heat ratio, but also a function of the expansion ratio and the load Q_{in}. The ratio of the Miller cycle thermal efficiency to an equivalent Otto cycle efficiency with the same compression ratio is plotted in Figure 2-9 for a range of compression ratios and λ values. For example, with $\lambda = 2$ and $r_c = 12$, the Miller cycle is about 20% more efficient than the Otto cycle.

The ratio of the Miller/Otto cycle imep is plotted as a function of λ in Figure 2-10. As λ increases, the imep decreases significantly, since the fraction of the displacement volume that is filled with the inlet fuel-air mixture decreases. This decrease in imep is a disadvantage of the Miller cycle. Supercharging of the inlet mixture is one technique that can be used to increase the imep.

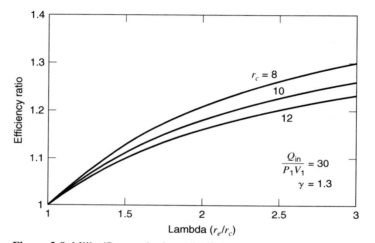

Figure 2-9 Miller/Otto cycle thermal efficiency ratio.

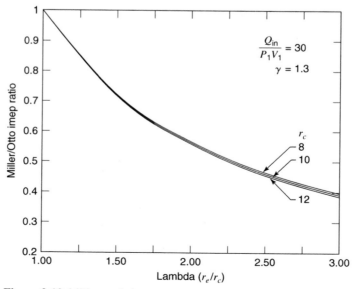

Figure 2-10 Miller cycle imep.

2.6   FINITE HEAT RELEASE

In the Otto and Diesel cycles the fuel is assumed to burn at rates which result in constant volume top dead center combustion, or constant pressure combustion, respectively. Actual engine pressure and temperature profile data do not match these simple models, and more realistic modeling, such as a finite heat release model, is required. A finite heat release model is a differential model of an engine power cycle in which the heat addition is specified as a function of the crank angle.

Heat release models can answer questions that the simple gas cycle models cannot address. For example, if one wants to know about the effect of spark timing or heat transfer on engine work and efficiency, a heat release model is required. If heat transfer is included, as is done in Chapter 8, then the state changes for compression and expansion are no longer isentropic, and cannot be expressed as simple algebraic equations.

A typical cumulative heat release or "burn fraction" curve for a spark ignition engine is shown in Figure 2-11. The figure plots the cumulative heat release fraction $x_b(\theta)$ versus the crank angle. The characteristic features of the heat release curve are an initial small slope region beginning with spark ignition, followed by a region of rapid growth, and then a more gradual decay. The three regions correspond to the initial ignition delay, a rapid burning region, and a burning completion region. This S-shaped curve can be represented analytically by a Weibe function (similar to a Weibull function)

$$x_b(\theta) = 1 - exp\left[-a\left(\frac{\theta - \theta_s}{\theta_d}\right)^n\right] \qquad (2.22)$$

where

θ = crank angle

θ_s = start of heat release

θ_d = duration of heat release

n = Weibe form factor

a = Weibe efficiency factor

The parameters a and n are adjustable parameters used to fit experimental data. The start of heat release is at $\theta = \theta_s$, with $f(\theta_s) = 1 - exp\,(0) = 0$. Since the cumulative heat release curve asymptotically approaches 1, the end of combustion is defined by an arbitrary limit, such as 90% or 99% complete combustion; i.e., $x_b = 0.90$ or 0.99. Corresponding values of the efficiency factor a are $a = 2.3$ and $a = 4.6$, respectively. Values of $a = 5$

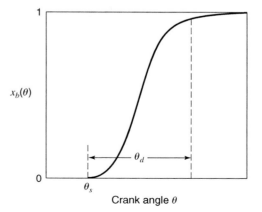

Figure 2-11 Cumulative heat release function.

and $n = 3$ have been reported to fit well with experimental data (Heywood, 1988). However, the parameter values also depend to some degree on the engine load and speed, as well as the particular type of engine. Discussion of the heat release curve for the compression ignition engine is deferred until Chapter 9.

The rate of heat release as a function of crank angle is obtained by differentiating the cumulative heat release Weibe function.

$$\frac{dQ}{d\theta} = Q_{in} \frac{dx_b}{d\theta} \tag{2.23}$$

$$= na \frac{Q_{in}}{\theta_d} (1 - x_b) \left(\frac{\theta - \theta_s}{\theta_d} \right)^{n-1}$$

EXAMPLE 2.1   *Heat Release Fractions*

Compare the heat release fraction curves for three types of combustion events. One event (1) has the heat release starting at $\theta_s = -40°$ after top dead center with a duration of $\theta_d = 20°$. The second event (2) has the heat release starting at $\theta_s = -20°$ atdc with a duration of $\theta_d = 40°$. The third event (3) has the heat release starting at $\theta_s = 0°$ atdc with a duration of $\theta_d = 60°$. For the heat release process, the value of the Weibe efficiency factor a is 5 and the value of the Weibe form factor n is 3. At what crank angle is half the heat released for each event?

SOLUTION   The three heat release curves are shown in Figure 2-12. By symmetry, the (1) event starting at $\theta_s = -40°$ has a half-heat release angle of about 30° before top dead center. The (2) event starting at $\theta_s = -20°$ has a half-heat release angle at top dead center. The (3) event starting at $\theta_s = 0°$ has a half-heat release angle at 30° atdc.

The finite heat release models are derived by incorporating the heat release equation, Equation 2.23, into the differential energy equation. The analysis assumes that the heat release occurs during the compression and expansion strokes. As shown in the following derivation, the resulting differential form of the energy equation does not have a simple analytical solution due to the heat release term, and so it is solved numerically.

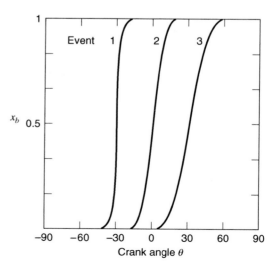

Figure 2-12 Heat release curves for Example 2.1.

The closed-system differential energy equation (note that work and heat interaction terms are not true differentials) for a small crank angle change, $d\theta$, is

$$\delta Q - \delta W = dU \tag{2.24}$$

since $\delta W = PdV$, and $dU = mc_v \, dT$

$$\delta Q - PdV = mc_v \, dT \tag{2.25}$$

Assuming ideal gas behavior,

$$PV = mRT \tag{2.26}$$

which in differential form is

$$m \, dT = \frac{1}{R}(PdV + VdP) \tag{2.27}$$

The energy equation is therefore

$$\delta Q - PdV = \frac{c_v}{R}(PdV + VdP) \tag{2.28}$$

per unit crank angle,

$$\frac{dQ}{d\theta} - P\frac{dV}{d\theta} = \frac{c_v}{R}\left(P\frac{dV}{d\theta} + V\frac{dP}{d\theta}\right) \tag{2.29}$$

Solving for the pressure, P,

$$\frac{dP}{d\theta} = -\gamma \frac{P}{V}\frac{dV}{d\theta} + \frac{\gamma - 1}{V}\left(\frac{dQ}{d\theta}\right) \tag{2.30}$$

This is a first order differential equation of the form $dP/d\theta = f(\theta,P,Q)$ for the cylinder pressure as a function of the crank angle, θ, pressure, P, and the heat release, Q. In order to integrate the differential energy equation an equation for the cylinder volume as a function of crank angle is needed. By reference to Chapter 1, the cylinder volume $V(\theta)$ is

$$V(\theta) = \frac{V_d}{r-1} + \frac{V_d}{2}[R + 1 - \cos\theta - (R^2 - \sin^2\theta)^{1/2}] \tag{2.31}$$

upon differentiation,

$$\frac{dV}{d\theta} = \frac{V_d}{2}\sin\theta[1 + \cos\theta(R^2 - \sin^2\theta)^{-1/2}] \tag{2.32}$$

where

V_d = displacement volume, $\dfrac{\pi}{4}b^2 s$

r = compression ratio

$R = 2l/s$

For the portion of the compression and expansion strokes with no heat release, where $\theta < \theta_s$ and $\theta > \theta_s + \theta_d$, $\delta Q/d\theta = 0$. With no heat release, the energy equation can be integrated and the isentropic algebraic pressure-volume relation recovered:

$$\frac{dP}{d\theta} = -\frac{P}{V}\frac{dV}{d\theta} \tag{2.33}$$

$$\frac{dP}{P} = -\gamma\frac{dV}{V} \tag{2.34}$$

$$PV^\gamma = \text{constant} \tag{2.35}$$

Table 2-1   Engine Parameters for Finite Heat Release Model

Q_{in}	Total heat addition (J)
θ_s	Start of heat release (degrees)
θ_d	Duration of heat release (degrees)
a	Weibe efficiency factor
n	Weibe form factor
T_1	Initial gas temperature (K)
P_1	Initial gas pressure (kPa)
M	Gas molecular weight (kg/kmol)
γ	Gas specific heat ratio (c_p/c_v)
r	Compression ratio
l	Connecting rod length (m)
b	Cylinder bore (m)
s	Cylinder stroke (m)
N	Engine speed (rpm)

The differential energy equation, Equation 2.30, is integrated numerically for the pressure using an integration routine such as fourth order Runge-Kutta integration. The integration starts at bottom dead center ($\theta = -180°$), with initial inlet conditions P_1, V_1, T_1, the gas molecular weight, M, and specific heat ratio, γ, given. The mass of gas, m, is $P_1 V_1 M / R_u T_1$. The integration proceeds degree by degree to top dead center and back to bottom dead center. Once the pressure is computed as a function of crank angle, the work and cylinder temperature can also be readily computed from $\delta W = PdV$ and the ideal gas law $T = PV/mR$.

The differential energy equation, Equation 2.30, can also be used "in reverse" to compute heat release curves from experimental measurements of the cylinder pressure. As discussed in Chapter 5, commercial combustion analysis software is available to perform such analysis in real time during an experiment.

The engine parameters required for solution of the cylinder pressure as a function of the crank angle are summarized in Table 2-1.

EXAMPLE 2.2   *Finite Heat Release*

A single cylinder Otto cycle engine is operated at full throttle. The engine has a bore of 0.1 m, stroke of 0.1 m, and connecting rod length of 0.15 m, with a compression ratio of 10. The initial cylinder pressure, P_1, and temperature, T_1, at bottom dead center are 1 bar and 300 K. The engine speed is 3000 rpm. The heat addition, Q_{in}, is 1800 J. Compare the effect of start of heat release at $\theta_s = -20$ and $\theta_s = 0°$ atdc on (a) the net work, the thermal efficiency of the cycle, and the mean effective pressure, and (b) the pressure, temperature, and work profiles versus crank angle. Assume that the ideal gas specific heat ratio is 1.4, the gas molecular weight is 29, the combustion duration is constant at 40°, and that the Weibe parameters are $a = 5$ and $n = 3$.

SOLUTION   The web pages accompanying the text contain an applet entitled "Simple Heat Release Applet," which can be used to compute engine performance, and compare the effect of changing combustion and geometric parameters. The applet computes gas cycle performance by numerically integrating Equation 2.30 for the pressure as a function of crank angle, then uses the ideal gas law to compute the temperature as a function of crank angle. The above engine parameters are entered into the "Simple Heat Release Applet" as

Simple Heat Release Applet

| Parameters | Result Table | Pressure Plot | Temperature Plot | Work Plot |

Calculate

Combustion Parameters	Engine 1	Engine 2	Geometric Parameters	Engine 1	Engine 2
Spark Timing	-20.0	0.0	s - stroke [mm]	100.0	100.0
Duration of combustion	40.0	40.0	b - bore [mm]	100.0	100.0
Weibe parameter a	5	5	l - connecting rod [mm]	150.0	150.0
Weibe parameter n	3	3	r - compression ratio	10	10
Inital Temperature [K]	300	300	Engine speed [rpm]	3000	3000
Inital Pressure [bar]	1	1			
Gas molecular weight	29				

Heat Addition

Heat Input (J)	1800
Gamma	1.4

Figure 2-13 Finite heat release applet input for Example 2.2.

Engine Performance Table

	Engine1	Engine2
Qin [J]	1800.0	1800.0
Max. Temperature (K)	2924.3	2572.5
Max. Pressure (kPa)	8730.0	4845.0
Mean Effective Pressure [bar]	13.29	12.23
Indicated Work [J]	1043.93	960.3
Indicated Power [kW]	26.1	24.01
Thermal Efficiency	0.58	0.533

Figure 2-14 Engine performance table for Example 2.2.

shown in Figure 2.13. Note that the start of heat release is $\theta_s = -20°$ for Engine 1 and $\theta_s = 0°$ for Engine 2, and all other parameters are the same for both engine conditions.

(a) As shown in Figure 2-14, as the start of heat release is changed from $\theta_s = -20$ to $\theta_s = 0°$, the work output and the mean effective pressure decrease, and the thermal efficiency decreases from 0.58 to 0.53.

(b) The pressure profiles are compared in Figure 2-15a. The pressure rise for Engine 1 (at $\theta_s = -20°$) is almost double that of Engine 2 (at $\theta_s = 0°$). The maximum pressure occurs at 10° after top dead center for Engine 1, and at about 25° after top dead center for

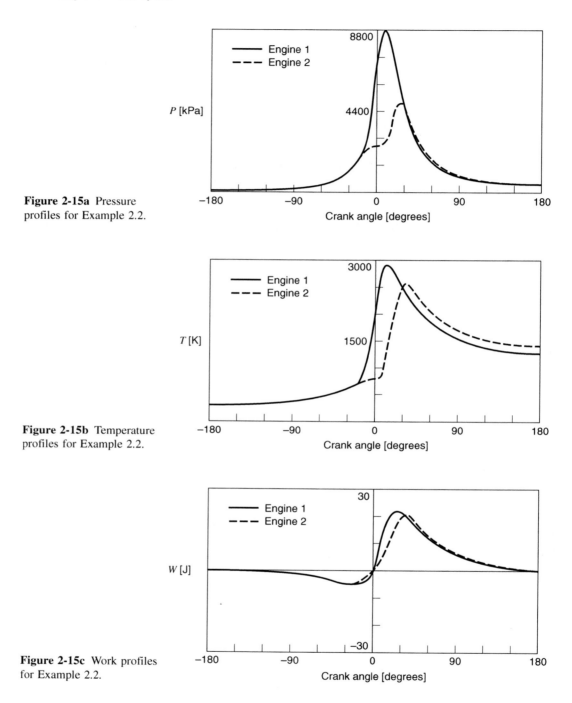

Figure 2-15a Pressure profiles for Example 2.2.

Figure 2-15b Temperature profiles for Example 2.2.

Figure 2-15c Work profiles for Example 2.2.

Engine 2. The temperature profiles are compared in Figure 2-15b. Engine 1 has a peak temperature of about 2900 K, almost 400 K above that of Engine 2. The work profiles are shown in Figure 2-15c. Note that the location of peak work δW, $\theta = 25°$ atdc for Engine 1 and 30° atdc for Engine 2, is not the same as the location of maximum pressure, since the incremental volume change $dV(\theta)$, see Equation 2.32, increases with increasing crank angle.

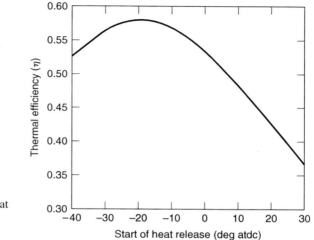

Figure 2-16 Effect of start of heat release on thermal efficiency for Example 2.2.

Table 2-2   Effect of Start of Heat Release on imep and Peak Pressure Crank Angle

Start of heat release θ_s (deg atdc)	Imep (bar)	Peak pressure crank angle θ_{max} (deg atdc)
−40	12.09	0
−30	12.96	5
−25	13.20	7
−20	13.29	10
−15	13.23	12
−10	13.02	17
−5	12.68	21
0	12.23	27
10	11.09	37
20	9.79	50
30	8.44	60

The results of Example 2.2 show that if the start of heat release begins too late, the heat release will occur in an expanding volume, resulting in lower combustion pressure, and lower net work. If the start of heat release begins too early during the compression stroke, the negative compression work will increase, since the piston is doing work against the expanding combustion gases. Using the engine specified in Example 2.2, the effect of the start of heat release on the thermal efficiency is shown in Figure 2-16, and the effect on imep and peak pressure is tabulated in Table 2-2. For this computation, the optimum start of heat release is about $\theta_s = -20°$.

In practice, the optimum spark timing also depends on the engine load, and is in the range of $\theta_s = -20°$ to $\theta_s = -5°$. The resulting location of the peak combustion pressure will typically be between 12° and 18° atdc.

2.7   IDEAL FOUR-STROKE PROCESS AND RESIDUAL FRACTION

The simple gas cycle models assume that the heat rejection process occurs at constant volume, and neglect the gas flow that occurs when the intake and exhaust valves are opened and closed. In this section, we use the energy equation to model the exhaust and intake

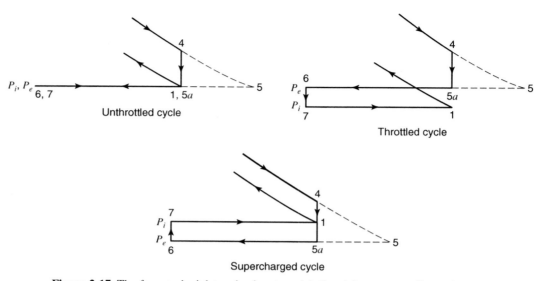

Figure 2-17 The four-stroke inlet and exhaust model. P_i = inlet pressure, P_e = exhaust pressure.

strokes. During the exhaust stroke, the exhaust valve is assumed to open instantaneously at bottom dead center and close instantaneously at top dead center. Similarly, during the intake stroke, the intake valve is assumed to open at the dead center and remain open until bottom dead center. The intake and exhaust valve overlap, that is, the time during which they are open simultaneously is assumed to be zero.

The intake and exhaust strokes are also assumed to occur adiabatically and at constant pressure. Constant pressure intake and exhaust processes occur only at low engine speeds. More realistic computations model the instantaneous pressure drop across the valves and further would account for the heat transfer which is especially significant during the exhaust. Such considerations are deferred to Chapters 7 and 8. Referring to Figure 2-17, the ideal intake and exhaust processes are as follows:

4 to 5a	Constant cylinder volume blowdown
5a to 6	Constant pressure exhaustion
6 to 7	Constant cylinder volume reversion
7 to 1	Constant pressure induction

Exhaust Stroke

The exhaust stroke has two processes: gas blowdown and gas displacement. At the end of the expansion stroke 3 to 4, the pressure in the cylinder is greater than the exhaust pressure. Hence, when the exhaust valve opens, gas will flow out of the cylinder even if the piston does not move. Typically the pressure ratio, P_4/P_e, is large enough to produce sonic flow at the valve so that the pressure in the cylinder rapidly drops to the exhaust manifold pressure, P_e, and the constant volume approximation is justified.

The remaining gas in the cylinder that has not flowed out through the exhaust valve undergoes an expansion process. If heat transfer is neglected, this unsteady expansion process can be modeled as isentropic. The temperature and pressure of the exhaust gases remaining in the cylinder are

$$T_5 = T_4\left(\frac{P_5}{P_4}\right)^{\frac{\gamma-1}{\gamma}} \tag{2.36}$$

$$P_5 = P_e \tag{2.37}$$

As the piston moves upward from bottom dead center, it pushes the remaining cylinder gases out of the cylinder. The cylinder pressure is assumed to remain constant at $P_5 = P_6 = P_4$. Since internal combustion engines have a clearance volume, not all of the gases will be pushed out. There will be exhaust gas left in the clearance volume, called residual gas. This gas will mix with the incoming fuel-air mixture in an Otto cycle, and the incoming air flow in a Diesel cycle.

The state of the gas remaining in the cylinder during the exhaust stroke can be found by applying the closed system first law to the cylinder gas from state 5 to state 6 as shown in Figure 2-18. The closed system control volume will change in shape as the cylinder gases flow out the exhaust port across the exhaust valve.

The energy equation is

$$Q_{5-6} - W_{5-6} = U_6 - U_5 \tag{2.38}$$

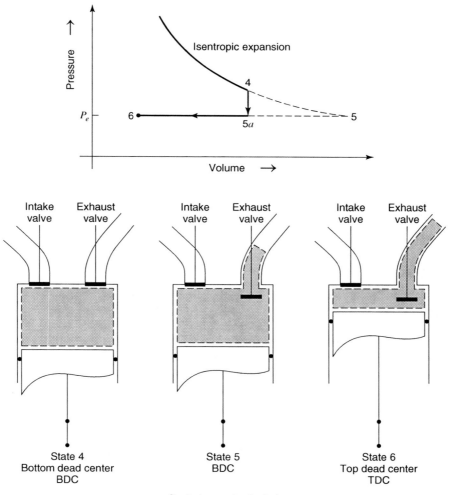

Control mass is shaded

Figure 2-18 The exhaust stroke (4 to 5 to 6) illustrating residual mass. Note that while the blowdown is assumed to occur at constant cylinder volume, the control mass is assumed to expand isentropically.

The work term is

$$W_{5-6} = P_e(V_6 - V_5) \tag{2.39}$$

and if the flow is assumed to be adiabatic, the first law becomes

$$U_6 + P_eV_6 = U_5 + P_eV_5 \tag{2.40}$$

or

$$h_6 = h_5 \tag{2.41}$$

$$T_e = T_6 = T_5 \tag{2.42}$$

Therefore, during an adiabatic exhaust stroke, the enthalpy and temperature of the exhaust gases remain constant as they leave the cylinder, and the enthalpy of the residual gas left in the cylinder clearance volume is constant.

The residual gas fraction, f, is the ratio of the residual gas mass, m_6, in the cylinder at the end of the exhaust stroke (state 6) to the mass, $m_4 = m_1 = m$, of the fuel-air mixture:

$$f = \frac{V_6/v_6}{V_4/v_4} = \frac{1}{r}\frac{v_4}{v_6} = \frac{1}{r}\frac{T_4}{T_e}\frac{P_e}{P_4} \tag{2.43}$$

Since

$$T_e = T_4\left(\frac{P_e}{P_4}\right)^{\frac{\gamma-1}{\gamma}} \tag{2.44}$$

the residual fraction is

$$f = \frac{1}{r}\left(\frac{P_e}{P_4}\right)^{\frac{1}{\gamma}} \tag{2.45}$$

For example, for a compression ratio of $r = 9$, $P_e = 101$ kPa, $P_4 = 500$ kPa, and $\gamma = 1.3$, $f = 1/9(101/450)^{1/1.3} = 0.035$. Typical values of the residual gas fraction, f, are in the 0.03 to 0.12 range. The residual gas fraction is lower in Diesel cycles than in Otto cycles, due to the higher compression ratio in Diesel cycles.

Intake Stroke

When the intake valve is opened, the intake gas mixes with the residual gas. Since the intake gas temperature is usually less than the residual gas temperature, the cylinder gas temperature at the end of the intake stroke will be greater than the intake temperature. In addition, if heat transfer is neglected, the flow across the intake valve, either from the intake manifold to the cylinder or the reverse, is at constant enthalpy.

There are three different flow situations for the intake stroke, depending on the ratio of inlet to exhaust pressure. If the inlet pressure is less than the exhaust pressure, the engine is throttled. In this case there is flow from the cylinder into the intake port when the intake valve opens. In the initial portion of the intake stroke, the induced gas is primarily composed of combustion products that have previously flowed into the intake port. In the latter portion of the stroke, the mixture flowing in is fresh charge, undiluted by any combustion products.

If the inlet pressure is greater than the exhaust pressure, the engine is said to be supercharged (turbocharging is a special case of supercharging in which a compressor driven by an exhaust turbine raises the pressure of atmospheric air delivered to an engine). In this case there is flow from the intake port into the engine until the pressure equilibrates.

In actual engines, because of valve overlap, there may be a flow of fresh mixture from the inlet to the exhaust port, which can waste fuel and be a source of hydrocarbon exhaust emissions. The third case is when inlet and exhaust pressures are equal; the engine is then said to be unthrottled.

The unsteady open system mass and energy equations can be used to determine the state of the fuel air mixture and residual gas combination at state 1, the end of the intake stroke. The initial state of the gas in the system at the beginning of the intake process is at state 6.

As discussed above, there is a flow of a gas mixture into or out of the cylinder when the intake valve is opened, depending on the relative pressure difference. The net gas flow into the cylinder control volume has mass, m_i, enthalpy, h_i, and pressure, P_i. As the piston moves downward, it is assumed that the cylinder pressure remains constant at the inlet pressure, P_i, which is consistent with experimental observations. For the overall process from state 6 to state 1 with the inlet flow at state i, the conservation of mass equation is

$$m_i = m_1 - m_6 \qquad (2.46)$$

The unsteady energy equation is

$$Q_{6-1} - W_{6-1} = -m_i h_i + m_1 u_1 - m_6 u_6 \qquad (2.47)$$

If heat transfer is neglected, $Q_{6-1} = 0$, and the work done by the gas is $W_{6-1} = P_1(V_1 - V_6)$, so

$$-P_i(V_1 - V_6) = -(m_1 - m_6)h_i + m_1 u_1 - m_6 u_6 \qquad (2.48)$$

Since $u_1 = h_1 - P_1 v_1$ and $u_6 = h_6 - P_e v_6$, we can express the energy equation in terms of enthalpy:

$$(P_i - P_e)m_6 v_6 = -(m_1 - m_6)h_i + m_1 h_1 - m_6 h_6 \qquad (2.49)$$

Solving for h_1:

$$h_1 = \frac{m_6}{m_1}\left[h_6 + \left(\frac{m_1}{m_6} - 1\right)h_i + (P_i - P_e)v_6 \right] \qquad (2.50)$$

Therefore, the enthalpy at the end of the intake stroke is not just the average of the initial and intake enthalpies, as would be the case for a steady flow situation, but also includes the flow work term.

The equation for the enthalpy at the end of the intake stroke, Equation 2.50, can also be expressed in terms of the residual gas fraction, f. From Equation 2.43,

$$m_6 = m_1 f \quad \text{and} \quad m_1 - m_6 = m_1(1 - f) \qquad (2.51)$$

and from the ideal gas law,

$$P_i v_6 = RT_6 \qquad (2.52)$$

Upon substitution of Eqs. (2.51) and (2.52) into Eq. (2.50),

$$h_1 = (1 - f)h_i + f h_6 - \left(1 - \frac{P_i}{P_e}\right)f R T_e \qquad (2.53)$$

If the reference enthalpy is chosen so that $h_i = c_p T_i$, then

$$T_1 = (1 - f)T_i + f T_e\left[1 - \left(1 - \frac{P_i}{P_e}\right)\left(\frac{\gamma - 1}{\gamma}\right) \right] \qquad (2.54)$$

For example, if $f = 0.05$, $P_i/P_e = 0.5$, $k = 1.35$, $T_i = 320$ K, and $T_e = 1400$ K, then $T_1 = 365$ K.

The volumetric efficiency of the inlet stroke for a gas cycle is given by

$$e_v = \frac{m_i}{\rho_i V_d} = 1 - \frac{P_e/P_i - 1}{\gamma(r - 1)} \tag{2.55}$$

During the intake process, the gas within the control volume does work since the piston is expanding the cylinder volume. During exhaust, work is done on the gas. The net effect during the intake and exhaust strokes is

$$W_{5a-1} = (P_i - P_e)V_d \tag{2.56}$$

The negative of that work is called pumping work since it is a loss of useful work for the throttled engine. The pumping mean effective pressure is defined as the pumping work per unit displacement volume:

$$\text{pmep} = P_e - P_i \tag{2.57}$$

The indicated mean effective pressure (imep) is defined as the work per unit displacement volume done by the gas during the compression and expansion stroke. The work per unit displacement volume required to pump the working fluid into and out of the engine during the intake and exhaust strokes is termed the pumping mean effective pressure (pmep). We shall call the work per unit displacement volume of the whole cycle the net indicated mean effective pressure (imep_{net}). In Chapter 6, we also include friction considerations. The following relations should be clear:

$$(\text{imep})_{\text{net}} = \text{imep} - \text{pmep} \tag{2.58}$$

$$\eta_{\text{net}} = \eta\left(1 - \frac{\text{pmep}}{\text{imep}}\right) \tag{2.59}$$

Prior to doing a gas cycle computation, the reader should be careful to distinguish the mass in quantities that are mass intensive. When the residual gas fraction is taken into account, the heat added is

$$Q_{\text{in}} = m_i q_{\text{in}} = m(1 - f)q_{\text{in}} \tag{2.60}$$

where q_{in} is the heat addition per unit mass of gas inducted (kJ/kg$_{\text{gas}}$). Now suppose that we compute the work done during compression as

$$w_{1-2} = c_v(T_1 - T_2) \tag{2.61}$$

In the metric system of units, this compression work is expressed as kJ/kg. Confusion can arise as to whether that is per kilogram of gas mixture in the cylinder (it is!) or per kilogram of gas inducted. The reader is advised to beware because here and in the literature the meaning is implicit in the context in which the symbol in question appears.

Four-Stroke Otto Gas Cycle Analysis

When we include the exhaust and intake strokes, we have two additional equations for the gas cycle analysis, the exhaust energy equation, and the intake energy equation. The two unknown parameters in these equations are the residual gas fraction, f, and the gas temperature at the end of the intake stroke, T_1. Since it is difficult to solve these two equations algebraically, the solution is found by iteration, as shown in this section. The overall cycle input parameters in this analysis are summarized in Table 2-3.

Table 2-3   Input Parameters for Four-Stroke Gas Cycle

T_i	inlet air or mixture temperature
P_e	exhaust pressure
r	compression ratio
P_i	inlet pressure
γ	ideal gas specific heat ratio
q_{in}	heat added per unit mass of gas induced

Since T_1 is dependent on the residual gas fraction, f, and the residual gas temperature, T_e, if values of f and T_e are assumed, the cycle calculation can proceed, and improved values of f and T_e calculated.

6, i-1: *Intake stroke*

$$T_1 = (1 - f)T_i + f[1 - (1 - P_i/P_e)^{(\gamma - 1)/\gamma}]T_e$$
$$P_1 = P_i$$

1-2: *Isentropic compression stroke*

$$P_2 = P_1(V_1/V_2)^\gamma = P_1 r^\gamma$$
$$T_2 = T_1 r^{\gamma - 1}$$

2-3: *Constant volume heat addition*

$$T_3 = T_2 + q_{in}(1 - f)/c_v$$
$$P_3 = P_2(T_3/T_2)$$

3-4: *Isentropic expansion stroke*

$$P_4 = P_3(1/r)^\gamma$$
$$T_4 = T_3(1/r)^{\gamma - 1}$$

4-5: *Isentropic blowdown*

$$T_5 = T_4(P_4/P_e)^{(1 - \gamma)/\gamma}$$
$$P_5 = P_e$$

5-6: *Constant pressure adiabatic exhaust stroke*

$$T_e = T_5$$
$$P_6 = P_5 = P_e$$
$$f = 1/r(P_6/P_4)^{1/\gamma}$$

EXAMPLE 2.3   *Four-Stroke Otto Cycle*

Compute the volumetric efficiency, net thermal efficiency, residual fraction, intake stroke temperature rise, and the exhaust stroke temperature drop of an engine that operates on the ideal four-stroke Otto cycle. The engine is throttled with an inlet pressure of $P_i = 50$ kPa and an inlet temperature of $T_1 = 300$ K. The exhaust pressure is $P_e = 100$ kPa. The compression ratio, $r = 10$. Assume a heat input of $q_{in} = 2500$ kJ/kg and $\gamma = 1.3$. Plot the volumetric efficiency, net thermal efficiency, and residual fraction as a function of the intake/exhaust pressure ratio for $0.3 < P_i/P_e < 1.5$.

SOLUTION   The web pages accompanying the text contain an applet entitled: *Four-Stroke Otto Gas Cycle* which iterates through the four-stroke Otto gas cycle equations to determine the pressures, temperatures, and other cycle parameters. The applet input/output page is

Four Stroke Gas Otto Cycle

Intake Pressure (kPa)	50
Intake Temperature (K)	300
Exhaust Pressure (kPa)	100
Compression ratio	10
Gamma	1.3
qin (kJ/kg gas)	2500
Compute Cycle	Enter

State	1	2	3	4
Pressure (kPa):	50.0	997.63	4888.88	245.02
Temperature (K):	329.2	656.9	3219.1	1613.4

Exhaust Temp. (K) =	1312.0	Volumetric Efficiency =	0.915
Residual Mass Fraction =	0.0502	Net Imep (kPa) =	668.75
Ideal Thermal Eff. =	0.499	Net Thermal Eff. =	0.464

Figure 2-19 Four-stroke gas cycle applet input/output for Example 2.3.

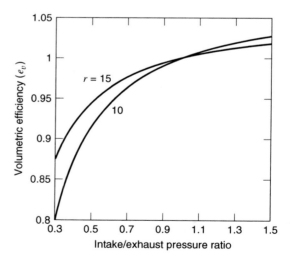

Figure 2-20 Four-stroke Otto gas cycle volumetric efficiency for Example 2.3.

shown in Figure 2-19. For the above conditions, the intake stroke temperature rise, $T_1 - T_i$, is about 30 K, and the exhaust blowdown temperature decrease, $T_4 - T_e$, is about 300 K. The volumetric efficiency, $e_v = 0.92$, the net thermal efficiency, $\eta_{net} = 0.46$, and the residual fraction, $f = 0.050$.

The volumetric efficiency (Equation 2.55), the residual fraction (Equation 2.45), and the net thermal efficiency (Equation 2.59) are plotted in Figures 2-20, 2-21, and 2-22, respectively, as a function of the intake/exhaust pressure ratio. As the pressure ratio decreases,

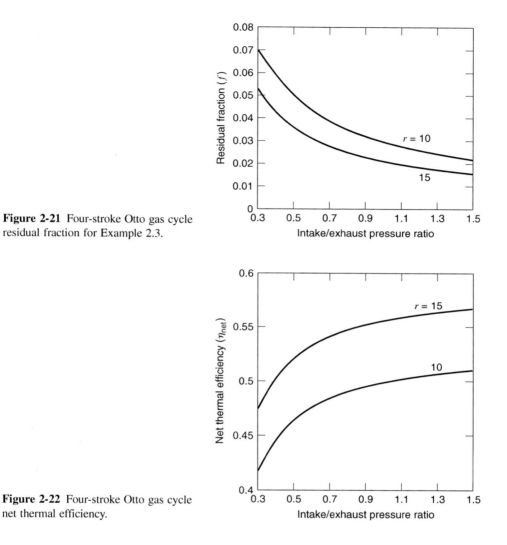

Figure 2-21 Four-stroke Otto gas cycle residual fraction for Example 2.3.

Figure 2-22 Four-stroke Otto gas cycle net thermal efficiency.

the volumetric efficiency and thermal efficiency decrease, and the residual fraction increases. The dependence of e_v on compression ratio is reversed for the throttled and supercharged conditions. In addition, the residual gas fraction increases. The increase in residual fraction is due to the decrease in the intake mass relative to the residual mass as the intake pressure is decreased.

2.8   DISCUSSION OF GAS CYCLE MODELS

Maximizing the mean effective pressure is important in engine design so that one can build a smaller, lighter engine to produce a given amount of work. As shown in Equation 2-8, there are evidently two ways to do this: (1) to increase the compression ratio, r, and (2) to increase the heat input, Q_{in}. However, there are practical limitations to these approaches. For spark ignition engines of conventional design, the compression ratio must be low enough to avoid engine knock; whereas for diesel engines increasing engine friction limits the utility of increasing compression ratio. Other more complicated factors influence the selection of compression ratio, especially constraints imposed by emission standards and, for some diesel engines, problems of startability.

One might expect that we can increase Q_{in} by increasing the fuel flow rate delivered to an engine. As we shall see in our studies of fuel-air cycles in Chapter 4, this is not always correct, for in fuel-rich mixtures not all of the fuel energy is used since there is not enough oxygen to burn the carbon monoxide to carbon dioxide nor the hydrogen to water. The fuel-air cycle predicts that the efficiency decreases the richer the mixture is beyond stoichiometric.

According to the gas cycles, and to the fuel-air cycles to be discussed later, the efficiency is greatest if heat can be added at constant volume:

$$\eta_{Otto} > \eta_{dual} > \eta_{Diesel} \tag{2.62}$$

Why then do we build engines that resemble constant pressure heat addition when we recognize that constant volume heat addition would be better? To illustrate how complicated that question is let us ask the following: Suppose that the maximum pressure in the cycle must be less than P_{max}. How should the heat be added to produce the required work? The answer is now

$$\eta_{Diesel} > \eta_{dual} > \eta_{Otto} \tag{2.63}$$

This can be demonstrated with the aid of the temperature-entropy diagram. If the Otto cycle and the Diesel cycle are drawn on such a diagram so that the work done in each cycle is the same, it can then be seen that the diesel is rejecting less heat and must therefore be the most efficient.

We close this chapter with discussion of an issue with gas cycle analysis: How does one choose γ and q_{in}? For hydrocarbon fuels one can choose

$$\gamma = 1.4 - 0.16 \phi \tag{2.64}$$

$$q_{in}\left(\frac{kJ}{kg_{mix}}\right) = \frac{m_f}{m} q_c = \frac{\phi F_s}{1 + \phi F_s} q_c \quad \text{for} \quad \phi \le 1$$

$$= \frac{\phi F_s}{1 + \phi F_s}[q_c - 3890(\phi - 1)] \quad \text{for} \quad \phi > 1 \tag{2.65}$$

where q_c is the heat of combustion of the fuel, $\phi = F/F_s$ the ratio of the actual fuel-air ratio F to the stoichiometric fuel-air ratio F_s. These three parameters are defined and used in the combustion analysis of Chapter 3. Equation 2.65 accounts for the fact that in rich combustion not all the heat of combustion is released. Equations 2.64 and 2.65 approximate the results of a more rigorous fuel-air combustion analysis instead of an equivalent heat release.

In fuel-air cycle analysis the equivalence ratio, ϕ is specified. The gas cycle equivalent is the specification of the heat added per unit mass of gas mixture inducted, q_{in}. The heat input to the gas cycle, q_{in}, can be estimated from Equations 2.64 and 2.65. The mass of gas inducted can be determined using the ideal gas law with known cylinder volume, inlet pressure and temperature, and residual fraction.

2.9 REFERENCES

HEYWOOD, J. (1988), *Internal Combustion Engine Fundamentals*, McGraw-Hill, New York.

TAYLOR, C. F. (1985), *The Internal Combustion Engine in Theory and Practice*, Vol. 1, MIT Press, Cambridge, Massachusetts.

VAN WYLEN, G. I. and R. E. SONNTAG (1994), *Fundamentals of Classical Thermodynamics*, Wiley, New York.

2.10   HOMEWORK

2.1 (a) Show for an Otto cycle that $T_3/T_2 = T_4/T_1$.

(b) Derive the Otto cycle efficiency equation, Equation 2.7.

2.2 Derive the Otto and Diesel cycle imep equation, Equation 2.8.

2.3 (a) Show that for a Diesel cycle $(T_3/T_2)^\gamma = T_4/T_1$.

(b) Derive Equation 2.12 and Equation 2.13.

2.4 For equal maximum temperature and work done, which cycle will be more efficient, the Diesel or Otto? The two cycles should have a common state corresponding to the start of compression.

2.5 What does the compression ratio of a Diesel cycle need to be to have the same thermal efficiency of an Otto cycle engine that has a compression ratio, $r = 9$?

2.6 Show that for the Otto and Diesel cycles as $r \to 1$, $\mathrm{imep}/P_1 \to Q_{in}(\gamma - 1)/P_1 V_1$ (use l'Hopital's rule).

2.7 A engine is modeled with a Dual cycle. The maximum pressure is to be 8000 kPa. The compression ratio is 17:1, the inlet conditions are 101 kPa and 320 K, and the nondimensional heat input $Q_{in}/P_1 V_1 = 30$. Find the thermal efficiency and the values of α and β.

2.8 (a) Derive the equation for the Miller cycle efficiency, Equation 2.20.

(b) Derive the equation for the Miller cycle imep, Equation 2.21.

2.9 For Otto and Miller cycles that have equal imep and compression ratios $(r_c = 10)$, what are the respective thermal efficiencies? Assume that the parameter, λ, is equal to 1.5 for the Miller cycle, the specific heat ratio $\gamma = 1.3$, and $Q_{in}/P_1 V_1 = 30$.

2.10 Develop a complete expansion cycle model in which the expansion stroke continues until the pressure is atmospheric. Derive an expression for the efficiency in terms of γ, $\alpha = V_4/V_3$, and $\beta = V_1/V_4$.

2.11 Using the heat release fraction applet,

(a) Plot the Wiebe heat release fraction curve for the following form factor values: $n = 1, 2, 3,$ and 4.

(b) At what crank angle is 0.10, 0.50, and 0.90 of the heat released?

For (a) and (b), assume that $a = 5$, the beginning of heat addition is $-10°$, and the duration of heat addition is $40°$.

2.12 Using the finite heat release applet, determine the effect of heat release duration on the net work, power, mean effective pressure, and thermal efficiency for a range of heat release duration of 40, 30, 20, 10, and 5°. Assume that the fuel is gasoline at stoichiometric conditions, the spark angle remains constant at $-10°$ atdc, $a = 5$, and $n = 3$. The engine bore and stroke are 0.095 m, the connecting rod length is 0.15 m, and the compression ratio is 9:1.

2.13 If a four cylinder unthrottled Otto cycle engine is to generate 100 kW at an engine speed of 2500 rpm, what should its bore and stroke be? Assume a square block with equal bore and stroke, and connecting rod length 1.5 × stroke. The fuel is gasoline at stoichiometric conditions, and standard atmospheric inlet conditions with a compression ratio of 10:1. The spark angle is $-10°$, the combustion duration is $40°$, $a = 5$, and $n = 3$. Hint: use the single cylinder finite heat release applet, and solve for a power output of 25 kW.

2.14   Develop a four-stroke Diesel cycle model (along the lines used in Example 2.3) with the following data: $r = 22$, $\gamma = 1.3$, $T_i = 300$ K, $P_i = 101$ kPa, $P_i/P_e = 0.98$, $M = 29$, and $q_{in} = 2090$ kJ/kg$_{gas}$.

2.15   Using the four-stroke gas Otto cycle applet, plot the effect of inlet throttling from 100 kPa to 25 kPa on the peak pressure, P_3. Assume the following conditions: $T_i = 300$ K, $r = 9$, $\gamma = 1.3$, and $q_{in} = 2400$ kJ/kg$_{gas}$.

2.16   If the engine in Example 2.3 were a four-cylinder, four-stroke engine with a 0.1 m bore and an 0.08 m stroke operating at 2000 rpm, how much indicated power (kW) would it produce? What if it were a two-stroke engine?

2.17   In Example 2.3, T_e is the exhaust temperature during the constant pressure exhaust stroke. It is not the same as the average temperature of the gases exhausted. Explain.

2.18   Derive Equation 2.65 for the $\phi < 1$ case.

Chapter 3

Fuel, Air, and Combustion Thermodynamics

3.1 INTRODUCTION

It has already been mentioned that an understanding of internal combustion engines will require a better thermodynamic model than that used in Chapter 2. In this chapter we review the thermodynamics of combustion and develop models suitable for application to internal combustion engines. The chapter starts with multicomponent ideal gas property models, followed by stoichiometry and computation of equilibrium combustion components and properties. The equilibrium combustion model is applied to adiabatic and isentropic processes. Finally, the heat of combustion and the adiabatic flame temperature are determined for a variety of fuel-air mixtures.

A few words about air are in order. The properties of air vary geographically, with altitude, and with time. We will assume that air is 21% oxygen and 79% nitrogen by volume, i.e., for each mole of O_2, there are 3.76 moles of N_2. Selected physical properties of air, oxygen, and nitrogen are given in Appendix A and B. Extension of our analyses to different air mixtures encountered in practice is straightforward. The most frequent differences accounted for are the presence of water and argon in air.

3.2 IDEAL GAS EQUATIONS OF STATE

The mass, m, of a mixture is the sum of the mass of all n components

$$m = \sum_{i=1}^{n} m_i \quad \text{(kg)} \tag{3.1}$$

The mass fraction, x_i, of any given species is defined as

$$x_i = m_i/m \tag{3.2}$$

and it should be clear that

$$\sum_{i=1}^{n} x_i = 1 \tag{3.3}$$

The internal energy U of a mixture is the sum of the internal energy u_i of all n components

$$U = \sum_{i=1}^{n} m_i u_i \quad \text{(kJ)} \tag{3.4}$$

The mass intensive internal energy u is

$$u = \sum_{i=1}^{n} x_i u_i \quad \text{(kJ/kg)} \tag{3.5}$$

Analogous relations for the enthalpy H are

$$H = \sum_{i=1}^{n} m_i h_i \quad \text{(kJ)} \tag{3.6}$$

$$h = \sum_{i=1}^{n} x_i h_i \quad \text{(kJ/kg)} \tag{3.7}$$

The total number of moles, N, of a mixture is the sum of moles of all n components

$$N = \sum_{i=1}^{n} n_i \tag{3.8}$$

and the mole fraction y_i of any given species is

$$y_i = \frac{n_i}{N} \tag{3.9}$$

The internal energy of a mixture can also be written as

$$U = \sum_{i=1}^{n} n_i u_i \quad \text{(kJ)} \tag{3.10}$$

where the molar intensive properties are denoted with an overbar. The molar intensive internal energy is thus

$$\bar{u} = \sum_{i=1}^{n} y_i \bar{u}_i \quad \text{(kJ/kmol)} \tag{3.11}$$

Likewise the enthalpy is

$$H = \sum_{i=1}^{n} n_i \bar{h}_i \quad \text{(kJ)} \tag{3.12}$$

$$\bar{h} = \sum_{i=1}^{n} y_i \bar{h}_i \quad \text{(kJ/kmol)} \tag{3.13}$$

Notice that capital letters have been adopted for extensive variables and lowercase letters are reserved for intensive variables. The molecular weight, M, of a mixture

$$M = \sum_{i=1}^{n} y_i M_i \quad \text{(kg/kmol)} \tag{3.14}$$

is the conversion factor required between molar intensive and mass intensive units. For example, the mass intensive (specific) gas constant R is related to the molar intensive (universal) gas constant R_u by

$$R = \frac{R_u}{M} \tag{3.15}$$

The familiar relationships between pressure P, temperature T, and volume V are

$$\begin{aligned} PV &= N R_u T \\ PV &= m R T \\ Pv &= R T \end{aligned} \tag{3.16}$$

The entropy S of a mixture is also the sum of the entropy of each component

$$S = \sum_{i=1}^{n} m_i \, s_i = \sum_{i=1}^{n} n_i \, \bar{s}_i \quad (\text{kJ/K}) \qquad (3.17)$$

but unlike enthalpy, where the enthalpy of a component is evaluated at the total pressure ($h_i = u_i + Pv_i$), the entropy of a component is evaluated at its partial pressure. The partial pressure of a component is

$$P_i = y_i \, P \quad (\text{kPa}) \qquad (3.18)$$

and the entropy s_i of any component is

$$s_i = s_i^o - R_i \ln (P_i/P_o) \quad (\text{kJ/kg K}) \qquad (3.19)$$

where s_i^o depends only on temperature and is the entropy of that component when $P_i = P_o$. Substitution of Equation 3.19 into Equation 3.17 yields the convenient relations

$$s = -R \ln (P/P_o) + \sum_{i=1}^{n} x_i (s_i^o - R_i \ln y_i) \quad \left(\frac{\text{kJ}}{\text{kg K}}\right) \qquad (3.20)$$

$$\bar{s} = -R_u \ln (P/P_o) + \sum_{i=1}^{n} y_i (\bar{s}_i^o - R_u \ln y_i) \quad \left(\frac{\text{kJ}}{\text{kmol K}}\right) \qquad (3.21)$$

Thermodynamic data for elements, combustion products, and many pollutants are available in a compilation published by the National Bureau of Standards, called the JANAF Tables (1971). For single component fuels, the data presented by Stull, Westrum, and Sinke (1969) is in the same format as that of the JANAF Tables. In addition to these two references, a compilation by Rossini (1953) is useful for hydrocarbon fuels at temperatures as high as 1500 K. Tabular data for ideal gases at $P = 1$ bar is given in Appendix B. The tables in Appendix B provide the molecular weight, enthalpy of formation ($\bar{h}_f^o$), and tabular data of the molar enthalpy and entropy.

The enthalpy of formation is the enthalpy of a substance at a specified state due to its composition. The standard reference state is $T = 298$ K and $P = 1$ bar. For compounds, the enthalpy of formation is the enthalpy required to form the compound from its elements in their stable state. The enthalpy of formation of the stable form of the elements hydrogen H_2, oxygen O_2, nitrogen N_2, and solid carbon $C(s)$ is assigned a value of zero at $T = 298$ K. The enthalpy of formation of CO_2 is $-393,522$ kJ/kmol or -8942 kJ/kg. The negative sign indicates that the enthalpy of CO_2 at the standard reference state is less than the enthalpy of C and O_2 at the same state. Similarly, the enthalpy of formation of H_2O vapor is $-241,826$ kJ/kmol or $-13,424$ kJ/kg.

EXAMPLE 3.1   *Properties of Gas Mixtures*

Compute the molecular weight, enthalpy (kJ/kg), and entropy (kJ/kg K) of a gas mixture at $P = 2000$ kPa and $T = 1000$ K. The constituents and their mole fractions are:

Species	y_i
CO_2	0.109
H_2O	0.121
N_2	0.694
CO	0.0283
H_2	0.0455

SOLUTION   Using the tabular data in Appendix B, the following table was prepared:

Species	y_i	M_i	$\bar{h}_f^o$	$\bar{h}_i - \bar{h}_f^o$	$\bar{s}_i^o$	$y_i(\bar{s}_i^o - R_u \ln y_i)$
CO_2	0.109	44.01	−393,522	33,397	269.30	31.36
H_2O	0.121	18.015	−241,826	26,000	232.74	30.28
N_2	0.694	28.013	0	21,463	228.17	160.05
CO	0.0283	28.01	−110,527	21,686	234.54	7.48
H_2	0.0455	2.016	0	20,663	166.22	8.73

molar enthalpy: $\bar{h} = \Sigma\, y_i\, \bar{h}_i = -52{,}047$ kJ/kmol
molar entropy: $\bar{s} = -R_u \ln (P/P_0) + \Sigma\, y_i\, (\bar{s}_i^o - R_u \ln y_i) = 213.5$ kJ/kmol K
molecular weight: $M = \Sigma\, y_i\, M_i = 27.3$

Therefore,

$$h = \bar{h}/M = -1906 \text{ kJ/kg}$$

and

$$s = \bar{s}/M = 7.82 \text{ kJ/kg K}$$

For computer calculations it is awkward to deal with tabular data. For this reason the specific heats are curve-fitted to polynomials by minimizing the least squares error. The function we will employ for any given species is

$$\frac{c_p}{R} = \frac{\bar{c}_p}{R_u} = a_1 + a_2 T + a_3 T^2 + a_4 T^3 + a_5 T^4 \tag{3.22}$$

It follows that the enthalpy and entropy at atmospheric pressure are

$$\frac{h}{RT} = \frac{\bar{h}}{R_u T} = a_1 + \frac{a_2}{2} T + \frac{a_3}{3} T^2 + \frac{a_4}{4} T^3 + \frac{a_5}{5} T^4 + \frac{a_6}{T} \tag{3.23}$$

$$\frac{s^o}{R} = \frac{\bar{s}^o}{R_u} = a_1 \ln T + a_2 T + \frac{a_3}{2} T^2 + \frac{a_4}{3} T^3 + \frac{a_5}{4} T^4 + a_7 \tag{3.24}$$

where a_6 and a_7 are constants of integration determined by matching the enthalpy and entropy at some reference temperature. Values of the curve-fit constants for several species of interest in combustion, CO_2, H_2O, N_2, O_2, CO, H_2, H, O, OH, and NO are given in Appendix C. Similar curve-fit coefficients for several fuels are also given in Appendix C. The mass-specific enthalpies of CO_2 and H_2O given by Equation 3.23 are plotted versus temperature in Figure 3-1. Note that the slope of the H_2O curve is steeper than that of the CO_2 curve, as a result of the greater specific heat of the water vapor.

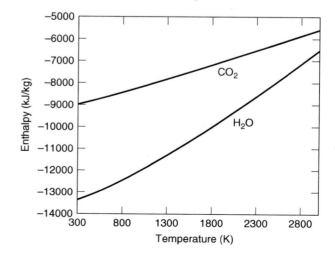

Figure 3-1 Enthalpy versus temperature curve-fits for CO_2 and H_2O.

3.3   LIQUIDS AND LIQUID-VAPOR-GAS MIXTURES

The thermodynamics involved with fuel injection, carburetion, water injection, and water condensation can be complicated. Fortunately, we can make some simplifications that are quite accurate for our intended use.

First, let us consider a pure substance in terms of its compressed liquid, saturated liquid, saturated vapor, and superheated vapor states. The simplifications that we will introduce are:

1. Compressed and saturated liquids are incompressible.
2. Saturated and superheated vapors are ideal gases.

For an incompressible substance, it can be shown that the internal energy and entropy depend only on temperature. Hence the approximation for compressed liquids can be

$$u = u_f(T) \quad \text{(kJ/kg)} \tag{3.25}$$
$$s = s_f(T) \quad \text{(kJ/kg K)} \tag{3.26}$$

where the notation $u_f(T)$ and $s_f(T)$ denote the internal energy and entropy of saturated liquid at the temperature T.

The enthalpy of a compressed liquid depends on pressure, and it is consistent with Equations 3.25 and 3.26 to assume that

$$h = h_{fo}(T) + (P - P_{\text{atm}})\, v \quad \text{(kJ/kg)} \tag{3.27}$$

where h_{fo} is the enthalpy of the compressed liquid at atmospheric pressure. The only property remaining to be prescribed is the specific volume. Let us choose it to be the specific volume of compressed liquid at atmospheric pressure as these data are readily available.

$$v = v_o(T_1) \quad \text{(m}^3\text{/kg)} \tag{3.28}$$

where T_1 is the initial temperature in the process being analyzed.

To introduce the enthalpy of vaporization into our equations of state for the liquids is convenient since these data are usually easier to find than the saturated liquid data. We then have

$$h = h_g - h_{fg} + (P - P_{\text{atm}})\, v \tag{3.29}$$

Unlike specific volume, data for the enthalpy of vaporization at saturation pressure are readily available. Hence we choose

$$h_{fg} = h_{fg}(T_1) \tag{3.30}$$

where, again, T_1 is the temperature at the start of the process being analyzed.

Typically

$$|P - P_{\text{atm}}|\, v << |h_{fg}| \quad \text{and} \quad |Pv| << |h| \tag{3.31}$$

so that in many cases we can write for liquids

$$h \approx h_g - h_{fg} \tag{3.32}$$
$$u \approx h \tag{3.33}$$

Table 3-1 gives the ideal gas molar enthalpy of formation, molar enthalpy of vaporization, the saturation pressure, and the specific volume of compressed liquid, for several liquid fuels and water at $T = 298$ K. Table 3-2 gives most of the same information for octane but as a function of temperature.

Table 3-1   Enthalpy of Formation, Enthalpy of Vaporization, Saturation Vapor Pressure, and Specific Volume of Some Liquid Fuels at $T = 298$ K

Formula	Name	$\bar{h}_f^o$ (kJ/mole)	$\bar{h}_{fg}$ (kJ/mole)	P_{sat} (bar)	v_o (m³/kg)
CH_3NO_2	Nitromethane	−74.73	38.37	0.050	0.879×10^{-3}
CH_4O	Methanol	−201.17	37.92	0.186	1.264×10^{-3}
C_2H_6O	Ethanol	−234.81	42.34	0.084	1.267×10^{-3}
$C_4H_{10}O$	Ethyl ether	−252.21	26.53	0.733	1.413×10^{-3}
C_5H_{12}	n-Pentane	−146.00	26.44	0.710	1.597×10^{-3}
C_6H_{14}	Hexane	−167.19	31.49	0.242	1.514×10^{-3}
C_6H_6	Benzene	+82.93	34.02	0.129	1.138×10^{-3}
C_7H_{17}	Gasoline	−267.12	38.51		1.449×10^{-3}
C_8H_{18}	Octane	−208.45	41.51	0.022	1.423×10^{-3}
C_8H_{18}	Isooctane	−224.14	35.11	0.071	1.445×10^{-3}
C_8H_{10}	Ethylbenzene	+29.29	42.01	0.013	1.153×10^{-3}
$C_{12}H_{26}$	n-Dodecane	−290.87	61.32	< 0.001	1.336×10^{-3}
$C_{14}H_{30}$	Tetradecane	−332.13	71.24	< 0.001	1.311×10^{-3}
$C_{14.4}H_{24.9}$	Diesel fuel	−100.00	74.08		1.176×10^{-3}
$C_{16}H_{34}$	Hexadecane (cetane)	−373.34	81.14	< 0.001	1.293×10^{-3}
$C_{19}H_{40}$	Nonadecane	−435.14	95.03	< 0.001	1.286×10^{-3}
H_2O	Water	−241.83	44.02	0.031	1.000×10^{-3}

Source: P_{sat} and v^o are from the *CRC Handbook of Chemistry and Physics* (1975–1976); $\bar{h}_{fg}$ is from N. B. Vargaflik (1975); and $\bar{h}_f^o$ is from Stull et al. (1969).

Table 3-2   Enthalpy of Vaporization, Saturation Vapor Pressure and Atmospheric Specific Volume of Octane

T (K)	$\bar{h}_{fg}$ (kJ/mole)	P_{sat} (bar)	v_o (m³/kg)
298	41.50	0.018	1.432×10^{-3}
325	39.99	0.075	1.479×10^{-3}
350	38.22	0.211	1.527×10^{-3}
375	36.45	0.503	1.579×10^{-3}
400	34.46	1.058	1.639×10^{-3}
425	32.35	2.003	1.708×10^{-3}
450	30.07	3.500	1.789×10^{-3}
475	27.49	5.707	1.890×10^{-3}
500	24.32	8.837	2.022×10^{-3}
550	20.35	13.21	2.214×10^{-3}

Source: Vargaftik (1975).

We will also deal with mixtures of gases in contact with a liquid phase. In these cases we assume:

1. The liquid contains no dissolved gases.
2. The gases are ideal.
3. At equilibrium the partial pressure of the liquid's vapor is equal to the saturation pressure corresponding to the mixture temperature.

Let χ denote the quality of the condensable substance, that is, the ratio of vapor to liquid and vapor. The enthalpy of the system is then

$$H = (1 - \chi) m_1 (h_{g,1} - h_{fg}) + \chi m_1 h_1 + \sum_{i=2}^{n} m_i h_i \qquad (3.34)$$

or

$$h = \sum_{i=1}^{n} x_i h_i - (1 - \chi) x_1 h_{fg} \qquad (3.35)$$

The indexing is chosen so that $i = 1$ corresponds to the condensable substance and $i = 2, \ldots, n$ corresponds to all other gases.

Likewise, the system volume is

$$V = (1 - \chi) m_1 (v_{g,1} - v_{fg}) + \chi m_1 v_g + \sum_{i=2}^{n} m_i v_i \qquad (3.36)$$

and the specific volume is

$$v = \sum_{i=1}^{n} x_i v_i - (1 - \chi) x_1 v_{fg} \qquad (3.37)$$

where the v_i are computed at the total pressure (*Amagat-Leduc Law of Additive Volumes*).

3.4 STOICHIOMETRY AND LOW TEMPERATURE COMBUSTION MODELING

Let us represent the chemical formula of a fuel as $C_\alpha H_\beta O_\gamma N_\delta$. A stoichiometric reaction is defined such that the fuel burns completely and the only products are carbon dioxide and water. Let us write the following stoichiometric reaction

$$C_\alpha H_\beta O_\gamma N_\delta + a_s(O_2 + 3.76 \, N_2) \rightarrow n_1 \, CO_2 + n_2 \, H_2O + n_3 \, N_2 \qquad (3.38)$$

and solve for a_s, the stoichiometric molar air-fuel ratio, and the moles $n_i \, (i = 1, 2, 3)$ that describe the product composition. Note that the stoichiometric reaction is expressed per mole of fuel. We know that atoms are conserved, so we can write

$$\begin{array}{ll} \text{C:} & \alpha = n_1 \\ \text{H:} & \beta = 2n_2 \\ \text{O:} & \gamma + 2a_s = 2n_1 + n_2 \\ \text{N:} & \delta + 2 \cdot 3.76 \cdot a_s = 2n_3 \end{array} \qquad (3.39)$$

Solution of these four equations gives

$$\begin{aligned} a_s &= \alpha + \frac{\beta}{4} - \frac{\gamma}{2} \\ n_1 &= \alpha \\ n_2 &= \frac{\beta}{2} \\ n_3 &= \frac{\delta}{2} + 3.76\left(\alpha + \frac{\beta}{4} - \frac{\gamma}{2}\right) \end{aligned} \qquad (3.40)$$

The stoichiometric air-fuel ratio A_s is

$$A_s = \frac{28.85 \, (4.76 \, a_s)}{(12.01 \, \alpha + 1.008 \, \beta + 16.00 \, \gamma + 14.01 \, \delta)} \qquad (3.41)$$

Table 3-3 Molecular Weights, Stoichiometric Ratios, and Combustion Product Mole Fractions

Fuel	Chemical Formula	M	A_s	F_s	a_s	y_{CO_2}	y_{H_2O}	y_{N_2}
Methane	CH_4	16.04	17.12	0.0584	2.00	0.095	0.190	0.715
Propane	C_3H_8	44.09	15.57	0.0642	5.00	0.116	0.155	0.729
Gasoline	C_7H_{17}	101.21	15.27	0.0655	11.25	0.121	0.147	0.732
Octane	C_8H_{18}	114.22	15.03	0.0665	12.50	0.125	0.141	0.734
Diesel	$C_{14.4}H_{24.9}$	198.04	14.30	0.0699	20.63	0.138	0.119	0.743
Eicosane	$C_{20}H_{40}$	280.52	14.69	0.0681	30.00	0.131	0.131	0.738
Methanol	CH_4O	32.04	6.43	0.1556	1.50	0.116	0.231	0.653
Ethanol	C_2H_6O	46.07	8.94	0.1118	3.00	0.123	0.184	0.693
Nitromethane	CH_3NO_2	61.04	1.69	0.5927	0.75	0.158	0.237	0.604
Hydrogen	H_2	2.02	34.06	0.0294	0.50	0.000	0.347	0.653
Acetylene	C_2H_2	26.04	13.19	0.0758	2.50	0.161	0.081	0.758
Ammonia	NH_3	17.03	6.05	0.1654	0.75	0.000	0.282	0.718
Cyanogen	C_2N_2	52.04	5.28	0.1895	2.00	0.174	0.000	0.826

Typical values are given in Table 3-3. Notice that for hydrocarbons ($\gamma = \delta = 0$) the stoichiometric air-fuel ratio by mass is not a particularly strong function of type. For typical hydrocarbon fuels the stoichiometric air-fuel ratio, A_s, is about 15. Table 3-3 also lists the molar stoichiometric air-fuel ratio, and the product mole fractions.

The mole and mass fraction of fuel in a stoichiometric fuel-air mixture are as follows:

$$y_s = \frac{1}{1 + 4.76\, a_s} \qquad x_s = \frac{1}{1 + A_s} \qquad (3.42)$$

The fuel-air equivalence ratio, ϕ, is defined as the stoichiometric air-fuel ratio, A_s, divided by the actual air-fuel ratio, A, or equivalently, the actual fuel-air ratio, F, divided by the stoichiometric fuel-air ratio, F_s:

$$\phi = \frac{A_s}{A} = \frac{F}{F_s} \qquad (3.43)$$

The equivalence ratio has the same value on a mole or mass basis. If $\phi < 1$ the mixture is called lean, if $\phi > 1$ the mixture is said to be rich, and if $\phi = 1$ the mixture is said to be stoichiometric.

EXAMPLE 3.2 *Molecular Weight of Gas Mixtures*

(a) What is the molecular weight M (kg/kmole) of a stoichiometric mixture of octane ($C_8 H_{18}$) and air?

(b) If the mixture fills a cylinder volume of $9.0 \times 10^{-4}\ m^3$ at a pressure $P = 100\ kPa$ and temperature $T = 300\ K$, what is the mass of the fuel-air mixture in the cylinder?

SOLUTION (a) The molecular weight of air is $M_a = 28.97$ kg/kmol. From Table 3-3, the molecular weight M_f of the octane is 114.22 kg/kmol, and the stoichiometric mass fuel-air ratio F_s is 0.0665.

$$y_f = \frac{n_f}{N} = \frac{n_f}{n_a + n_f} = \frac{m_f/M_f}{\dfrac{m_a}{M_a} + \dfrac{m_f}{M_f}}$$

$$= \left(1 + \frac{1}{F_s \dfrac{M_a}{M_f}} \right)^{-1} = 0.0166$$

$$y_a = 1 - y_f = 0.9834$$

$$M = y_a M_a + y_f M_f = 30.39 \text{ kg/kmol}$$

(b)
$$m = \frac{PVM}{R_u T} = \frac{(100)(9.0 \times 10^{-4})(30.39)}{(8.314)(300)}$$

$$= 1.097 \text{ g}$$

So for an automotive engine, there is typically a gram of combustible mixture in the cylinder.

At low temperatures ($T < 1000$ K, such as in the exhaust) and carbon to oxygen ratios less than one, the overall combustion reaction for any equivalence ratio can be written

$$C_\alpha H_\beta O_\gamma N_\delta + \frac{a_s}{\phi}(O_2 + 3.76 \, N_2) \rightarrow \tag{3.44}$$
$$n_1 \, CO_2 + n_2 \, H_2O + n_3 \, N_2 + n_4 \, O_2 + n_5 \, CO + n_6 \, H_2$$

For reactant C/O ratios greater than one we would have to add solid carbon C(s) and several other species to the product list, as we shall see. This equation assumes the dissociation of molecules is negligible, and a more general case is treated in Section 3.6.

For lean combustion ($\phi < 1$) products at low temperature, we will assume no product CO and H_2, i.e., $n_5 = n_6 = 0$. In this case atom balance equations are sufficient to determine the product composition since there are four equations and four unknowns.

For rich combustion we will assume that there is no product O_2, i.e., $n_4 = 0$. In this case there are five unknowns, so we need an additional equation to supplement the four atom balance equations. We assume equilibrium considerations between the product species CO_2, H_2O, CO, and H_2 determine the product composition. This reaction is termed the water-gas reaction:

$$CO_2 + H_2 \rightleftharpoons CO + H_2O \tag{3.45}$$

with the equilibrium constant providing the fifth equation:

$$K(T) = \frac{n_2 \, n_5}{n_1 \, n_6} \tag{3.46}$$

The equilibrium constant K(T) equation is a curve fit of JANAF Table data for $400 < T < 3200$ K:

$$\ln K(T) = 2.743 - \frac{1.761}{t} - \frac{1.611}{t^2} + \frac{0.2803}{t^3} \quad \left(t = \frac{T}{1000} \right) \tag{3.47}$$

Solutions for both rich and lean cases are given in Table 3-4. In the rich case, the parameter n_5 is given by the solution of the quadratic equation

$$n_5 = \frac{-b_1 + \sqrt{b_1^2 - 4a_1c_1}}{2a_1} \tag{3.48}$$

Table 3-4   Low Temperature Combustion Products

Species	n_i	$\phi \leq 1$	$\phi > 1$
CO_2	n_1	α	$\alpha - n_5$
H_2O	n_2	$\beta/2$	$\beta/2 - d_1 + n_5$
N_2	n_3	$\delta/2 + 3.76\, a_s/\phi$	$\delta/2 + 3.76\, a_s/\phi$
O_2	n_4	$a_s(1/\phi - 1)$	0
CO	n_5	0	n_5
H_2	n_6	0	$d_1 - n_5$

The a_1, b_1, c_1 coefficients are given by

$$a_1 = 1 - K$$
$$b_1 = \frac{\beta}{2} + K\alpha - d_1(1 - K)$$
$$c_1 = -\alpha d_1 K \tag{3.49}$$
$$d_1 = 2\, a_s\left(1 - \frac{1}{\phi}\right)$$

In reciprocating engines there is residual gas mixed with the fuel and air. We need to determine the composition of such a fuel-air-residual gas mixture for analysis of the compression stroke and later the unburned mixture ahead of the flame. The residual gas is assumed to be at a low enough temperature so that the species relations in Table 3-4 specify its composition.

The fuel-air-residual gas mixture will contain both reactants and products. Let us rewrite the combustion equation as

$$n_0' C_\alpha H_\beta O_\gamma N_\delta + n_4' O_2 + n_3' N_2 \rightarrow$$
$$n_1'' CO_2 + n_2'' H_2O + n_3'' N_2 + n_4'' O_2 + n_5'' CO + n_6'' H_2 \tag{3.50}$$

where

n_i' = reactant coefficient
n_i'' = product coefficient

Adopting similar notation for other symbols, it should be clear that for a mixture of residual gas and premixed fuel-air

$$x_i = (1 - f)\, x_i' + f x_i'' \qquad i = 0, 6 \tag{3.51}$$

The mole fractions y_i are given by

$$y_i = (1 - y_r)\, y_i' + y_r\, y_i'' \qquad i = 0, 6 \tag{3.52}$$

and the residual mole fraction, y_r, is

$$y_r = \left[1 + \frac{M''}{M'}\left(\frac{1}{f} - 1\right)\right]^{-1} \tag{3.53}$$

With the composition of the fuel-air-residual gas mixture known, the thermodynamic properties of the mixture are found by application of the property relations of Section 3.2. This is illustrated by the following example.

EXAMPLE 3.3   *Low Temperature Combustion*

What is the mole fraction (ppm) of CO produced when C_8H_{18} is burned at $\phi = 1.2$ and $T = 1000$ K?

SOLUTION   An equivalence ratio of $\phi = 1.2$ is a rich mixture, so the product concentration of O_2 is assumed to be zero, i.e., $n_4 = 0$. The calculation proceeds as follows:

$$\alpha = 8, \beta = 18, \gamma = \delta = 0$$
$$a_s = \alpha + \beta/4 - \gamma/2 = 12.5$$
$$d_1 = 2\,a_s(1 - 1/\phi) = 4.167$$
$$t = T/1000 = 1$$
$$\ln K = 2.743 - 1.761/t - 1.611/t^2 + 0.2803/t^3 = -0.34 \Rightarrow K = 0.705$$
$$a_1 = 1 - K = 0.295$$
$$b_1 = \beta/2 + \alpha K - d_1 (1 - K) = 13.41$$
$$c_1 = -\alpha d_1 K = -23.50$$
$$n_5 = [-b_1 + (b_1^2 - 4a_1c_1)^{1/2}]/2a_1 = 1.690$$
$$n_1 = \alpha - n_5 = 6.310$$
$$n_2 = \beta/2 - \delta + n_5 = 6.523$$
$$n_3 = \delta/2 + 3.76\,a_s/\phi = 39.167$$
$$n_6 = d_1 - n_5 = 2.477$$
$$N = \Sigma n_i = 56.167$$

The reaction equation is

$$C_8H_{18} + 10.42\,(O_2 + 3.76\,N_2) \rightarrow 6.310\,CO_2 + 6.523\,H_2O + 39.167\,N_2$$
$$+ 1.690\,CO + 2.477\,H_2$$

and

$$y_5 = n_5/N = 0.03009 \cong 30{,}000 \text{ ppm}$$

3.5   GENERAL CHEMICAL EQUILIBRIUM

In general, we often consider a combustion problem that has many product species. The fuel is initially mixed with air with an equivalence ratio ϕ. After combustion, the products of reaction are assumed to be in equilibrium at temperature T and pressure P. The composition and thermodynamic properties of the products are to be determined. The overall combustion reaction per mole of fuel is:

$$C_\alpha H_\beta O_\gamma N_\delta + \frac{a_s}{\phi}(O_2 + 3.76\,N_2) \rightarrow$$

$$n_1\,CO_2 + n_2\,H_2O + n_3\,N_2 + n_4\,O_2 + n_5\,CO + n_6\,H_2 \qquad (3.54)$$
$$+ n_7\,H + n_8\,O + n_9\,OH + n_{10}\,NO + n_{11}\,N + n_{12}\,C(s)$$
$$+ n_{13}\,NO_2 + n_{14}\,CH_4 + \cdots$$

The condition for equilibrium is usually stated in terms of thermodynamic functions such as the minimization of the Gibbs or Helmholtz free energy or the maximization of entropy. If temperature and pressure are used to specify a thermodynamic state, the Gibbs free energy is most easily minimized since temperature and pressure are its fundamental variables. For a product mixture of n species, the Gibbs free energy is

$$g = \sum_{j=1}^{n} \mu_j n_j \quad (kJ) \tag{3.55}$$

where the chemical potential, μ_j, of species j is defined by

$$\mu_j = \left(\frac{\partial g}{\partial n_j}\right)_{T,P,n_{i\neq j}} \quad (kJ/kmol) \tag{3.56}$$

The equilibrium state is determined by minimizing the Gibbs free energy subject to the constraints imposed by atom conservation, that is,

$$b'_i = \sum_{j=1}^{n} a_{ij} n_j \qquad i = 1, ..., l \tag{3.57a}$$

or

$$b_i - b'_i = 0 \qquad i = 1, ..., l \tag{3.57b}$$

where l is the number of atom types, a_{ij} is the number of atoms of element i in species j, b'_i is the number of atoms of element i in the reactants, and

$$b_i = \sum_{j=1}^{n} a_{ij} n_j \tag{3.58}$$

is the number of atoms of element i in the products.

Defining a term G to be

$$G = g + \sum_{i=1}^{l} \lambda_i (b_i - b'_i) \tag{3.59}$$

where λ_i are Lagrangian multipliers, the condition for equilibrium becomes

$$\delta G = \sum_{j=1}^{n} \left(\mu_j + \sum_{i=1}^{l} \lambda_i a_{ij}\right) \delta n_j + \sum_{i=1}^{l} (b_i - b'_i) \, \delta\lambda_i = 0 \tag{3.60}$$

Treating the variations δn_j and $\delta\lambda_i$ as independent

$$\mu_j + \sum_{i=1}^{l} \lambda_i a_{ij} = 0 \quad j = 1, ..., n \tag{3.61}$$

For ideal gases

$$\mu_j = \mu_j^o + R_u T \ln (n_j/N) + R_u T \ln (P/P_o) \tag{3.62}$$

so that

$$\frac{\mu_j^o}{R_u T} + \ln (n_j/N) + \ln (P/P_o) + \sum_{i=1}^{l} \pi_i a_{ij} = 0 \quad j = 1, ..., n \tag{3.63}$$

where

$$\pi_i = \lambda_i/R_u T \tag{3.64}$$

To determine the equilibrium composition using the Lagrange multiplier approach, we have to solve a set of $n + l + 1$ equations. For a given temperature and pressure (T, P) Equation 3.63 is a set of n equations for the n unknown n_j, l unknown π_i, and N.

Equation 3.57b provides an additional l equation and we close the set with

$$N = \sum_{j=1}^{n} n_j \qquad (3.65)$$

Once the composition of the products has been determined, we can now compute the thermodynamic properties of the equilibrium mixture. Recall that any two of the independent properties T, P, H, S, U, and V specify the thermodynamic state. For example, for constant pressure combustion, the enthalpy is known instead of the temperature. For this case we include an equation for the known enthalpy to our set of equations,

$$H = \sum_{j=1}^{n} n_j \overline{h}_j \qquad (3.66)$$

For an isentropic compression or expansion, or expansion to a specified pressure, the entropy is given instead of enthalpy or temperature. In this case we have

$$S = \sum_{j=1}^{n} n_j(\overline{S}_j^o - R_u \ln(n_j/N) - R_u \ln(P/P_o)) \qquad (3.67)$$

Finally, if in any case specific volume rather than pressure is known, then we have to minimize the Helmholtz free energy. In this case a similar analysis (Gordon and McBride, 1971) shows that Equation 3.63 is replaced by

$$\frac{\mu_j^o}{R_u T} + \ln(n_j/N) + \ln\left(\frac{RT}{P_o v}\right) + \sum_{i=1}^{l} \pi_i a_{ij} = 0 \quad j = 1, ..., n \qquad (3.68)$$

For constant volume combustion, the internal energy is known, so we include

$$U = \sum_{j=1}^{n} n_j (\overline{h}_j - R_u T) \qquad (3.69)$$

For an isentropic expansion or compression to a specified volume v we include

$$S = \sum_{j=1}^{n} n_j\left(\overline{s}_j^o - R_u \ln\left(\frac{n_j}{N}\right) - R_u \ln\left(\frac{RT}{P_o v}\right)\right) \qquad (3.70)$$

A summary of the appropriate sets of equations to solve for given thermodynamic variables follows in Table 3-5.

Solution of these problems for practical application requires numerical iteration on a computer. Fortunately, there are now several computer programs available. Thermodynamic properties can be computed using a NASA program called TRANS72 (Svehla and McBride, 1973) for hydrocarbon-air mixtures. The STANJAN program (Reynolds, 1986), which is PC based, uses the Lagrange multiplier approach for determination of the equilibrium mole fractions.

Table 3-5   Equations Required for Property Calculation

Given properties	Equations required
T, P	(3.57), (3.63), (3.65)
H, P	(3.57), (3.63), (3.65), (3.66)
S, P	(3.57), (3.63), (3.65), (3.67)
T, V	(3.57), (3.65), (3.68)
U, V	(3.57), (3.65), (3.68), (3.69)
S, V	(3.57), (3.65), (3.68), (3.70)

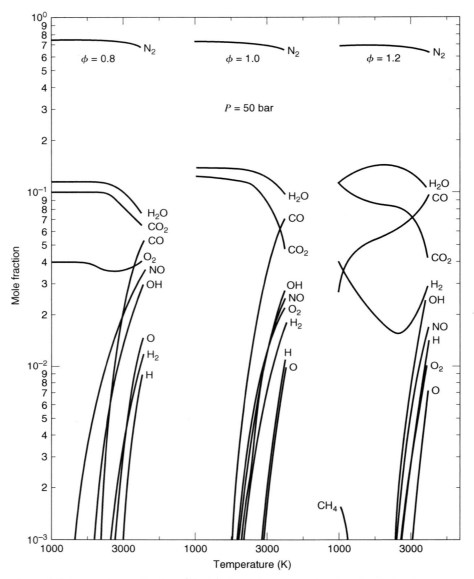

Figure 3-2 Composition of octane (C_8H_{18})-air mixtures at equilibrium for different temperatures at $\phi = 0.8$, 1.0, and 1.2.

Results illustrating composition shifts with temperature and equivalence ratio are given in Figure 3-2 and 3-3 for the combustion of C_8H_{18} at $P = 50$ bar. Composition as a function of temperature is shown in Figure 3-2. The largest mole fractions are N_2, H_2O, and CO_2. At this pressure, the composition predicted using Table 3-4 is a good approximation for temperatures less than about 2000 K. At lower pressures, dissociation is even greater, so that at atmospheric pressure, Table 3-4 is valid for temperatures less than about 1500 K. As the reaction temperature is increased above 1500 K, there is an exponential rise in product species such as CO, NO, OH, O_2, O, H_2, and H. For lean ($\phi < 1$) conditions, the O_2 fraction is relatively insensitive to temperature. For rich conditions, the H_2 mole fraction first decreases, then increases with increasing temperature.

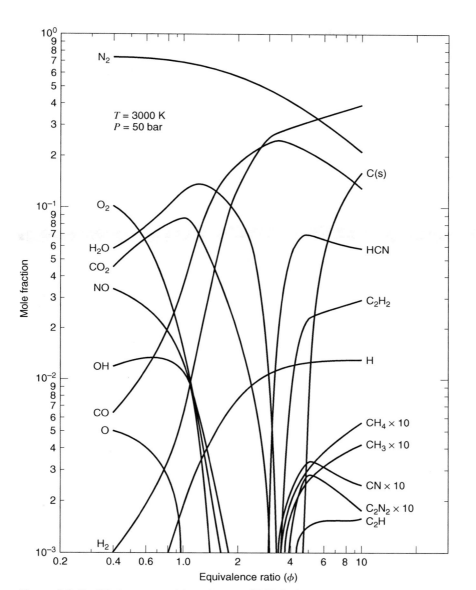

Figure 3-3 Equilibrium composition of octane (C_8H_{18})-air mixture at $T = 3000$ K, $P = 50$ bar.

Notice that at high temperatures there is a significant amount of nitric oxide (NO). If any gas in an engine cylinder is raised to these high temperatures, that gas will tend toward equilibrium at a rate determined by chemical kinetics. Since the chemistry for most species that contribute to the thermodynamic properties is fast enough relative to engine time scales, in many cases local equilibrium may be assumed. Nitric oxide, however, is significant even though its concentrations are relatively low because it is an air pollutant. Unlike the species of thermodynamic importance, its chemistry is not fast enough to assume that it is in equilibrium concentrations. Likewise, once formed, its chemistry freezes during the expansion stroke so that even in the low temperature exhaust gases nitric oxides are found. This will be discussed more fully when we deal with emissions.

Composition as a function of equivalence ratio is illustrated in Figure 3-3. The mole fraction behavior relative to equivalence ratio is complex. The results show the general trends expected from Figure 3-2 and Table 3-4. The product species CO and H_2 generally increase with equivalence ratio, while the O_2, NO, OH, and O mole fractions decrease.

If the equivalence ratio is greater than about 4, the product species list becomes quite large and includes solid carbon, C(s); hydrogen cyanide, HCN; acetylene, C_2H_2; and methane, CH_4. Thus, if anywhere in the cylinder there are fuel air pockets where $\phi > 3$, such as in diesel or stratified charge engines, there will be a tendency for these species to form. Similar to nitric oxides, they may freeze when mixed with leaner pockets or when the temperature drops, so these species can appear in the exhaust. With diesel engines, the maximum power is limited by the appearance of solid carbon (smoke and soot) in the exhaust even though the engine is running lean.

3.6 CHEMICAL EQUILIBRIUM USING EQUILIBRIUM CONSTANTS

This section presents a solution for the properties of equilibrium combustion products based on an equilibrium constant method applied by Olikara and Borman (1975) to the gas phase products of combustion of hydrocarbon fuels. Use of equilibrium constants is based on the minimization of the Gibbs free energy of the gas. The equilibrium constant method is simpler than the Lagrange multiplier approach when considering restricted species lists. However, the equilibrium constant method requires that the equilibrium reactions, such the water-gas reaction given by Equation 3.45, be specified.

Therefore, a more complete reaction calculation needs to be performed first to determine the significant product species to include in the equilibrium constant analysis. Inspection of Figures 3-2 and 3-3 shows that if $\phi < 3$, the only species of importance because of dissociation are O, H, OH, and NO. The species list in Equation 3.54 can be terminated at $i = 10$; that is, we need to consider only 10 species. Let us consider the following reaction:

$$C_\alpha H_\beta O_\gamma N_\delta + \frac{a_s}{\phi}(O_2 + 3.76\,N_2) \rightarrow$$

$$\begin{aligned}
&n_1\,CO_2 + n_2\,H_2O + n_3\,N_2 \\
&+ n_4\,O_2 + n_5\,CO + n_6\,H_2 + n_7\,H \\
&+ n_8\,O + n_9\,OH + n_{10}\,NO
\end{aligned} \tag{3.71}$$

Atom balancing yields the following four equations:

$$\text{C:}\quad \alpha = (y_1 + y_5)N \tag{3.72}$$

$$\text{H:}\quad \beta = (2y_2 + 2y_6 + y_7 + y_9)N \tag{3.73}$$

$$\text{O:}\quad \gamma + \frac{2a_s}{\phi} = (2y_1 + y_2 + 2y_4 + y_5 + y_8 + y_9 + y_{10})N \tag{3.74}$$

$$\text{N:}\quad \delta + \frac{3.76\,a_s}{\phi} = (2y_3 + y_{10})N \tag{3.75}$$

where N is the total number of moles. By definition, the following can be written

$$\sum_{i=1}^{10} y_i - 1 = 0 \tag{3.76}$$

We now introduce six gas phase equilibrium reactions. These reactions include the dissociation of hydrogen, oxygen, water, carbon dioxide, and equilibrium OH and NO

Table 3-6   Equilibrium Constant Curve-Fit Coefficients for Equation 3.83

i	A_i	B_i	C_i	D_i	E_i
1	0.432168E + 00	−0.112464E + 05	0.267269E + 01	−0.745744E − 04	0.242484E − 08
2	0.310805E + 00	−0.129540E + 05	0.321779E + 01	−0.738336E − 04	0.344645E − 08
3	−0.141784E + 00	−0.213308E + 04	0.853461E + 00	0.355015E − 04	−0.310227E − 08
4	0.150879E − 01	−0.470959E + 04	0.646096E + 00	0.272805E − 05	−0.154444E − 08
5	−0.752364E + 00	0.124210E + 05	−0.260286E + 01	0.259556E − 03	−0.162687E − 07
6	−0.415302E − 02	0.148627E + 05	−0.475746E + 01	0.124699E − 03	−0.900227E − 08

formation:

$$\frac{1}{2}\,H_2 \rightleftharpoons H \qquad K_1 = \frac{y_7 P^{1/2}}{y_6^{1/2}} \tag{3.77}$$

$$\frac{1}{2}\,O_2 \rightleftharpoons O \qquad K_2 = \frac{y_8 P^{1/2}}{y_4^{1/2}} \tag{3.78}$$

$$\frac{1}{2}\,H_2 + \frac{1}{2}\,O_2 \rightleftharpoons OH \qquad K_3 = \frac{y_9}{y_4^{1/2} y_6^{1/2}} \tag{3.79}$$

$$\frac{1}{2}\,O_2 + \frac{1}{2}\,N_2 \rightleftharpoons NO \qquad K_4 = \frac{y_{10}}{y_4^{1/2} y_3^{1/2}} \tag{3.80}$$

$$H_2 + \frac{1}{2}\,O_2 \rightleftharpoons H_2O \qquad K_5 = \frac{y_2}{y_4^{1/2} y_6 P^{1/2}} \tag{3.81}$$

$$CO + \frac{1}{2}\,O_2 \rightleftharpoons CO_2 \qquad K_6 = \frac{y_1}{y_5 y_4^{1/2} P^{1/2}} \tag{3.82}$$

The unit of pressure in the six equations (Equations 3.77 to 3.82) is the atmosphere. Olikara and Borman (1975) have curve fitted the equilibrium constants $K_i(T)$ to JANAF Table data for $600 < T < 4000$ K. Their expressions are of the form

$$\log_{10} K_i(T) = A_i \ln (T/1000) + \frac{B_i}{T} + C_i + D_i T + E_i T^2 \tag{3.83}$$

where T is in Kelvin. The equilibrium constant curve-fit coefficients are listed in Table 3-6. Given pressure P, temperature T, and equivalence ratio ϕ, Equations 3.72 to 3.82 will yield 11 equations for the 11 unknowns: the 10 unknown mole fractions y_i and the unknown total product moles N. The set of 11 equations are nonlinear and solved by Newton-Raphson iteration. With the product mole fraction composition known, one can proceed to compute the thermodynamic properties of interest: enthalpy, entropy, specific volume, and internal energy.

EXAMPLE 3.4   *Equilibrium Combustion Mole Fraction*

What are the mole fractions and mixture properties resulting from the combustion of a gasoline mixture at a pressure of 3000 kPa, temperature of 3000 K, and a fuel-air equivalence ratio of 1.2?

SOLUTION   The web pages contain an applet *Equilibrium Combustion Solver* that computes the product mole fractions and properties for five fuels given the pressure, temperature, and the fuel-air equivalence ratio. The fuels are gasoline (C_7H_{17}), diesel ($C_{14.4}H_{24.9}$), methanol (CH_3OH), methane (CH_4), and nitromethane (CH_3NO_2). The above information is entered into the applet *Equilibrium Combustion Solver* as shown in Figure 3-4. The resulting mole

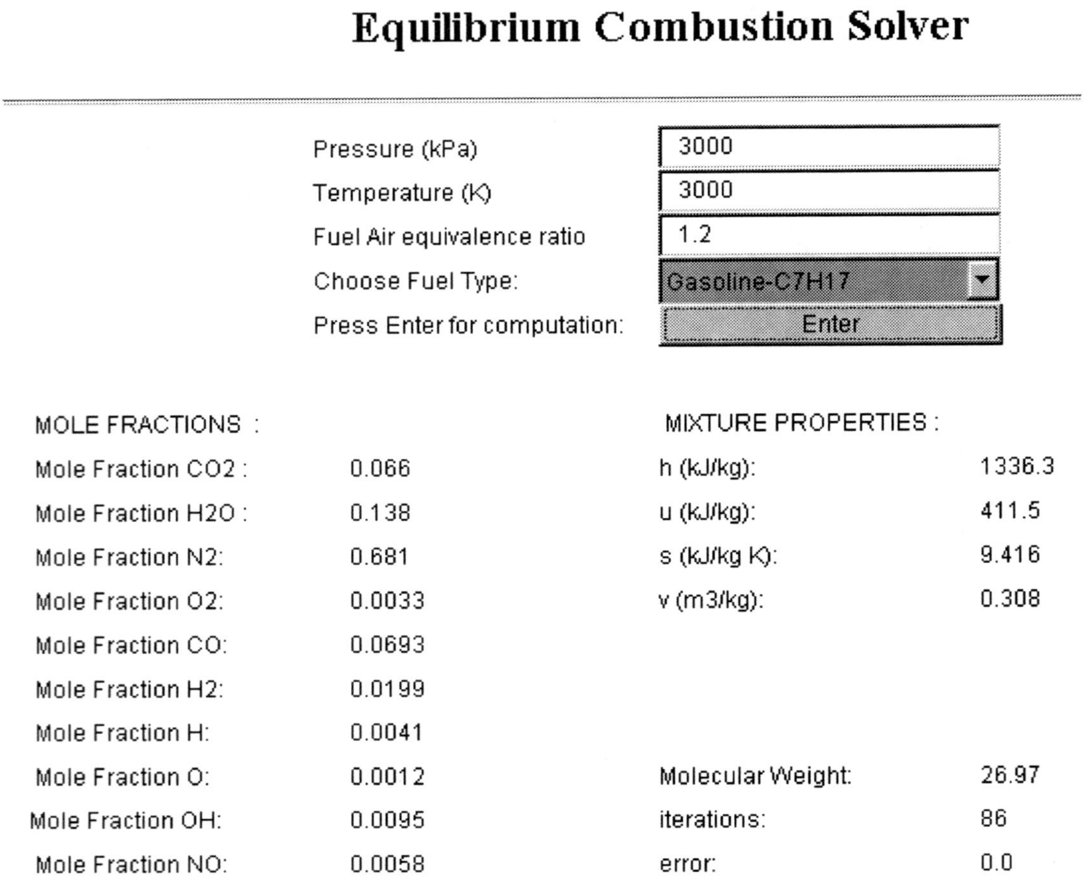

Equilibrium Combustion Solver

Pressure (kPa)	3000
Temperature (K)	3000
Fuel Air equivalence ratio	1.2
Choose Fuel Type:	Gasoline-C7H17 ▼
Press Enter for computation:	Enter

MOLE FRACTIONS :

		MIXTURE PROPERTIES :	
Mole Fraction CO2 :	0.066	h (kJ/kg):	1336.3
Mole Fraction H2O :	0.138	u (kJ/kg):	411.5
Mole Fraction N2:	0.681	s (kJ/kg K):	9.416
Mole Fraction O2:	0.0033	v (m3/kg):	0.308
Mole Fraction CO:	0.0693		
Mole Fraction H2:	0.0199		
Mole Fraction H:	0.0041		
Mole Fraction O:	0.0012	Molecular Weight:	26.97
Mole Fraction OH:	0.0095	iterations:	86
Mole Fraction NO:	0.0058	error:	0.0

Figure 3-4 Equilibrium combustion solver applet.

fractions and properties are also indicated in Figure 3-4. Note that the equilibrium mole fractions calculated with the simple equilibrium constant model compare well with the mole fractions of Figure 3-2 computed by the more general Lagrange multiplier method.

The *Equilibrium Combustion Solver* applet can be used to compute general trends for fuel-air combustion that are not immediately obvious. For example, the effect of temperature on enthalpy of the combustion products for three different equivalence ratios is shown in Figure 3-5 for the combustion of gasoline at a pressure of 101.3 kPa. Note that the lowest value of enthalpy occurs at a stoichiometric equivalence ratio, and as the equivalence ratio is made lean or rich, below 2500 K, the enthalpy increases. This behavior is also shown in Figure 3-6, a plot of the enthalpy of the combustion products of methanol versus equivalence ratio at pressures of 101 kPa and 2000 kPa. The enthalpy is a minimum at near stoichiometric conditions, as on either side of stoichiometric, the combustion is incomplete. If the mixture is lean, there is an excess of unburnt oxygen. If the mixture is rich, there will be unburnt carbon monoxide. A minimal value of enthalpy implies that the specific heat of the combustion products is also a minimum, which will maximize the adiabatic flame temperature, discussed in Section 3.7.

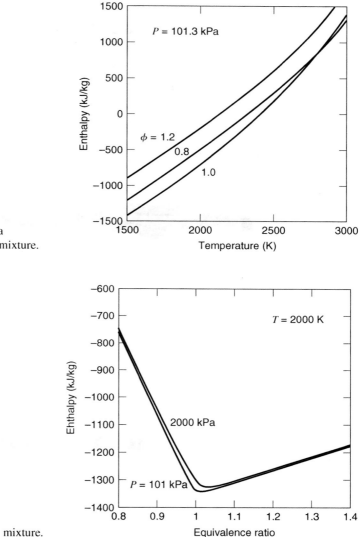

Figure 3-5 Enthalpy of combustion products of a gasoline/air equilibrium mixture.

Figure 3-6 Enthalpy of combustion products for methanol/air equilibrium mixture.

3.7   COMBUSTION AND THE FIRST LAW

With chemical equilibrium modeling we are able to predict the equilibrium state that results from burning a fuel-air mixture as a function of initial conditions, such as pressure, temperature, equivalence ratio, and residual fraction. In this and the next section, we apply the first law of thermodynamics to fuel-air combustion processes, and compute energy interactions between the fuel-air mixture and the environment. We discuss constant pressure and constant volume combustion to illustrate the principles, and introduce the heat of combustion and the adiabatic flame temperature.

Let us first consider the case in which combustion occurs at constant pressure. Suppose that the fuel, air, and residual gases are premixed to a homogeneous state and burned in a combustion system. Application of the first law to this combustion process leads to

$$q = h_p - h_r \quad \text{(kJ/kg)} \tag{3.84}$$

where the subscript p represents products, and the subscript r represents reactants. The enthalpy of the products is equal to the enthalpy of the reactants plus any heat transferred to the system.

The heat of combustion, q_c, of a fuel is defined as the heat transferred out of a system per unit mass or mole of fuel when the initial and final states are at the same temperature and pressure, typically $T_o = 298$ K and $P_o = 101$ kPa. Furthermore, the combustion is assumed to be complete, with the fuel burning to carbon dioxide and water. Since the products are at low temperature ($T_o < 1000$ K), the analyses that led to Table 3-4 for the lean or stoichiometric case can be used to compute the heat of combustion. In this case, it can be shown that Equation 3.84 becomes

$$-q_c = \alpha h_{CO_2} + \frac{\beta}{2} \left[h_{H_2O} - (1 - \chi) h_{fg, H_2O} \right] - h_{fuel} \quad \text{(kJ/kg)} \tag{3.85}$$

Two values of q_c are recognized: The lower heat of combustion is defined as the state where all of the water in the products is vapor (the quality is $\chi = 1$), and the higher heat of combustion is defined as the state where all of the water in the products is liquid ($\chi = 0$).

The higher heat of combustion and stoichiometric adiabatic flame temperature of several fuels are given in Table 3-7. The heat of combustion is used primarily in two ways: (1) in some cases, such as in the gas cycles of Chapter 2 or when solving reacting Navier-Stokes equations, it is desirable to relax the rigor of the thermodynamics by using the heat of combustion to define an equivalent heat release; and (2) for practical fuels, discussed in Chapter 10, the enthalpy at $T_o = 298$ K can be determined inexpensively by measurement of the heat of combustion.

An analogous discussion could be presented for constant-volume combustion where

$$q = u_p - u_r \quad \text{(kJ/kg)} \tag{3.86}$$

Table 3-7 Higher Heat of Combustion and Stoichiometric Adiabatic Flame Temperature of Some Fuels at $P = 1.0$ atm, $T = 298$ K, $f = 0.0$

FUEL		$q_{c,298}$ (MJ/kg)	$T_{f, \phi = 1.0}$ (K)
C_2N_2 (g)	Cyanogen	21.0	2596
H_2 (g)	Hydrogen	141.6	2383
NH_3 (g)	Ammonia	22.5	2076
CH_4 (g)	Methane	55.5	2227
C_3H_8 (g)	Propane	50.3	2268
C_8H_{18} (l)	Octane	47.9	2266
$C_{15}H_{32}$ (l)	Pentadecane	47.3	2269
$C_{20}H_{40}$ (g)	Eicosane	47.3	2291
C_2H_2 (g)	Acetylene	49.9	2540
$C_{10}H_8$ (s)	Naphthalene	40.3	2328
CH_4O (l)	Methanol	22.7	2151
C_2H_6O (l)	Ethanol	29.7	2197
CH_3NO_2 (l)	Nitromethane	11.6	2545

However, as a rule of thumb, when the heat of combustion or adiabatic flame temperature is referred to without qualification, constant pressure combustion is implied.

Given the heat transferred per unit mass from a control volume, and the combustion pressure, we can use Equation 3.84 to solve for the product enthalpy h_p. With two thermodynamic variables, P and h, known, the other properties such as the temperature, specific volume, internal energy, and the like of the products can also be computed. For adiabatic combustion $(q = 0)$, the resultant product temperature is called the adiabatic flame temperature.

The equilibrium constant method of Section 3.6 is formulated with the assumption that the product pressure and temperature are known. Since the product temperature is generally unknown in first law combustion calculations, iteration with an initial temperature estimate is required.

EXAMPLE 3.5   *Adiabatic Flame Temperature*

A stoichiometric mixture of gasoline C_7H_{17}, air, and residual gas is burned at constant pressure. Give that $T_1 = 298$ K, $P_1 = 101.3$ kPa, and $f_1 = 0.10$, compute the adiabatic flame temperature.

SOLUTION   The text contains an applet *Adiabatic Flame Temperature* that computes the adiabatic flame temperature and product mixture properties using Newton-Raphson iteration for a set of fuels and for a given pressure, initial temperature, fuel-air equivalence ratio,

Adiabatic Flame Temperature

Pressure (kPa)	101.3
Initial Temperature (K)	298
Fuel Air equivalence ratio	1.1
Residual mass fraction	0.1
Choose Fuel Type:	Gasoline-C7H17 ▼
Press Enter for computation:	Enter

THERMODYNAMIC PROPERTY :	VALUE
Adiabatic Flame Temperature (K):	2093
Pressure (kPa):	101.3
Volume (m3/kg):	6.162
Enthalpy (kJ/kg):	-446
Internal Energy (kJ/kg):	-1070

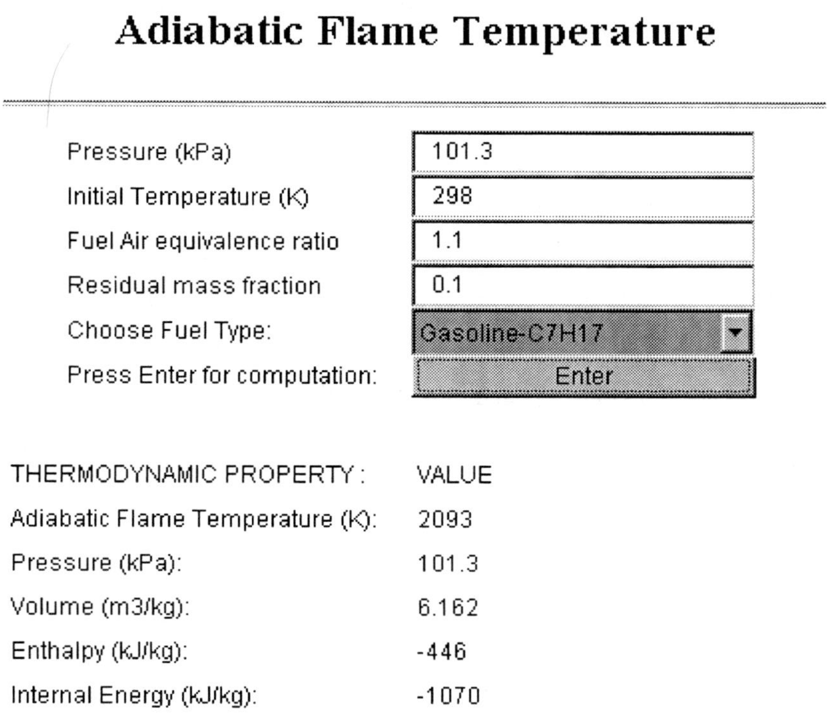

Figure 3-7 Use of adiabatic flame temperature applet (Example 3-4).

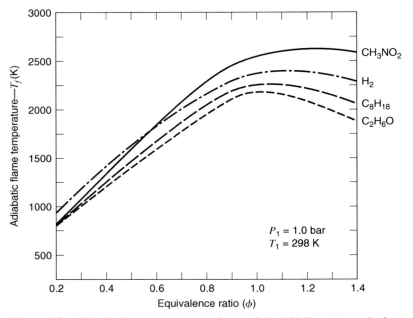

Figure 3-8 Adiabatic flame temperature of some fuels initially at atmospheric pressure and temperature ($f = 0.0$).

and initial residual mass fraction. The fuels are gasoline (C_7H_{17}), diesel ($C_{14.4}H_{24.9}$), methanol (CH_3OH), methane (CH_4), and nitromethane (CH_3NO_2). The above information is entered into the applet as shown in Figure 3-7. The resulting adiabatic flame temperature is $T = 2094$ K. Associated properties are also shown in Figure 3-7.

The stoichiometric adiabatic flame temperature of several fuels burned in air is given in Table 3-7. The adiabatic flame temperatures for several of these fuels burned at $T_1 = 298$ K, $P_1 = 1$ bar, and no residual fraction, are shown in Figure 3-8 as functions of equivalence ratio. There is little dependence on fuel type among the hydrocarbons, which have adiabatic flame temperatures of about 2250 K. The adiabatic flame temperatures are maximum near stoichiometric. This is consistent with the effect of equivalence ratio on enthalpy shown in Figures 3-5 and 3-6, since mixtures with a relatively lower specific heat will undergo a larger temperature change for a given heat release. Additional calculations show that the adiabatic flame temperature decreases with increasing residual fraction and slightly increases with pressure.

3.8   ISENTROPIC PROCESSES

One component of engine cycle analysis is the computation of the change of state due to an isentropic compression or expansion to a specified pressure or specific volume. With a known change of state, the first law can be used to determine the work transfer, given by Equation 3.87:

$$w_{1\text{-}2} = h_2 - h_1 \tag{3.87}$$

For a mixture of ideal gases that chemically reacts to changing constraints, such as volume or pressure, simple relationships between the initial and final state cannot be derived and we resort to the computer. The equilibrium constant calculations discussed in Sections 3.4 and 3.6 compute the properties of a mixture of gases given the temperature and pressure. For an isentropic change of state to an unknown temperature, iteration is required.

EXAMPLE 3.6   *Isentropic Fuel-Air Processes*

A gasoline fuel-air mixture initially at $T_1 = 300$ K, $P_1 = 101.3$ kPa, and $\phi = 0.8$ is compressed isentropically to $P_2 = 2020$ kPa. What is T_2 and the work w_{1-2}? Assume the control volume is closed.

SOLUTION   Using the applet *Equilibrium Combustion Solver* the properties at state 1 are

$$T_1 = 300 \text{ K}$$
$$P_1 = 101.3 \text{ kPa}$$
$$h_1 = -2364 \text{ kJ/kg}$$
$$u_1 = -2451 \text{ kJ/kg}$$
$$s_1 = 7.040 \quad \text{kJ/kg K}$$
$$v_1 = 0.862 \quad \text{m}^3/\text{kg}$$

Since the compression is isentropic $s_2 = s_1 = 7.040$ kJ/kg K. Using a bracketing procedure with the applet we find at $T_2 = 660.5$ K and $P_2 = 2020$ kPa that $u_2 = -2156$ kJ/kg. Therefore,

$$-w_{1-2} = u_2 - u_1$$
$$= -2156 - (-2451)$$

and
$$w_{1-2} = -295 \text{ kJ/kg}$$

3.9   REFERENCES

CRC Handbook of Chemistry and Physics (1975-1976), 56th ed., CRC Press, Cleveland, Ohio.

GORDON, S. and B. MCBRIDE (1971), "Computer Program for Calculation of Complex Chemical Equilibrium Compositions, Rocket Performance, Incident and Reflected Shocks, and Chapman-Jouquet Detonations," NASA SP-273.

JANAF Thermochemical Tables (1971), 2nd ed., National Bureau of Standards Publications, NSRDS-N35 37, Washington D.C.

OLIKARA, C. and G. L. BORMAN (1975), "A Computer Program for Calculating Properties of Equilibrium Combustion Products with Some Applications to I.C. Engines," SAE paper 750468.

REYNOLDS, W. (1986), "The Element Potential Method for Chemical Equilibrium Analysis: Implementation in the Interactive Program STANJAN," M.E. Dept., Stanford Univ.

ROSSINI, F. D. (1953), *Selected Values of Physical and Thermodynamic Properties of Hydrocarbons and Related Compounds,* Carnegie Press, Pittsburgh.

STULL, D. R., E. F. WESTRUM, Jr., and G. C. SINKE (1969), *The Chemical Thermodynamics of Organic Compounds,* Wiley, New York.

SVEHLA, R. A. and B. H. MCBRIDE (1973), "Fortran IV Computer Program for Calculation of Thermodynamic and Transport Properties of Complex Chemical Systems," NASA TND-7056.

VARGAFTIK, N. B. (1975), *Tables on the Thermophysical Properties of Liquids and Gases,* Wiley, New York.

3.10   HOMEWORK

3.1   Using Table C.1 (Appendix), verify that the enthalpy and entropy in Example 3.1 are correct.

3.2   Why does Equation 3.20 contain y_i?

3.3   What are the molecular weight, enthalpy (kJ/kg), and entropy (kJ/kg K) of a gas mixture at $P = 1000$ kPa and $T = 500$ K, if the mixture contains the following species and mole fractions?

Species	y_i
CO_2	0.10
H_2O	0.15
N_2	0.70
CO	0.05

3.4   A system whose composition is given below is in equilibrium at $P = 101$ kPa and $T = 298$ K. What are the enthalpy (kJ/kg), specific volume (m³/kg), and quality of the mixture?

Species	y_i
CO_2	0.125
H_2O	0.141
N_2	0.734

3.5   What are the composition, enthalpy, and entropy of the combustion products of methanol, CH_3OH, at $\phi = 1.0$, $T = 1200$ K, and $P = 101$ kPa?

3.6   If a lean ($\phi = 0.8$) mixture of methane is burned at a temperature of 1500 K and pressure of 500 kPa, what are the mole fractions of the products, and the product enthalpy and entropy?

3.7   Plot the product equilibrium mole fractions as a function of equivalence ratio ($0.5 < \phi < 2$) resulting from the combustion of methane at 5000 kPa and 2500 K. (Use the model of Section 3.6.)

3.8   Derive Equations 3.52 and 3.53.

3.9   At what equivalence ratio for octane-air mixtures does the carbon to oxygen ratio of the system equal one? Why is this of interest?

3.10   At what temperature is the concentration of H_2 a minimum for the combustion of gasoline and air at $\phi = 1.2$ at 5000 kPa?

3.11   At what temperature does the mole fraction of NO reach 0.01 for the equilibrium products resulting from the combustion of gasoline and air at $\phi = 1.0$ at 5000 kPa?

3.12   Why are Equations 3.84 and 3.86 not valid for molar intensive variables?

3.13   Compute the higher, lower, and equilibrium heats of combustion for methanol CH_3OH (l). The equilibrium computation determines the quality of the water in the products at atmospheric pressure.

3.14   The heat of combustion could have been defined without requiring complete conversion to carbon dioxide and water. What would the lower heat of combustion be for the case $\phi = 1.4$, fuel = C_8H_{18} (l) octane, $T_o = 298$ K, $P_o = 1$ atm? Assume the water quality is zero, and that $K = 9.95 \times 10^{-6}$.

3.15 What is the residual mass fraction required to reduce the adiabatic flame temperature of gasoline, diesel, methane, methanol, and nitromethane below 2000 K? Assume $\phi = 1.0$ at 101 kPa and 298 K.

3.16 Plot the adiabatic flame temperature of gasoline as a function of pressure $(50 < P < 5000 \text{ kPa})$ for $T = 298$ K, $\phi = 1.0$, and $f = 0.1$.

3.17 Equilibrium combustion products at $\phi = 1.0$ of methane are expanded isentropically from $T_1 = 2000$ K, $P_1 = 1000$ kPa to a pressure of 100 kPa. Find the final temperature and the work done.

3.18 Equilibrium combustion products of gasoline are expanded isentropically by a volume ratio of 10:1. For $\phi = 1.0$ and an initial state of $T_1 = 3000$ K, $P_1 = 5000$ kPa, find the final state (T_2, P_2) and the work done.

Chapter 4

Fuel-Air Cycles

4.1 INTRODUCTION

We now combine thermodynamic processes with the fuel-air equations of state to form a fuel-air cycle analysis of the efficiency and work produced by an internal combustion engine. A fuel-air cycle model includes the effect of the change in composition of the fuel-air mixture as a result of combustion. The groundwork for introducing fuel-air cycles was laid in Chapter 2, where basic thermodynamic processes were presented; and in Chapter 3, where the required equations of state were given.

4.2 COMPARISON OF FIRST AND SECOND LAW EFFICIENCY

The appropriate definition of efficiency for any of the gas cycles presented in Chapter 2 is clear, since the efficiency for a gas cycle is defined as the fraction of an "equivalent heat release" converted to work. When the analysis takes into account that the fuel is burned rather than heat being released, it is usually assumed that the efficiency used in the analysis is the first law efficiency. The first law efficiency for a control volume (c.v.) in this case is defined as the ratio of the net work done per unit mass of fuel inducted to the heat of combustion of the fuel.

$$\eta_I = w_{c.v.}/q_c \qquad (4.1)$$

It is instructive to examine internal combustion engine efficiency from the perspective of the second law of thermodynamics. The second law definition of efficiency is the ratio of the net work done to the maximum possible work

$$\eta_{II} = W_{c.v.}/W_{max} \qquad (4.2)$$

Following Obert (1973), the maximum possible work is found from application of the first and second law to the control volume shown in Figure 4-1.

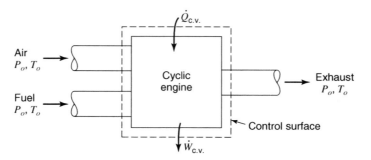

Figure 4-1 A control volume for analyzing maximum work that a cyclic engine can produce by burning a fuel.

$$\dot{Q}_{\text{c.v.}} - \dot{W}_{\text{c.v.}} = \left(\frac{dE}{dt}\right)_{\text{c.v.}} + \sum_{\text{out}} \dot{m} h - \sum_{\text{in}} \dot{m} h \quad (\text{kW}) \tag{4.3}$$

Let us integrate over one period of the engine's cycle

$$Q_{\text{c.v.}} - W_{\text{c.v.}} = \sum_{\text{out}} mh - \sum_{\text{in}} mh \quad (\text{kJ}) \tag{4.4}$$

The maximum work is obtained only if the process is reversible, in which case the second law applied to the control volume is

$$Q_{\text{c.v.}} = T_o \left(\sum_{\text{out}} ms - \sum_{\text{in}} ms \right) \tag{4.5}$$

Note that the only way in which the reversible heat transfer of Equation 4.5 can occur between an engine and its surroundings is via an intervening Carnot engine. Upon substitution of Equation 4.5 into Equation 4.3, and substituting the heat of combustion for the enthalpy change of the fuel-air mixture (assuming inlet and outlet T and P are 298 K and 1 bar), the maximum work is

$$W_{\text{max}} = \sum_{\text{in}} mh - \sum_{\text{out}} mh + T_o \left(\sum_{\text{out}} ms - \sum_{\text{in}} ms \right) \tag{4.6}$$

$$= m_f q_c + T_o \left(\sum_{\text{out}} ms - \sum_{\text{in}} ms \right) \quad (\text{kJ})$$

The available energy of combustion a_c is defined as the maximum work per unit mass of fuel

$$a_c = \frac{W_{\text{max}}}{m_f} = q_c + T_o \left(\sum_{\text{out}} ms - \sum_{\text{in}} ms \right) \Big/ m_f \quad (\text{kJ/kg}_{\text{fuel}}) \tag{4.7}$$

so the second law efficiency can be expressed as

$$\eta_{\text{II}} = \frac{W_{\text{c.v.}}}{m_f \, a_c} \tag{4.8}$$

The difference between the available energy of combustion, $\Delta(H - T_o S)$, and the heat of combustion, ΔH, is that the available energy of combustion takes into account the change in entropy due to changes in composition of reactants. Note that the maximum work is attained only if the exhaust is in equilibrium at the state T_o, P_o; therefore the exhaust water quality should be evaluated by setting the partial pressure of the vapor equal to the saturation vapor pressure at $T = T_o$.

Figure 4-2 compares the available energy of combustion with the heat of combustion at $T_o = 298$ K, $P_o = 1.013$ bar. It is evident from Figure 4-2 that more energy is available per unit mass of fuel if an engine is fueled lean than if it is fueled rich. In the rich case, there is significant carbon monoxide and hydrogen in the exhaust. Thus, not all of the heat is released and the exhaust gases could, in principle, be used as a fuel for some other engine. In practice, however, those gases are usually exhausted to the atmosphere and the energy is wasted. Equation 4.8 is appropriate only when the exhaust of the engine is used as fuel for some other device. In the more usual case, a definition that reflects the waste of energy by running rich is called for. For this reason, we will base our second law efficiency on the maximum available energy of combustion which occurs for very lean equivalence ratios.

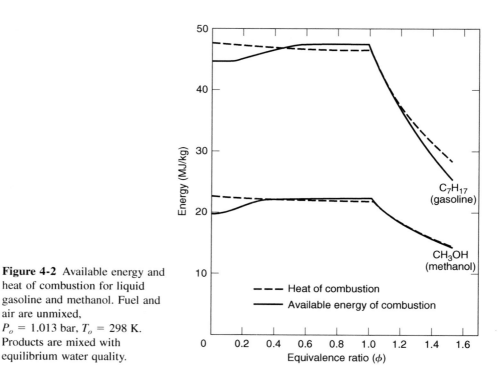

Figure 4-2 Available energy and heat of combustion for liquid gasoline and methanol. Fuel and air are unmixed, $P_o = 1.013$ bar, $T_o = 298$ K. Products are mixed with equilibrium water quality.

Thus, letting

$$a_o = a_{c,\phi=0.01} \tag{4.9}$$

the second law efficiency becomes

$$\eta_{II} = \frac{W_{c.v.}}{m_f a_o} = \frac{W_{c.v.}(1 + \phi F_s)}{\phi F_s (1 - f) a_o} \tag{4.10}$$

To a certain extent, the definition of the second law efficiency is equivocal. It seems impractical to take into account the small amount of work that can in principle be realized because the exhaust composition is different from that of the atmosphere. As shown in homework problem 4-1, the second law efficiency approaches the first law efficiency in the limit of no combustion, since the exhaust composition would be no different than the atmosphere. The arguments presented could be extended to recognize that more energy is available on a cold day than on a hot day. Thus, rather than $T_o = 298$ K, one should evaluate a_0 at the temperature equal to that of the coldest day ever recorded at Antarctica since, in principle, one could run all engines down there on the coldest of days. This concept is easily dismissed if one recognizes the practical constraints.

The choice of $\phi = 0.01$ in Equation 4.9 is also equivocal. Figure 4-1 suggests taking the limit of $\phi \to 0$ to evaluate a_o. Strictly speaking, that limit does not exist, for it would imply that thermodynamic states exist with no bound in volume. Indeed, if one tries to evaluate the limit one will find it blows up. Thus, $\phi = 0.01$ was chosen as being close enough to zero for practical purposes. Indeed, it is unlikely that any engine would be operated significantly leaner than this.

Table 4-1 provides a list of the enthalpy of formation, the absolute entropy at 298 K, the maximum available energy of combustion, and the lower heat of combustion for various gaseous and liquid fuels. The difference between the enthalpy of formation for a given

Table 4-1 Maximum Available Energy of Combustion[1], a_o, Compared with the Lower Heat of Combustion, q_c

FUEL		h_f^o (mJ/kmol)	$\bar{s}_{298}^o$ (kJ/kmol K)	q_c (MJ/kg)	a_o (MJ/kg)
C_2N_2 (g)	Cyanogen	309.1	241.5	21.06	21.29
H_2 (g)	Hydrogen	0.0	130.6	119.95	119.52
NH_3 (g)	Ammonia	−45.7	192.6	18.61	20.29
CH_4 (g)	Methane	−74.9	186.2	50.01	52.42
C_3H_8 (g)	Propane	−103.9	269.9	46.36	49.16
C_7H_{17} (l)	Gasoline[2]	−305.6	345.8	44.51	47.87
C_8H_{18} (l)	Octane	−249.5	360.8	44.43	47.67
C_8H_{18} (l)	Isooctane	−259.3	328.0	44.35	47.67
$C_{14.4}H_{24.9}$ (l)	Diesel[2]	−174.0	525.9	42.94	45.73
$C_{15}H_{32}$ (l)	Pentadecane	−428.9	587.5	43.99	47.22
C_2H_2 (g)	Acetylene	226.7	200.8	48.22	48.58
C_6H_6 (l)	Benzene	48.91	173.0	40.14	42.14
$C_{10}H_8$ (s)	Naphthalene	78.1	166.9	38.86	40.84
CH_4O (l)	Methanol	−239.1	126.8	19.91	22.68
C_2H_6O (l)	Ethanol	−277.2	160.7	26.82	29.71
CH_3NO_2 (l)	Nitromethane	−113.1	171.8	10.54	12.43
C (s)	Graphite	0.0	5.7	32.76	33.70
$C_{176}H_{144}O_8N_3$ (s)	Good coal[2]	−10000.0	3000.0	31.57	33.57

[1]Based on equilibrium water quality, lean combustion at $\phi = 0.01$, $T_o = 298K$, $P_o = 1.013$ bar and unmixed reactants.
[2]Estimated for typical fuel.

liquid fuel in Table 4-1 and the value given for the same fuel in a gaseous state in Table 3-1 is the enthalpy of vaporization, h_{fg}. The exhaust H_2O is assumed to be in a gaseous state. Notice that for the most part there is little difference between the lower heat of combustion q_c and the maximum available energy a_o, so that for fuel-air cycle modeling there is little difference between the first and the second law efficiencies. In this chapter, the thermal efficiency of fuel-air cycles is computed on a first law basis.

4.3 OTTO CYCLE

The Otto cycle models engines whose combustion is rapid enough that it occurs at constant volume near top dead center. It is generally most applicable to homogeneous charge spark ignition engines. The basic processes of a fuel-air Otto cycle necessary to compute the efficiency and the indicated mean effective pressure are:

1 to 2   Isentropic compression of fuel, air, and residual gases
2 to 3   Adiabatic, constant volume combustion
3 to 4   Isentropic expansion of equilibrium combustion products

The inputs to an Otto fuel-air cycle are the compression ratio r, the fuel-air equivalence ratio ϕ, the residual mass fraction f, the fuel type, and the initial temperature T_1 and pressure P_1. Using the fuel-air models developed in Chapter 3, it is possible to compute the properties at states 1, 2, 3, and 4. Since the combustion process is assumed to be adiabatic and constant volume, $u_3 = u_2$, and the increase in T and P is due to the change in chemical composition from an unburned fuel air mixture to an equilibrium combustion product mixture. The work of the fuel-air Otto cycle is

$$w_{net} = (u_3 - u_4) - (u_2 - u_1) = u_1 - u_4 \quad (kJ/kg_{mixture}) \tag{4.11}$$

and the imep is

$$\text{imep} = \frac{w_{net}}{v_1 - v_2} \quad (kPa) \tag{4.12}$$

The first law efficiency is given by Equation 4.1, and the second law efficiency is given by Equation 4.8.

EXAMPLE 4.1   *Fuel-air Otto cycle*

Compute the state properties, work, imep, and thermal efficiency of a fuel-air Otto cycle with the following initial conditions: gasoline fuel with $P_1 = 101.3$ kPa, $T_1 = 298$ K, $\phi = 0.8, f = 0.1$, and a compression ratio of 10.

SOLUTION   The web applet *Fuel-air Otto cycle* is shown in Figure 4-3. Using Newton-Raphson iteration, the applet computes the mixture properties at the four states, as well as the work,

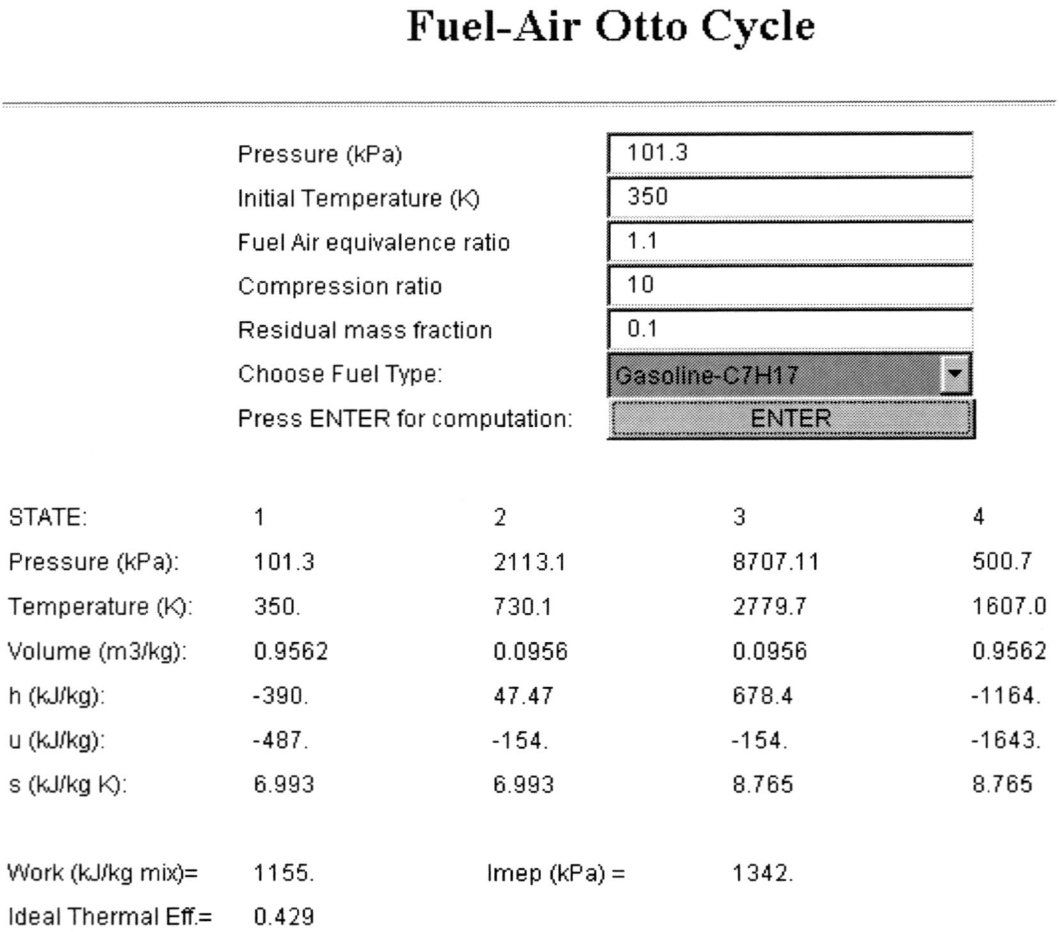

Fuel-Air Otto Cycle

Pressure (kPa)	101.3
Initial Temperature (K)	350
Fuel Air equivalence ratio	1.1
Compression ratio	10
Residual mass fraction	0.1
Choose Fuel Type:	Gasoline-C7H17 ▼
Press ENTER for computation:	ENTER

STATE:	1	2	3	4
Pressure (kPa):	101.3	2113.1	8707.11	500.7
Temperature (K):	350.	730.1	2779.7	1607.0
Volume (m3/kg):	0.9562	0.0956	0.0956	0.9562
h (kJ/kg):	-390.	47.47	678.4	-1164.
u (kJ/kg):	-487.	-154.	-154.	-1643.
s (kJ/kg K):	6.993	6.993	8.765	8.765

Work (kJ/kg mix)=	1155.	Imep (kPa) =	1342.
Ideal Thermal Eff.=	0.429		

Figure 4-3 Fuel-air Otto cycle applet.

imep, and thermal efficiency. The maximum temperature and pressure are $T_3 = 2780$ K and $P_3 = 8707$ kPa. Note that the mixture entropy is constant for the compression from state 1 to 2 and the expansion from state 3 to 4. The work produced is 1155 kJ/kg, the imep is 1342 kPa, and the thermal efficiency is 0.43.

Results obtained using the *Fuel-air Otto cycle* model for different equivalence ratios and different compression ratios are given in Figure 4-4 and 4-5. Some important conclusions are:

1. Indicated efficiency increases with increasing compression ratio, is maximized by lean combustion, and is practically independent of the initial temperature and initial pressure. In actual engines, maximum efficiency occurs at stoichiometric or slightly lean; excessive dilution of the charge with air degrades the combustion.

2. Indicated mean effective pressure increases with increasing compression ratio, is maximized slightly rich of stoichiometric, and increases linearly with the initial density (i.e., imep $\sim P_1$ and imep $\sim 1/T_1$).

3. For a given compression ratio, the peak pressure is proportional to the indicated mean effective pressure.

4. Peak temperatures in the cycles are largest for equivalence ratios slightly rich of stoichiometric.

Taylor (1985) presents similar results for the fuel octane. In fact, the results shown are characteristic of most hydrocarbon fuels. It is of interest to explore the influence of

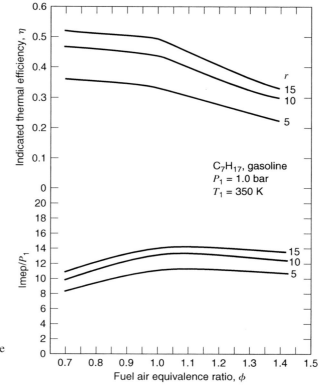

Figure 4-4 Effect of equivalence ratio on Otto fuel-air cycle characteristics.

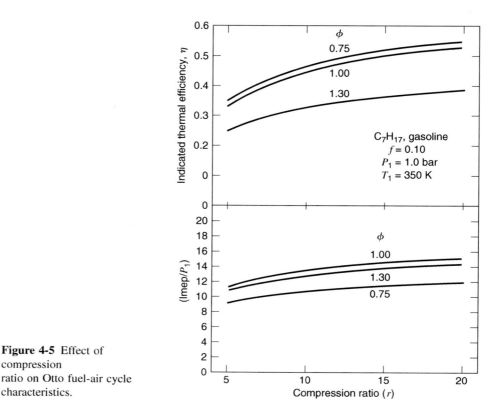

Figure 4-5 Effect of compression ratio on Otto fuel-air cycle characteristics.

fuel properties for some alternative fuels as we look to the future. Table 4-2 presents results obtained for two different compression ratios and five different fuels. Notice that there is very little difference among hydrocarbons. According to this analysis, diesel fuel would be just as good as gasoline in a homogeneous charge spark ignition engine; in reality, of course, knock would be a problem. Note that nitromethane is an excellent choice for a racing fuel, as it has the largest imep of the fuels in Table 4-2.

Because one technique for emission control is exhaust gas recirculation, it is also of interest to examine the influence of the residual fraction. By pumping exhaust gas into

Table 4-2   Effect of Fuel Type on Otto Fuel-Air Cycle

Fuel[1]	Chemical Formula	r	η_t	imep (bar)
Gasoline	C_7H_{17}	10	0.44	13.3
		15	0.495	14.4
Diesel	$C_{14.4}H_{24.9}$	10	0.44	13.7
		15	0.495	14.9
Methane	CH_4	10	0.44	12.2
		15	0.496	13.1
Methanol	CH_3OH	10	0.43	13.1
		15	0.48	14.2
Nitromethane	CH_3NO_2	10	0.39	21.0
		15	0.43	23.1

[1]$\phi = 1.0, f = 0.10, P_1 = 1.0$ bar, $T_1 = 350$K.

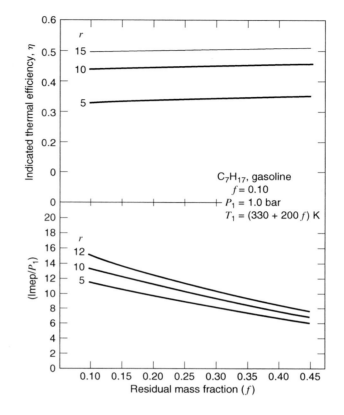

Figure 4-6 Effect of residual fraction on Otto fuel-air cycle characteristics.

the intake manifold and mixing it with the fuel and air, one has, in essence, increased the residual gas fraction. Although the exhaust gas so recirculated is cooled before introduction into the induction system, it is still considerably warmer than the inlet air. Therefore, we will increase the inlet temperature in our computations simultaneously to examine the overall effect. To illustrate, assume $T_1 = (330 + 200f)$ in degrees Kelvin. The results obtained for gasoline are given in Figure 4-6. Notice that the efficiency increases slightly with increasing dilution of the charge by residual gas. Notice too that imep falls with increasing f; it falls because the residual gas displaces fuel-air and because it warms the fuel-air, thereby reducing the charge density.

The indicated efficiency and mean effective pressure of actual engines are determined in practice by measuring the cylinder pressure as a function of cylinder volume and integrating $\int PdV$ to find the work. It is also possible to measure the residual fraction and charge density trapped within the cylinder. Otto fuel-air cycles, with P_1, T_1, and f chosen such that the charge density and entropy are matched to the conditions in the cylinder at the time of inlet valve closing, over-predict the work and efficiency by a factor of only 1.1 to 1.2.

Stated differently, the ratio of the actual efficiency to that of the Otto fuel-air cycle is

$$\eta/\eta_{\text{Otto}} = 0.8 \text{ to } 0.9$$

Based on the analysis done in Chapter 2, this is to be expected. The difference is primarily due to heat loss, but also to mass loss and the finite burning rate. A small part of the discrepancy can be also attributed to opening the exhaust valve prior to bottom dead center to provide for the finite rate of blowdown.

For a given engine operated at optimum spark timing, this ratio is nearly independent of the fuel-air equivalence ratio, the inlet temperature, the inlet pressure, the exhaust gas recirculation, and the engine speed. All the trends predicted by the Otto fuel-air cycle are, in fact, observed in practice. As will be shown later in this chapter, all of the conclusions drawn from Figures 4-4 to 4-6 apply to actual engines, provided that they are operated at optimum spark timing.

4.4   FOUR-STROKE OTTO CYCLE

It is also instructive to consider the Otto fuel-air cycle with the ideal inlet and exhaust processes. In this case, the variables T_1, P_1, and f are no longer the independent variables. Instead, the intake pressure P_i, the exhaust pressure P_e, and the intake temperature T_i are the independent variables.

The additional processes, introduced in section 2.7, are reiterated here:

4 to 5   Constant cylinder volume blowdown
5 to 6   Constant pressure exhaustion
6 to 7   Constant cylinder volume reversion
7 to 1   Constant pressure induction

The blowdown is considered to be isentropic as far as the control mass is concerned. One solves for the temperature T_5 by requiring that $S_5 = S_4$ and $P_5 = P_e$. Application of the first law to the control mass during exhaust leads to the conclusion that

$$h_6 = h_5 \qquad P_6 = P_5$$
$$T_6 = T_5 \qquad V_6 = V_5$$

These are still valid conclusions even though now we are treating the exhaust gas as equilibrium combustion products.

The residual fraction, given by Equation 2.46, is restated here as

$$f = \frac{1}{r}\frac{v_4}{v_6} \tag{4.13}$$

The energy equation applied to the cylinder control volume during intake is given in Equation 2.47. Note that Equation 2.54 is no longer valid since it assumes constant specific heats. In this case Equation 2.54 is replaced by

$$h_1 = f[h_6 + (P_i - P_e)v_6] + (1 - f)h_i \tag{4.14}$$

and, of course, it is still true that if the pressure drop across the intake valves is neglected

$$P_1 = P_i \tag{4.15}$$

The volumetric efficiency and pumping work are

$$e_v = \frac{m_i}{\rho_i V_d} = \frac{r(1 - f)v_i}{(r - 1)v_1} \tag{4.16}$$

$$\text{pmep} = P_e - P_i \tag{4.17}$$

Finally, the net imep and thermal efficiency are

$$(\text{imep})_{\text{net}} = \text{imep} - \text{pmep} \tag{4.18}$$

$$\eta_{\text{net}} = \eta\left(1 - \frac{\text{pmep}}{\text{imep}}\right) \tag{4.19}$$

The inputs to a four-stroke Otto fuel-air cycle are the compression ratio r, the fuel-air equivalence ratio ϕ, the intake pressure P_i, the exhaust pressure P_e, the intake temperature T_i, and the fuel type. Using the fuel-air models developed in Chapter 3, it is possible to compute the properties at states 1, 2, 3, and 4. As in the four-stroke gas cycle, analysis of the four-stroke fuel-air cycle requires iteration.

EXAMPLE 4.2   *Four-stroke fuel-air Otto cycle*

Compute the state properties, volumetric efficiency, residual fraction, net imep, and net thermal efficiency of a throttled four-stroke fuel-air Otto cycle with the following intake conditions: gasoline fuel with $P_i = 50$ kPa, $P_e = 105$ kPa, $T_i = 300$ K, $\phi = 0.8$, and a compression ratio of 10.

SOLUTION   The web applet four-stroke fuel-air Otto cycle is shown in Figure 4-7. The applet assumes that the fuel is gasoline, and using Newton-Raphson iteration, computes the mixture properties at the four cycle states, as well as the volumetric efficiency, residual fraction, net imep, and net thermal efficiency.

Four Stroke Fuel-Air Otto Cycle

Gasoline (C7H17)

Intake Pressure (kPa)	50.	
Intake Temperature (K)	300	
Exhaust Pressure (kPa)	105.	
Compression Ratio	10	
Fuel-air Equivalence Ratio	0.8	
Compute Cycle	Enter	

State	1	2	3	4
Pressure (kPa):	50.0	1075.89	3809.12	217.11
Temperature (K):	350.4	753.1	2547.3	1457.5
Volume (m3/kg):	1.9552	0.1955	0.1955	1.9552
h (kJ/kg):	-203.1	154.23	688.71	-953.57
u (kJ/kg):	-398.07	-56.14	-56.06	-1378.08
s (kJ/kg K):	7.201	7.201	8.683	8.683

Exhaust Temp.(K) =	1243.7		Volumetric Efficiency =	0.894
Residual Mass Fraction =	0.0567		Net Imep (kPa) =	556.91
Ideal Thermal Eff. =	0.469		Net Thermal Eff. =	0.423

Figure 4-7 Four-stroke fuel-air Otto cycle applet.

The temperature rise of the inlet fuel-air mixture is about 50 K when mixed with the $f = 0.0567$ residual fraction. The maximum temperature and pressure are $T_3 = 2547$ K and $P_3 = 3809$ kPa. The exhaust temperature T_e is 1244 K. The volumetric efficiency is 0.895, net imep is 557 kPa, and the net thermal efficiency is 0.42.

Results obtained by varying the intake to exhaust pressure ratio and the compression ratio are given in Figures 4-8 and 4-9. The net efficiency and the net indicated mean effective pressure are each seen to be a strong function of the pressure ratio. The advantage of turbocharging and the disadvantage of throttling are clear. For pressure ratios corresponding to supercharging, the curves are not representative, for one would have to also account for the work to drive the compressor. Notice that throttling also hurts the volumetric efficiency, mainly because of an increase in the residual fraction. The residual fraction decreases with increasing compression ratio, as one would expect.

The modeling of the intake and exhaust portion of the Otto cycle is not nearly as good as the compression, combustion, and expansion portion of the cycle. This is because of the assumptions of isobaric intake and exhaust processes and the neglect of heat transfer. Neglect of the heat transfer causes the residual fraction to be under-predicted by a factor on the order of two. In Chapter 7, it will be shown that the processes are isobaric only at very low piston speeds; consequently, at high piston speeds, the pumping mean effective pressure can be in considerable error and can even have the wrong sign for super- or turbo-charged engines.

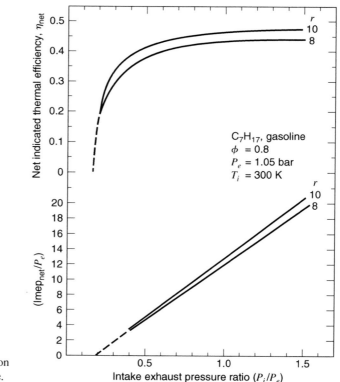

Figure 4-8 Effect of intake/exhaust pressure ratio on four-stroke Otto fuel-air cycle.

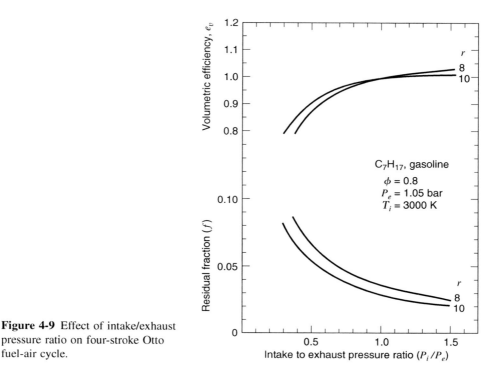

Figure 4-9 Effect of intake/exhaust pressure ratio on four-stroke Otto fuel-air cycle.

4.5   FUEL-INJECTED LIMITED-PRESSURE CYCLE

This cycle models diesel engines and fuel-injected stratified charge engines in which the fuel is injected at the time it is intended to burn. The processes are:

1 to 2   Isentropic compression of air and residual gas
2 to 2.5   Constant volume, adiabatic fuel injection, and combustion
2.5 to 3   Constant pressure, adiabatic fuel injection, and/or combustion
3 to 4   Isentropic expansion

These engines, in general, are fueled overall lean. In this case, the air-residual gas mixture is equivalent to equilibrium combustion products at an equivalence ratio given by

$$\phi_{12} = \frac{f\phi}{1 + (1 - f)\phi F_s} \tag{4.20}$$

where the residual fraction, f, is the ratio of the residual mass to the cylinder gas mass prior to fuel injection. The thermodynamic state during compression can then be determined with the equivalence ratio ϕ_{12} used as an argument.

The details of the fuel injection and combustion are of no concern at this level of modeling. We need only assume that at state 3 the gases in the cylinder are equilibrium combustion products at the overall fuel-air equivalence ratio. To specify the state, we know that

$$P_3 \leq P_{\text{limit}} \tag{4.21}$$

and we apply the energy equation to the process 2 to 3.

$$\Delta U = m_3 u_3 - m_2 u_2 = m_f h_f - P_3(m_3 v_3 - m_3 v_2) \tag{4.22}$$

It follows that

$$\frac{m_2}{m_3} = \frac{m_a + m_r}{m_a + m_r + m_f} = \frac{1}{1 + (1 - f)\,\phi F_s} \tag{4.23}$$

$$\frac{m_f}{m_3} = \frac{m_3 - m_2}{m_3} = \frac{(1 - f)\,\phi F_s}{1 + \phi F_s (1 - f)} \tag{4.24}$$

and the enthalpy at state 3 is

$$h_3 = u_3 + P_3 v_3 = \frac{u_2 + P_3 v_2 + (1 - f)\,\phi F_s\, h_f}{1 + (1 - f)\,\phi F_s} \tag{4.25}$$

The pressures during fuel injection P_f are high enough that Equation 3.27 should be used in lieu of Equation 3.32 to evaluate the fuel enthalpy. Hence,

$$h_{\text{fuel}} = h_{fo} + v_o (P_{\text{fuel}} - P_o) \tag{4.26}$$

where the subscript zero denotes conditions at atmospheric pressure ($P_o = 1.01325$ bar).

In doing a computation, one should first assume that the combustion and fuel injection are entirely at constant volume. If the resultant P_3 satisfies Equation 4.21, then indeed the process is at constant volume. However, if Equation 4.21 is not satisfied, then $P_3 = P_{\text{limit}}$, and one solves Equation 4.25 to find the state 3.

The expansion occurs to a specific volume at state 4 different from that at state 1 because of the fuel injected. It can be shown that expansion must satisfy the following constraints:

$$s_4 = s_3 \tag{4.27}$$

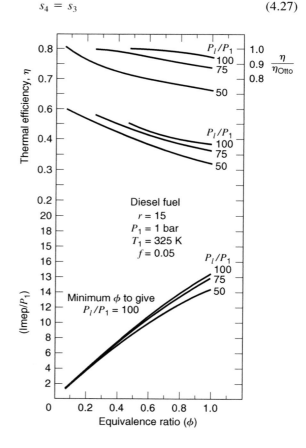

Figure 4-10 Effect of equivalence ratio on limited pressure fuel injected fuel-air cycle.

$$v_4 = rv_3 = r\frac{m_2}{m_3}v_2 = \frac{rv_2}{1 + (1-f)\,\phi F_s} \tag{4.28}$$

To evaluate the work w_{cv}, it is convenient to split it into three parts, w_{12}, w_{23}, and w_{34}, due to the change in mass and energy by fuel injection. The work components expressed per unit mass after fuel injection are

$$w_{c.v.} = w_{12} + w_{23} + w_{34} \tag{4.29}$$

$$w_{12} = \frac{W_{12}}{m_{12}} \cdot \frac{m_2}{m_3} = \frac{u_1 - u_2}{1 + (1-f)\,\phi F_s} \tag{4.30}$$

$$w_{23} = P_3\left[v_3 - \frac{v_2}{1 + (1-f)\,\phi F_s}\right] \tag{4.31}$$

$$w_{34} = u_3 - u_4 \tag{4.32}$$

The efficiency and imep are given by

$$\eta = \frac{w_{c.v.}\,m_3}{m_f\,q_c} = \frac{w_{c.v.}\left[1 + (1-f)\,\phi F_s\right]}{\phi F_s\,q_c\,(1-f)} \tag{4.33}$$

$$\text{imep} = \frac{w_{c.v.}\left[1 + \phi F_s(1-f)\right]}{v_1 - v_2} \tag{4.34}$$

Results obtained using the above modeling for different compression and equivalence ratios are given in Figures 4-10 and 4-11. Important conclusions are:

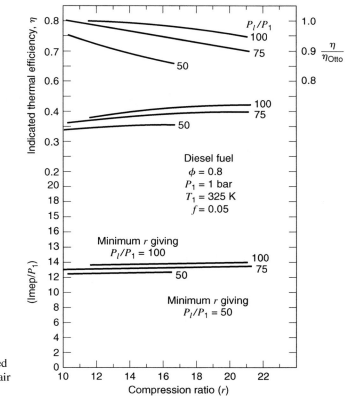

Figure 4-11 Effect of compression ratio on limited pressure fuel injected fuel-air cycle.

- The efficiency decreases with increased equivalence ratio.
- The imep increases with equivalence ratio.
- The efficiency and imep are a weaker function of compression ratio relative to an Otto cycle.
- Both efficiency and imep increase with increasing limit pressure.
- Even in the absence of heat and mass loss, the ratio η/η_{Otto} may be as low as 0.85.

The constraint on peak pressure results in the efficiency and imep being insensitive to compression ratio. In practice, the ratio η/η_{Otto} for diesel and fuel injected stratified engines is more sensitive to the particular design and the operating conditions than it is for homogeneous-charge spark ignition engines. Thus, a greater range of indicated efficiencies exists among engines made by different manufacturers and among engines of different sizes. Divided chamber engines usually have a smaller ratio than open chamber engines partly because of throttling losses through the throat between chambers, but mainly because of a greater heat loss.

The fuel-air cycle adequately models conventional spark ignition engines, but is not as useful for an engine as heterogeneous as a typical diesel engine. Diesel engine fuel-flow rates are limited by the appearance of solid carbon in exhaust that did not burn to carbon monoxide or carbon dioxide. This occurs even though the engine is running lean and is not predicted by fuel-air cycles. A more sophisticated model is required. These exist but are beyond the scope of this text.

4.6 COMPARISON OF FUEL-AIR CYCLE WITH ACTUAL SPARK IGNITION CYCLES

Since the efficiency of an actual engine must be less than the efficiency of its equivalent Otto fuel-air cycle, the fuel-air cycle is a convenient standard for comparison. With reference to Figure 4-12, an equivalent fuel-air cycle is constructed by matching the temperature, pressure, and composition (and thereby entropy) at some reference point after closing of the intake valve and prior to firing of the spark plug. Since the actual process is nearly isentropic, the compression curves of the two cycles nearly coincide.

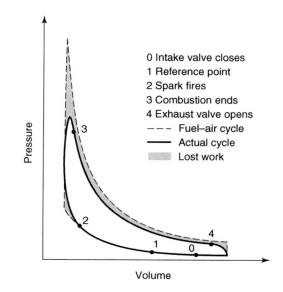

Figure 4-12 A comparison of an actual cycle with its equivalent fuel-air cycle.

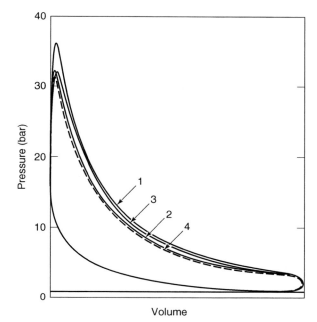

Figure 4-13 Effect of fuel-air equivalence ratio on *P–V* diagram. Reprinted from *The Internal Combustion Engine in Theory and Practice*, Vol. 1, by C. Fayette Taylor, with permission of MIT Press.

CURVE	ϕ	θ_s (deg atdc)	θ_d (deg)	bmep (bar)	imep (bar)	η/η_{Otto}
1	1.17	−15	33	—	7.7	0.85
2	0.80	−23	39	—	6.4	0.83
3	1.80	−20	39	—	7.0	0.83
4	0.74	−33	58	—	6.3	0.80

CFR engine, b = 82.6 mm, s = 114.3 mm, r = 7, 1200 rpm, P_i = 0.95 bar, P_e = 1.02 bar, T_i = 355 K.

Soon after the onset of combustion, the pressure of the actual cycle starts rising above that of the fuel-air cycle. Because the combustion actually is not at constant volume, the peak pressure is considerably less than that predicted by the fuel-air cycle. The expansion curve 3 to 4 is polytropic in character; measurements show that the entropy decreases during expansion, primarily due to heat transfer to the coolant. At point 4 the exhaust valve opens, and soon after, the pressure falls rapidly to the exhaust pressure. The cross-hatched area represents "lost work" that can mainly be attributed to the following

- Heat loss
- Mass loss
- Finite burn rate
- Finite blowdown rate

These losses result in the actual efficiency being less than that of the equivalent fuel-air cycle by a factor ranging from 0.8 to 0.9. Figures 4-13 through 4-18 show measured pressure-volume diagrams as functions of several engine variables: the equivalence ratio, spark timing, engine speed, compression ratio, and inlet pressure. The results strictly apply only to the engine tested but the trends revealed and orders of magnitude are typical of most engines. With the exception of Figure 4-14, all the data are obtained using optimum spark timing. Important points to notice are:

- The indicated mean effective pressure is maximized slightly rich of stoichiometric, and increases with increasing compression ratio and inlet pressure.

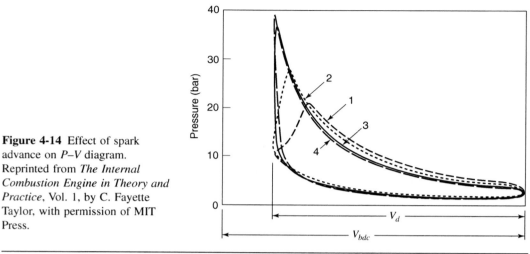

Figure 4-14 Effect of spark advance on *P–V* diagram. Reprinted from *The Internal Combustion Engine in Theory and Practice*, Vol. 1, by C. Fayette Taylor, with permission of MIT Press.

CURVE	θ_s (deg atdc)	θ_d (deg)	bmep (bar)	imep (bar)	η/η_{Otto}
1	0	40	5.0	6.0	0.73
2	−13	40	5.7	7.5	0.82
3	−26	38	5.8	7.5	0.82
4	−37	39	5.0	6.9	0.74

CFR engine, b = 82.6 mm, s = 114.3 mm, r = 6, ϕ = 1.13, P_i = 0.99 bar, P_e = 1.02 bar, T_i = 328 K, 1200 rpm.

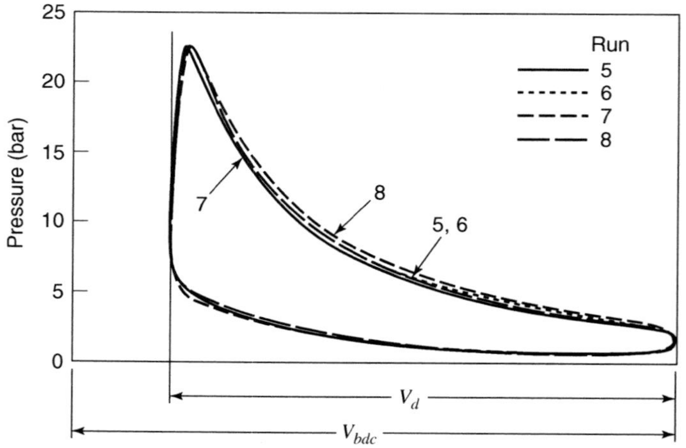

Figure 4-15 Effect of speed on *P–V* diagram, constant delivery ratio. From *The Internal Combustion Engine in Theory and Practice*, Vol. 1, by C. Fayette Taylor, with permission of MIT Press.

CURVE	P_i (bar)	RPM	θ_s (deg atdc)	θ_d (deg)	bmep (bar)	imep (bar)	η/η_{Otto}
5	0.76	900	−18	36	3.90	5.89	0.842
6	0.79	1200	−19	39	3.77	5.94	0.848
7	0.95	1500	−22	40	3.80	6.07	0.865
8	0.98	1800	−18	38	3.59	6.14	0.877

CFR engine, b = 82.6 mm, s = 114.3 mm, r = 6, T_i = 339 K, ϕ = 1.13. P_e = 1.02 bar.

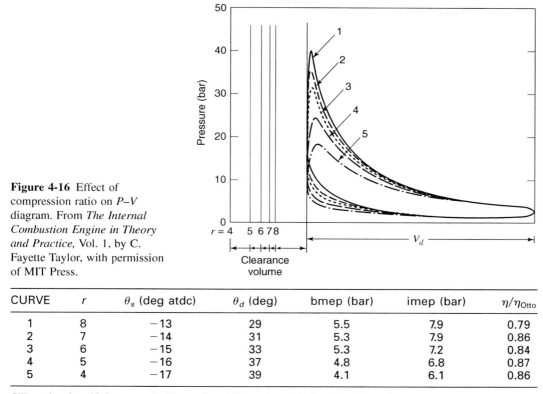

Figure 4-16 Effect of compression ratio on P–V diagram. From *The Internal Combustion Engine in Theory and Practice,* Vol. 1, by C. Fayette Taylor, with permission of MIT Press.

CURVE	r	θ_s (deg atdc)	θ_d (deg)	bmep (bar)	imep (bar)	η/η_{Otto}
1	8	−13	29	5.5	7.9	0.79
2	7	−14	31	5.3	7.9	0.86
3	6	−15	33	5.3	7.2	0.84
4	5	−16	37	4.8	6.8	0.87
5	4	−17	39	4.1	6.1	0.86

CFR engine, b = 82.6 mm, s = 114.3 mm, P_i = 0.95 bar, P_e = 1.04 bar, T_i = 339 K, ϕ = 1.13, 1200 rpm.

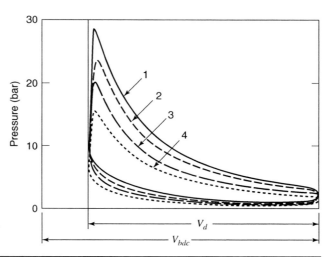

Figure 4-17 Effect of inlet pressure on P–V diagram. From *The Internal Combustion Engine in Theory and Practice,* Vol. 1, by C. Fayette Taylor, with permission of MIT Press.

CURVE	P_i (bar)	θ_s (deg atdc)	θ_d (deg)	bmep (bar)	imep (bar)	η/η_{Otto}
1	0.95	−19	36	5.1	7.3	0.86
2	0.81	−20	38	4.0	6.3	0.87
3	0.68	−26	42	2.7	5.0	0.89
4	0.54	−28	44	1.6	3.7	0.82

CFR engine, b = 82.6 mm, s = 114.3 mm, r = 6, P_e = 1.0 bar, T_i = 339 K, 1200 rpm, ϕ = 1.13.

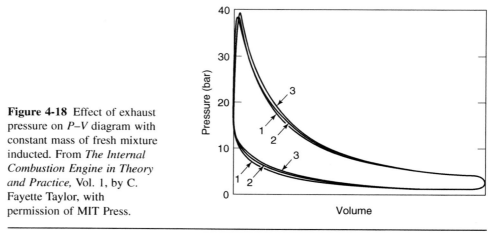

Figure 4-18 Effect of exhaust pressure on P–V diagram with constant mass of fresh mixture inducted. From *The Internal Combustion Engine in Theory and Practice,* Vol. 1, by C. Fayette Taylor, with permission of MIT Press.

CURVE	P_i (bar)	P_e (bar)	θ_s (deg atdc)	θ_d (deg)	bmep (bar)	imep (bar)	η/η_{Otto}
1	0.92	0.51	−18	24	—	7.7	0.83
2	0.95	0.95	−18	25	—	7.7	0.83
3	0.99	1.52	−18	25	—	7.7	0.83

CFR engine, $b = 82.6$ mm, $s = 114.3$ mm, $r = 7$, 1200 rpm, $\dot{m}_a = 4.70$ g/s, $\phi = 1.17$, $T_i = 356$ K.

- The ratio η/η_{Otto} is on the order of 0.85, and varies insignificantly with engine operating variables, at most decreasing slightly with increasing compression ratio.
- The combustion duration, θ_d, is on the order of 35°, decreases with increasing compression ratio or inlet pressure, and is minimum at a slightly rich equivalence ratio.
- The optimum spark advance, θ_s, increases with combustion duration and with increased engine speed.
- The imep increases with engine speed, while bmep decreases, which, as we will see in Chapter 6, is caused by increased friction.

The ratio η/η_{Otto} from these data shows a weak dependence on compression ratio and an even weaker dependence on inlet temperature. As the compression ratio increases, the peak temperature increases, increasing the wall heat transfer to the coolant. The ratio is also independent of equivalence ratio and the exhaust to inlet pressure ratio. Some of these data are compared with data for other engines in Figure 4-19. The coincidence in the or-

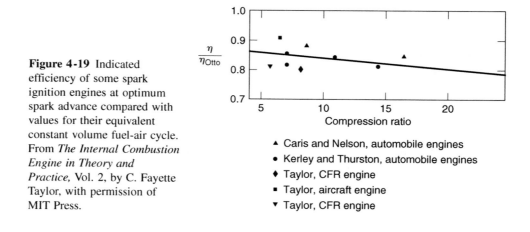

Figure 4-19 Indicated efficiency of some spark ignition engines at optimum spark advance compared with values for their equivalent constant volume fuel-air cycle. From *The Internal Combustion Engine in Theory and Practice,* Vol. 2, by C. Fayette Taylor, with permission of MIT Press.

▲ Caris and Nelson, automobile engines
● Kerley and Thurston, automobile engines
◆ Taylor, CFR engine
■ Taylor, aircraft engine
▼ Taylor, CFR engine

ders of magnitude among the various engines lends credence to the earlier assertion that the results shown in Figures 4-13 through 4-18 are typical of most engines.

An important implication of these results is that there is slightly greater potential for improving the efficiency of spark ignition engines by increasing their theoretical efficiency rather than by reducing their losses. To illustrate, suppose that by reducing the heat loss or increasing the burn rate one could increase η/η_{Otto} from 0.80 to 0.90. The efficiency might be 0.32 instead of 0.29. On the other hand, suppose that research results showed that the compression ratio could be increased to 20. The fuel-air cycle efficiency would increase to about 0.46, and if η/η_{Otto} were still 0.8, the actual efficiency would now be 0.37. There is greater potential with this approach because the second law of thermodynamics does not limit the choice of variables that fix the theoretical efficiency but it does limit the gains that can be realized once the parameters that specify the fuel-air cycle are fixed.

4.7   COMPARISON OF FUEL-AIR CYCLE WITH ACTUAL COMPRESSION IGNITION CYCLES

Diesel engines are designed to limit both the rates of pressure rise and the maximum pressures to satisfy durability, noise, and emissions considerations. Therefore, a convenient standard appears to be the equivalent limited pressure (lp) fuel-air cycle, and indeed this was the choice of Taylor (1985) in his book. Figures 4-20 and 4-21 compare actual pressure-volume diagrams with their ideal limited-pressure counterparts. The engine used in Figure 4-20 is a single cylinder prechamber diesel engine. The engine used in Figure 4-21 is a two-stroke open chamber diesel engine. As in the spark ignition engine, the losses are

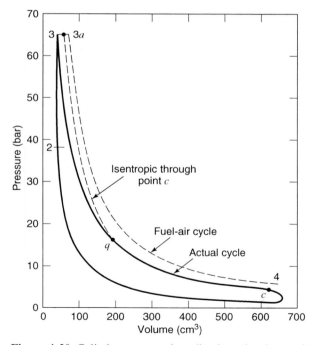

Figure 4-20 Cylinder pressures in a diesel prechamber engine compared with equivalent limited pressure fuel-air cycle: $f = 0.255$, imep $= 7.72$ bar, $\eta/\eta_{lp} = 0.614$. Single-cylinder comet-head diesel engine: $r = 14.5$, $b = 82.6$ mm, $s = 114.3$ mm, N $= 1000$ rpm, $\phi = 0.9$, optimum injection timing, air consumption $= 4.93$ g/s, $P_i = 0.99$ bar, $P_e = 1.01$ bar, $T_i = 560$ K. From *The Internal Combustion Engine in Theory and Practice,* Vol. 1, by C. Fayette Taylor, with permission of MIT Press.

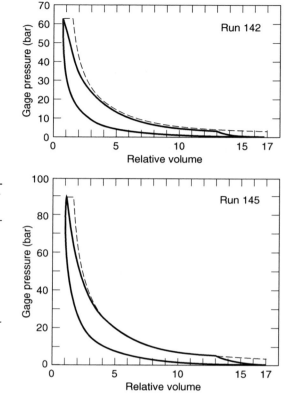

Figure 4-21 Actual open-chamber diesel cycles compared with equivalent limited pressure fuel–air cycles.

	Run Number	
	142	145
ϕ	0.56	0.59
f	0.08	0.06
m'_a (trapped), g	1.07	1.36
imep, bar	7.10	9.86
η/η_{lp}	0.85	0.84
P_i, bar	1.19	1.66
P_e, bar	1.02	1.02

Two stroke, $b = 108$ mm, $s = 150$ mm, $r = 17$, 1600 rpm. From *The Internal Combustion Engine in Theory and Practice,* Vol. 1, by C. Fayette Taylor, with permission of MIT Press.

attributed to heat and mass loss, the finite blowdown rate, and combustion occurring at less than the maximum pressure. Note that the ratio $\eta/\eta_{lp} = 0.61$ is less for the prechamber engine $(\eta/\eta_{lp} = 0.84 - 0.85)$ than it is for the direct-injection, two-stroke engine.

There are two problems with using the limited-pressure fuel-air cycle as a standard. The first is that an engine that can operate at a higher peak pressure and still satisfy the constraints imposed by durability, noise, and emissions considerations is a better engine and ought to be recognized as such. The second issue is that for some engines, it is not possible to construct an equivalent limited-pressure fuel-air cycle because the losses are so great that the peak pressure is less than would be achieved via isentropic compression alone.

An example of such a case is provided by the cycle shown in Figure 4-22. The compression ratio $(r = 22)$ of this diesel prechamber engine is considerably higher than those considered earlier for the spark ignition engines given in Figure 4-19. If the mixture in the cylinder at closure of the inlet valve were compressed isentropically to the volume at top center, the resultant pressure would exceed the observed peak pressure. Therefore, an equivalent limited-pressure fuel-air cycle cannot be constructed. As indicated in Figure 4-22, when this engine is compared to an equivalent fuel-injected Otto cycle, the ratio η/η_{Otto} is only 0.55. That is not to say that the engine is inefficient; recognize that at a compression ratio of 22 its equivalent Otto cycle efficiency is quite high.

We conclude by noting that if the ratio of η/η_{Otto} is a measure of how well an engine of a given compression ratio is developed, it appears that gasoline engines are more highly developed than diesel engines. This suggests that there is more potential for payoff from research and development on losses in diesel engines than there is on losses in spark ignition engines.

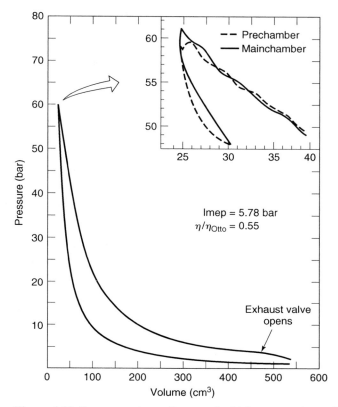

Figure 4-22 Pressure–volume diagram of a high compression ratio diesel prechamber engine. Multicylinder engine with a Ricardo Mark V6 combustion chamber: $b = 88$ mm, $s = 85$ mm, $\varepsilon = 0.266$, $r = 22.0$, $N = 1600$ rpm, $\phi = 0.57$, $e_v = 0.854$. Data courtesy D. Siegla and R. Talder, General Motors.

4.8   REFERENCES

OBERT, E. F. (1973), *Internal Combustion Engines and Air Pollution,* Harper & Row, New York.

TAYLOR, C. F. (1985), *The Internal Combustion Engine in Theory and Practice,* MIT Press, Cambridge, Massachusetts.

4.9   HOMEWORK

4.1   Show that Equation 4.8 for the second law efficiency reduces to Equation 4.1 for the first law efficiency when there is only heat transfer to the system instead of combustion.

4.2   Derive Equation 4.10 for the second law efficiency.

4.3   Derive Equation 4.16 for the volumetric efficiency.

4.4   With reference to Figure 4-2, explain why the heats of combustion at $\phi = 0.2$ and $\phi = 1.2$ are less then those at $\phi = 1.0$.

4.5   Compute a_0, the maximum energy of combustion, for liquid gasoline C_7H_{17} based on equilibrium water quality, lean combustion at $\phi = 0.01$, and unmixed reactants.

4.6   Plot the indicated thermal efficiency versus compression ratio (vary r from 4 to 20 in steps of 4) for a methane fuel-air Otto cycle. Compare the results with the gas Otto cycle. Assume $P_1 = 101.3$ kPa, $T_1 = 350$ K, and $\phi = 1.0, f = 0.15$, and $\gamma = 1.3$.

4.7   What compression ratio is required to have an imep of 1500 kPa with a methanol fuel-air Otto cycle, assuming $P_1 = 101.3$ kPa, $T_1 = 350$ K, $f = 0.05$, and $\phi = 0.95$?

4.8   What value of the equivalence ratio will maximize the imep for a gasoline fuel-air Otto cycle with a compression ratio of 10? Assume $P_1 = 101.3$ kPa, $T_1 = 350$ K.

4.9   Exhaust gas recirculation (EGR) is used in spark ignition engines to reduce the peak combustion temperature and the concentration of NO_x. EGR can be modeled in a fuel-air cycle by varying the residual fraction. What residual fraction is need to reduce T_3 to 2250 K in a gasoline engine with the following conditions: $r = 10$, $\phi = 0.95$, $P_1 = 101.3$ kPa, and $T_1 = 350$ K? If the original residual fraction was $f = 0.05$, what is the change in the imep, and why?

4.10   Plot the effect of supercharging on the volumetic efficiency of a four-stroke gasoline fuel-air Otto cycle model. Vary the inlet pressure from 50 to 130 kPa in 20 kPa steps, and assume $T_i = 300$ K, $\phi = 1.0$, and $P_e = 105$ kPa.

4.11   As the inlet pressure is throttled from 101.3 to 50 kPa, what is the change in the volumetric efficiency, imep, and residual fraction of a four-stroke gasoline fuel-air Otto cycle engine? Assume $T_i = 300$ K, $\phi = 0.9$, $r = 11$, and $P_e = 105$ kPa.

4.12   Derive Equations 4.20, 4.24, and 4.33.

Chapter **5**

Engine Testing and Control

5.1 INTRODUCTION

This chapter has two purposes: (1) to introduce engine instrumentation and analysis and (2) to discuss engine control hardware and software. One needs to instrument the engine to determine the value of engine parameters such as the engine torque, engine speed, fuel flow rate, air flow rate, emissions, cylinder pressure, residual fraction, coolant temperature, oil temperature, and the spark or fuel injection timing.

Some measurements are rather straightforward and require little, if any, explanation. The coolant temperature is easily measured by insertion of a thermocouple or thermistor into the coolant. Some of the measurements require analysis to obtain the desired result. For example, the air-fuel equivalence ratio is determined by measuring the composition of the exhaust gases and performing an analysis of the combustion equations.

The advent of digital microprocessors and advanced sensors has allowed the use of sophisticated engine control schemes to improve fuel economy and to reduce emissions. Adequate control of devices such as high speed fuel injectors is only possible with computer based control systems.

5.2 DYNAMOMETERS

The dynamometer is a device that provides an external load to the engine, and absorbs the power from the engine, as shown in Figures 5-1 and 5-2. The earliest dynamometers were brakes that used mechanical friction to absorb the engine power, hence the power absorbed was called the "brake horsepower."

The types of dynamometers currently used are hydraulic or electric. A hydraulic dynamometer or water brake is constructed of a vaned rotor mounted in a casing mounted to the rotating engine shaft. A continuous flow of water is maintained through the casing. The power absorbed by the rotor is dissipated in fluid friction as the rotor shears through the water. Adjusting the level of water in the casing varies the torque absorbed.

There are a number of different kinds of electric dynamometers. These include direct current, regenerative alternating current, and eddy current. The power absorbed in an electric dynamometer is converted into electrical energy, either as power or eddy currents. The electricity can then be dissipated as heat by resistance heating and transferred to a cooling water or air stream. In direct current or regenerative alternating-current machines, the electricity generated can be used, and transformers are available that allow it to supplement a power system. Historically, direct-current machines have offered the greatest testing flexibility but at the greatest cost.

Engine dynamometers can further be classified depending upon whether or not they also have the capability to motor an engine, that is, spin an engine not producing power as the starter motor of an automobile engine does. Hydraulic dynamometers cannot motor an engine. Strictly speaking, neither can eddy-current machines, but because they are often configured into a package with an electric motor to run an engine, for practical purposes the distinction is moot.

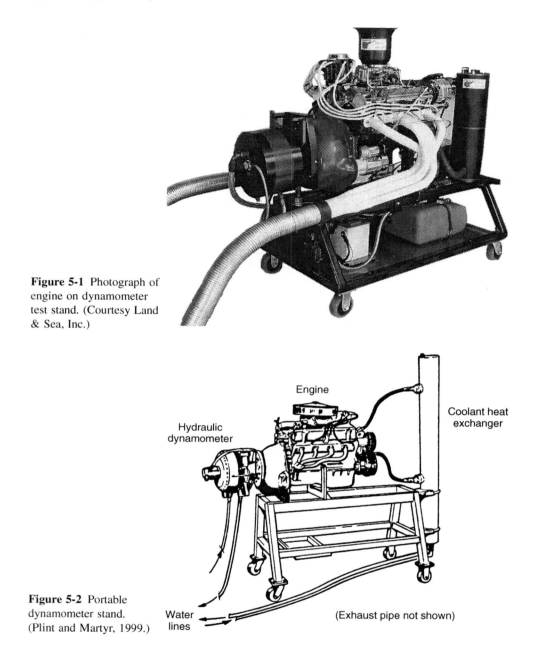

Figure 5-1 Photograph of engine on dynamometer test stand. (Courtesy Land & Sea, Inc.)

Figure 5-2 Portable dynamometer stand. (Plint and Martyr, 1999.)

The method most commonly employed to measure torque is shown in Figure 5-3. The dynamometer is supported by trunnion bearings and restrained from rotation only by a strut connected to a load cell. Whether the dynamometer is absorbing or providing power, a reaction torque is applied to the dynamometer. Hence, if the force applied by the strut is F, then the torque applied to the engine is

$$\tau = FR_0 \tag{5.1}$$

where R_0 is defined in Figure 5-3. The load cell measures the force F. For calibration, lever arms are located at R_1 and R_2 for hanging known weights.

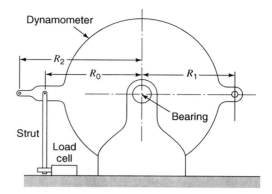

Figure 5-3 Torque measurement using a cradle mounted dynamometer.

Since the work done in rotating the engine's crankshaft through one revolution, or 2π radians, is $2\pi\tau$, it follows that for two- and four-stroke engines, respectively

$$\text{mep} = \frac{2\pi\tau}{V_d} \text{ (two-stroke)} \tag{5.2}$$

$$\text{mep} = \frac{4\pi\tau}{V_d} \text{ (four-stroke)} \tag{5.3}$$

If the engine is absorbing energy, then the brake mean effective pressure, bmep, is determined. If the engine is being motored, then the motoring mean effective pressure, mmep, is determined (which, as explained in Chapter 6, is a rough measure of the friction losses in an engine).

With an appropriate control system, the dynamometer can be used to control either engine speed or torque. For control of engine speed, the dynamometer applies whatever load is required to maintain that speed. For example, if the engine being tested were a spark ignition engine, then the response of the dynamometer to the operator increasing the inlet manifold pressure by opening the throttle would be to increase the load (resistance to turning or applied torque) to maintain the speed.

With torque control selected, the dynamometer maintains a fixed load. For example, the dynamometer's response to opening the throttle of a spark ignition engine would be to maintain a constant applied torque. In this case, the engine speed would increase to a point where the friction mean effective pressure in the engine would have increased by an amount equal to the increase in the net indicated mean effective pressure.

The engine speed is measured with optical or electrical techniques. One optical technique uses a disk with holes mounted on the revolving engine shaft. A light emitting diode is mounted on one side of the disk and a phototransistor is mounted on the other side. Each time a hole on the disk passes by the optical sensor, a pulse of light impinges on the phototransistor, which generates a periodic signal, the frequency of which is proportional to engine speed. Two electrical methods of engine speed measurement are introduced in Section 5.7.

Since about one third of the input fuel energy to the engine ends up as heat transfer to the coolant, a cooling tower or radiator is required for the dynamometer stand. The cooling tower will control the coolant temperature. A complete test stand has, in addition, provisions to control the fuel and air temperature, the atmospheric pressure, and the air humidity.

5.3   FUEL AND AIR FLOW MEASUREMENT

An old, but accurate and simple, way to measure the cumulative fuel flow to an engine is to locate the fuel supply on a weighing bridge and time the period required to consume a certain weight of fuel. The essence of such a system is shown in Figure 5-4. This method works equally well for both liquid and gaseous fuels. For liquid fuels, a pipette and stopwatch can be used, a method used to calibrate fuel flow meters.

A small positive displacement turbine can be installed in the fuel line as an electronic fuel flow transducer. Basically, the rotational speed of the turbine is proportional to the fuel flow rate. These transducers are also convenient in terms of minimizing bulk in the test cell, maximizing safety, and maintaining a clean fuel system. Unfortuantely, they measure a volumetric flow rate instead of a mass flow rate, and the calibration is weakly dependent upon the fuel viscosity. Thus, in practice, calibration curves have to be established as a function of the fuel temperature (and possibly pressure) and new ones generated if the fuel type is varied. The calibration curve needs to span the nominal range of fuel flow rates that can be as large as 50:1.

At considerably greater cost than the turbine-type flow meters, there are other types, such as the "Flowtron" meter, and the Coriolis flow meter. The "Flowtron" meter is the hydraulic equivalent of the Wheatstone bridge circuit. The bridge comprises four matched orifices and a recirculating pump. The external fuel flow through the meter generates a pressure imbalance that is proportional to the mass flow rate. Coriolis flow meters pass the fuel flow through a vibrating U-tube, as shown in Figure 5-5. The Coriolis force $(2\overline{\omega} \times \overline{U})$ acting on the flow will generate a twisting moment in the tube since the flow reverses direction as it passes through the U-tube. A strain gage mounted on the tube measures the magnitude of the twist, which is proportional to the flow rate. The accuracy and repeatability of the Flowtron and Coriolis meters is excellent, better than $\pm 0.5\%$ or less.

The return flow from fuel injection systems needs to be taken into account in fuel flow measurements, since it is of the same magnitude as the fuel flow through the fuel injectors. One approach is to cool the return fuel to the temperature of the supply fuel, and connect it to the supply fuel downstream of the fuel flow meter.

Air flow to engines cannot be measured with the same precision as the fuel flow. There are two main reasons for this: (1) the instrumentation available is at best accurate to within about $\pm 1\%$; and (2) it is harder to ensure that all of the air delivered to the engine is metered or retained. Air can leak into and out of the cylinder of engines, for example, through the valve guides. As a result, as recommended by Stone (1989), it is wise to measure air flow not only directly, but also by inference from exhaust gas analysis. A discussion of air flow meters follows. A discussion of air flow measurement via exhaust gas analysis is given in Section 5.4.

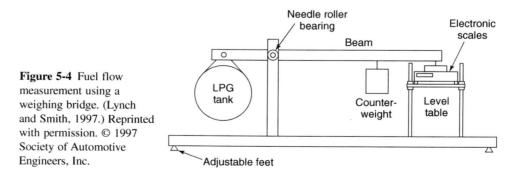

Figure 5-4 Fuel flow measurement using a weighing bridge. (Lynch and Smith, 1997.) Reprinted with permission. © 1997 Society of Automotive Engineers, Inc.

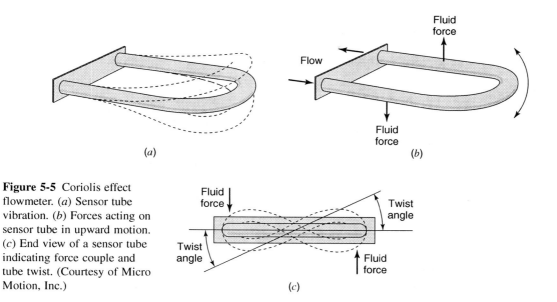

Figure 5-5 Coriolis effect flowmeter. (*a*) Sensor tube vibration. (*b*) Forces acting on sensor tube in upward motion. (*c*) End view of a sensor tube indicating force couple and tube twist. (Courtesy of Micro Motion, Inc.)

A common problem in measuring air flow is that the flow is unsteady or periodic, however, the available meters are usable only in a steady flow. A similar problem can be encountered in measuring fuel flow. The severity of the problem decreases with an increasing number of cylinders sharing a common intake manifold and with an increase in the volume between the meter and the intake ports. The volume acts as a fluidic capacitor to damp out fluctuations at the meter. The flow at the meter is smoother with multicylinder engines than with single-cylinder engines because the cylinders are out of phase with one another; thus as the peaks and valleys in the flow rates to individual cylinders are superimposed, the flow at the meter becomes smooth.

A solution for the worst case, that is, for a single-cylinder engine, that can be applied to steady-state engine testing is illustrated in Figure 5-6. All of the air to be delivered to

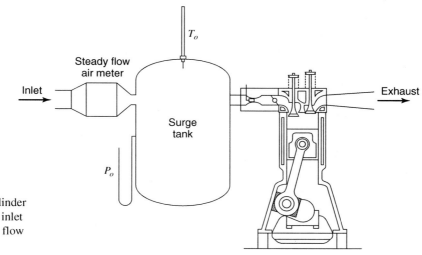

Figure 5-6 Single-cylinder engine equipped with inlet surge tank and steady flow air meter.

the engine is metered by a steady-state air flow meter located upstream of a surge tank. Kastner (1947) recommends that the volume of the surge tank be at least about 250 times the displacement volume of the engine.

The various types of air flow meters that can be used include the following:

ASME Orifice Often employed as a secondary standard to calibrate other meters; flow rate depends on the square root of the pressure drop across the orifice, so a range of orifice sizes is used to cover the air flow rate range.

Laminar Flow Meter A bundle of tubes (not necessarily round in cross section) sized so that the Reynolds number in each is well within the laminar regime; flow rate depends linearly on the pressure drop across the meter.

Critical Flow Nozzle A venturi in which the flow is choked; the flow rate is then linearly dependent upon the delivery pressure (an external compressor is thus required) and independent of the pressure in the surge tank.

Turbine Meter The air flow rate is linearly dependent upon the rotational speed of the turbine.

Hot Wire Meter A hot wire anemometer is inserted into the flow to measure the center-line velocity; the air flow rate is proportional to center-line velocity.

No matter which of the various methods is used, subsidiary measurements of temperature and pressure have to be made. Key components of systems employed using these various meters are identified in Figure 5-7. The calibration constants of the meters are a function of the Reynolds number of the flow through the meters.

For transient engine testing, only the hot wire meter can be used, as it can measure the instantaneous mass flow rate; it can also be used in steady-state testing where the required air box is viewed as a nuisance. It must, however, be used with care because it is possible that in some engines a flow reversal will occur and the meter does not know whether the flow is going forward or backward.

Correction factors are used to adjust measured data to standard atmospheric temperature and pressure conditions. The specific correction procedures are usually included in the laboratory practice manual.

5.4 EXHAUST GAS ANALYSIS

Electronic instruments that are easy to use and reliable are available for several of the exhaust constituents of interest to us including carbon dioxide, carbon monoxide, hydrocarbons, oxygen, and nitrogen oxides. Many laboratories, especially those studying or testing emissions, have a set of these instruments mounted together with a suitable sample handling system. We will briefly explain how the instruments for the species mentioned operate and then we will look at how experimental data can be used to compute the fuel-air ratio.

Carbon Dioxide and Carbon Monoxide

Nondispersive infrared analyzers (NDIR) are used for carbon dioxide and carbon monoxide. They can also be used for methane, hexane, nitric oxide, sulfur dioxide, ethylene, and water. The principle of operation of the infrared analyzer is based on the infrared absorption spectrum of gases. For the most part gases are transparent to electromagnetic radiation.

However, at certain frequencies in the infrared spectrum, the energy associated with a photon coincides with that required to change a molecule from one quantized energy level to another. At those frequencies a gas will absorb radiation. The concentration of a given compound in a gas mixture can then be determined from the absorption characteristics. As shown in Figure 5-8, carbon dioxide absorbs at about 4.2 μm; whereas carbon monoxide absorbs at about 4.6 μm. Thus, by using a radiation analyzer with a sensitivity as shown, one can detect carbon dioxide in a sample without interference from any carbon monoxide that may also be present.

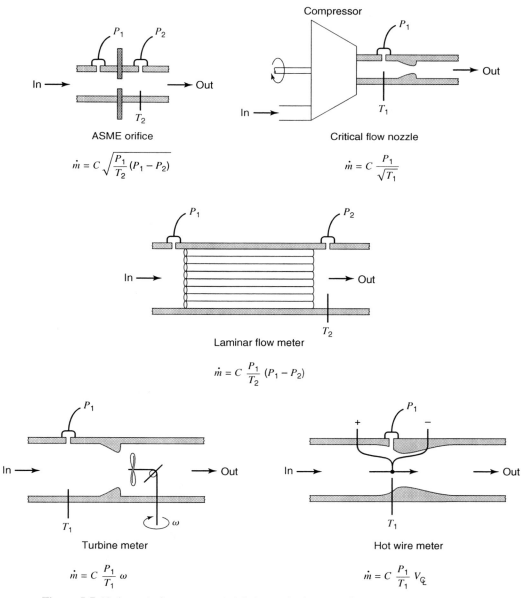

ASME orifice

$$\dot{m} = C \sqrt{\frac{P_1}{T_2}(P_1 - P_2)}$$

Critical flow nozzle

$$\dot{m} = C \frac{P_1}{\sqrt{T_1}}$$

Laminar flow meter

$$\dot{m} = C \frac{P_1}{T_2}(P_1 - P_2)$$

Turbine meter

$$\dot{m} = C \frac{P_1}{T_1}\omega$$

Hot wire meter

$$\dot{m} = C \frac{P_1}{T_1}V_{\mathcal{C}}$$

Figure 5-7 Various air flow meters and their required pressure/temperature measurements.

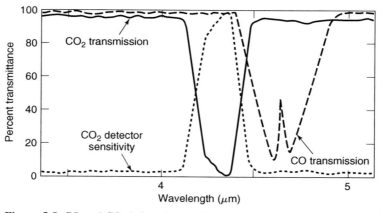

Figure 5-8 CO and CO_2 infrared transmittance spectra.

The operation of an infrared analyzer is shown in Figure 5-9. The analyzer passes infrared radiation through two cells; one a reference cell containing a nonabsorbing background gas and the other a sample cell containing a continuous flowing sample. The detector is filled with the component gas of interest to absorb infrared radiation transmitted through the two cells. The detector will absorb less radiation on the right than on the left because of the attenuation in the sample cell causing a diaphragm to deflect in proportion to the difference in the rates of energy absorption. Since the deflection will depend on the component density in the sample stream, the amount of deflection can be sensed and displayed on an electric meter calibrated to read in units of concentration. Notice that by filling the detector with the component of interest, one automatically obtains the desired sensitivity so as to eliminate interference from other components.

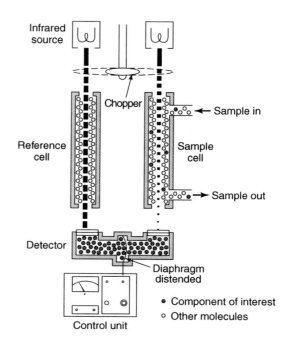

Figure 5-9 Infrared analyzer schematic. (Courtesy of Beckman Instruments, Inc.)

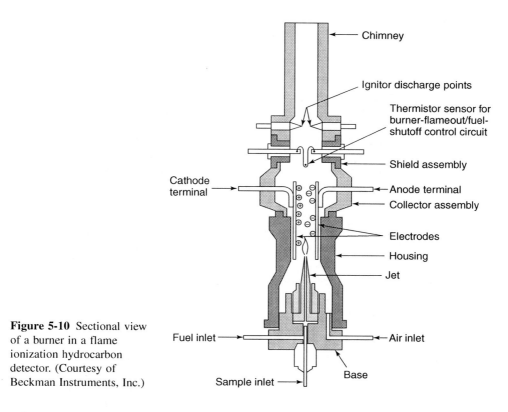

Figure 5-10 Sectional view of a burner in a flame ionization hydrocarbon detector. (Courtesy of Beckman Instruments, Inc.)

Hydrocarbons

Hydrocarbon detection is performed with a flame ionization detector (FID). Introduction of hydrocarbon molecules into a hydrogen-air flame produces, in a complex process, electrons and positive ions. By burning a sample exhaust gas in an electric field, positive ions are produced in an amount proportional to the number of carbon atoms introduced into the flame. An example of such a burner is shown in Figure 5-10. The sample is mixed with the hydrogen and helium and burned in a diffusion flame. The combustion products pass between electrodes producing an ion current. The hydrocarbon concentration is proportional to the ion current.

The magnitude of the current depends somewhat on the molecular structure of the hydrocarbon being detected. The characteristic response of a given molecular structure normalized by the response to methane is given in Table 5-1.

Table 5-1 FID Characteristic Response

Molecular structure	Approximate response
Methane	1.0
Alkanes	1.0
Aromatics	1.0
Alkenes	0.95
Alkynes	1.3
Carbonyl radical	0
Nitrile radical	0.3

According to Table 5-1, the following concentrations would all read approximately 1% on the meter:

1.00% of CH_4, methane

0.1% of $C_{10}H_{22}$, decane

0.132% of C_8H_{16}, octene

0.385% of C_2H_2, acetylene

The flame ionization detector gives no information about the type of hydrocarbons in the exhaust or their average hydrogen to carbon ratio. In recognition of this latter point, it is preferred to report measurements as ppmC (particles per million carbon) rather than as ppm CH_4 or C_3H_8 or $C_\alpha H_\beta$ equivalent.

Hydrocarbon speciation is performed with a Fourier transform infrared analyzer (FTIR), a gas chromatograph, or mass spectrometer. The FTIR operates on the same principle as the NDIR, but also computes a Fourier transform of the infrared absorption spectrum of the gas mixture. It is useful for the detection of methanol and formaldehyde. A gas chromatograph uses a solid or solid-liquid column to separate the hydrocarbon species. Detection limits for gas chromatographs used with a FID are on the order of 10 parts per billion carbon (ppbC).

Nitrogen Oxides

Nitrogen oxides (NO_x) are measured with a chemiluminescence detector (CLD). Chemiluminescence is the process of photon emission during a chemical reaction. When nitric oxide (NO) reacts with ozone (O_3), chemiluminescence from an intermediate product nitric dioxide (NO_2) occurs during the reaction. The amount of nitric oxide present is proportional to the number of photons produced.

A chemiluminescence reactor model is shown in Figure 5-11. The exhaust gas sample is first passed through a catalyst to convert nitric dioxide (NO_2) to nitric oxide (NO) prior to delivery to the reactor. The reactor has an exhaust gas sample port, an ozone inlet port, and an outlet port. The photons produced are measured with a photomultiplier. An optical filter is used to filter out photons from non-NO_2 chemiluminescence reactions that produce photons outside the wavelength band between 0.60 and 0.66 μm.

To simplify the reaction analysis, the reactor is assumed to be perfectly stirred, so the concentration of reactants is uniform throughout the reactor. The chemical reactions

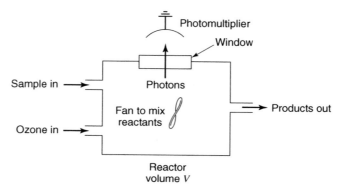

Figure 5-11 Model representation of the reactor in a chemiluminescent nitric oxide analyzer.

involved in this process are

$$NO + O_3 \rightarrow NO_2{}^* + O_2 \qquad (5.4)$$
$$NO_2{}^* \rightarrow NO_2 + photon \qquad (5.5)$$
$$NO_2{}^* + M \rightarrow NO_2 + M \qquad (5.6)$$

The asterisk in Equations 5.4 to 5.6 denotes NO_2 in an electronically excited state and M is a symbol chemists use to denote any molecule in the system. In Equation 5.4 the nitric oxide reacts with ozone to produce electronically excited nitric dioxide. The excited nitric oxides can be deactivated by emission of a photon or by collision with any other molecule, as indicated by Equations 5.5 and 5.6.

The conservation equation for excited nitric dioxide is:

$$\frac{d}{dt}(NO_2{}^*) = k_1(NO)(O_3) - k_2(NO_2{}^*) - k_3(NO_2{}^*)(M) - \dot{V}_f(NO_2{}^*) = 0 \qquad (5.7)$$

Equation 5.7 indicates that the rate at which excited nitric dioxide $NO_2{}^*$ is produced by reaction Equation 5.4 in the reactor is balanced by the chemical disappearance rates in reaction Equations 5.5 and 5.6 and the rate at which it flows out of the reactor. The parentheses in Equation 5.7 denote the concentrations in units of mol/m^3, and $\dot{V}_f$ is the volumetric flow rate of products leaving the reactor. The k's are the rate coefficients for the three reactions, Equations 5.4 to 5.6. When the system is in steady state, the rate of change of concentration of any species in the reactor is zero. The steady-state concentration of $NO_2{}^*$ is therefore

$$(NO_2{}^*) = \frac{k_1(NO)(O_3)}{k_2 + k_3(M) + \dot{V}_f} \qquad (5.8)$$

At steady state, the rate at which photons leave the system is equal to the rate at which they are produced in reaction 5.5. Therefore, the photon intensity, I, measured by the photomultiplier is

$$I = k_2(NO_2{}^*) = \frac{k_2 k_1(NO)(O_3)}{k_2 + k_3(M) + \dot{V}_f} \qquad (5.9)$$

The rate of change of concentration of nitric oxide is assumed to be zero, so

$$\frac{d}{dt}(NO) = +k_1(NO)(O_3) - \dot{V}_f(NO)_{Sample} = 0 \qquad (5.10)$$

Upon substitution of Equation 5.10 into Equation 5.9, the photon intensity is

$$I = \frac{k_2 \dot{V}_f(NO)_{Sample}}{k_2 + k_3(M) + \dot{V}_f} \qquad (5.11)$$

If the reactor is operated such that the ozone flow rate is large compared to the sample flow rate, then

$$\dot{V}_f = \dot{V}_{f, O_3} \qquad (5.12)$$
$$(M) = (O_3) = P/RT$$

Fixing the reactor temperature fixes the chemical rate constants k_1, k_2, and k_3, and fixing the pressure fixes the ozone concentration. Therefore, as indicated by Equation 5.11, for a given volumetric flow rate of ozone and exhaust gas sample, the photon intensity is proportional to the concentration of nitric oxide in the entering sample stream.

Oxygen

Both paramagnetic and polarographic analyzers are in common use for measuring oxygen concentration. In the latter, a Teflon® membrane separates the sample gas stream from a sensor that consists of a gold cathode and a silver anode immersed in potassium chloride gel. The rate at which oxygen diffuses through the membrane is proportional to its partial pressure in the same stream. At the cathode, the following reaction takes place

$$O_2 + 2H_2O + 4e^- \rightarrow 4OH^-$$

and at the anode

$$4Ag + 4Cl^- \rightarrow 4AgCl + 4e^-$$

There is a current flow because electrons are produced at the anode and consumed at the cathode. Because the oxygen concentration in the gel is small, this flow is proportional to the oxygen concentration in the sample.

Particulates

There are a number of techniques used to characterize and measure particulate emissions. These include light absorption, filter discoloration, and measurement of the total mass of particulates trapped on a filter paper. The exhaust particle size distribution can be measured using aerosol instruments such as the scanning mobility particle sizer (Wang and Flagan, 1990).

The absorption-type smoke meter uses the principle of light absorption by particles. A pump is used to draw undiluted exhaust gas into a measuring chamber that has a light source at one end and a photodiode at the opposite end. The attenuation of the beam of light by the exhaust is proportional to the particle concentration. The filter-type smoke meter draws a metered amount of exhaust gases through a filter paper. The blackening of the filter paper is compared against a "Bacharach grey scale."

Standards, such as SAE J 1280, that employ direct mass measurement also specify the use of a dilution tunnel in order to simulate the exhaust conditions near a vehicle. The particulates leaving the exhaust pipe are at a relatively high temperature and concentration in the outlet exhaust flow. These gases cool during the mixing process with the atmosphere, and the associated condensation and agglomeration processes will change the structure and density of the particulates in the exhaust gases. Dilution tunnels are used to standardize this near field (< 3m) mixing process.

A dilution tunnel is shown in Figure 5-12. The tunnel is typically 0.3 m in diameter. By flowing dilution air at a constant speed, typically 10 m/s, through a converging-

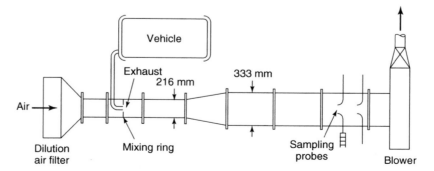

Figure 5-12 Exhaust gas dilution tunnel (SAE J 1280 Dynamic Dilution System B). Reprinted with permission. © 1992 Society of Automotive Engineers, Inc.

diverging nozzle, the venturi effect can be used to remove exhaust gas from the exhaust pipe. Mini-dilution tunnels with a 25-mm diameter have also been developed. Downstream of the nozzle the exhaust is well mixed with the dilution air. The diluted exhaust gas is sampled and drawn through Teflon®-coated glass fiber paper filters. The total particulate mass is trapped by the filter found by the increase in weight of the sample filter.

The carbon dioxide concentration can be measured in both the engine exhaust and the diluted sample in order to compute the dilution ratio, the ratio of dilute mixture flow rate to exhaust gas flow rate. The dilution ratio is typically about 10:1.

Fuel-air Equivalence Ratio

To solve for the fuel-air equivalence ratio from exhaust gas analysis we essentially invert the analysis that led to Table 3-4, but use a slightly different set of approximations. Let us write the combustion reaction as

$$C_\alpha H_\beta O_\gamma N_\delta + \frac{a_s}{\phi}(O_2 + 3.76\,N_2) \rightarrow n_1\,CO_2 + n_2\,H_2O$$
$$+ n_3\,N_2 + n_4\,O_2 + n_5\,CO + n_6\,H_2 + n_7\,CH_z\,(g) \tag{5.13}$$

where

$$z = \beta/\alpha \tag{5.14}$$

In this equation $CH_z(g)$ represents the gaseous hydrocarbons that a flame ionization detector records. The parameter z is the average hydrogen to carbon ratio of the hydrocarbons and is unknown. In engines that function properly, the exhaust gas contains negligible hydrocarbons, as far as atom balancing is concerned. They are included in the analysis because they are important in engines that misfire. A carbon and oxygen balance on this equation leads to an equation for the equivalence ratio

$$\phi = \frac{2\left(1 + \frac{1}{4}\frac{\beta}{\alpha} - \frac{1}{2}\frac{\gamma}{\alpha}\right)(y_1 + y_5 + y_7)}{2y_1 + y_2 + 2y_4 + y_5} \tag{5.15}$$

Notice that if we had an instrument to measure the mole fraction of water (y_2) in the exhaust gas (generally we do not), then the equivalence ratio could be determined with no further analysis or approximations. A complication arises in that most emission instruments do not function if water condenses in them. A common way to handle this is to condense the water from the sample prior to delivery to the instruments. With no water vapor in the sample, we then say the concentration is "dry," as opposed to "wet." The dry concentration depends on the amount of water condensed out. Denoting a dry concentration with a superscript zero, we can relate the dry concentration of any species i to the wet concentration by

$$y_i^0 = \frac{y_i}{1 - y_2} \tag{5.16}$$

Of course there is no physical significance to "dry water" ($i = 2$). It is mathematically convenient, however, to speak of dry water defined by Equation 5.16. In terms of dry concentrations, the equivalence ratio is

$$\phi = \frac{2\left(1 + \frac{1}{4}\frac{\beta}{\alpha} - \frac{1}{2}\frac{\gamma}{\alpha}\right)(y_1^0 + y_5^0 + y_7^0)}{2y_1^0 + y_2^0 + 2y_4^0 + y_5^0} \tag{5.17}$$

There is no change in the function since the wet water term $(1 - y_2)$ factors out. The water concentration is found from a hydrogen atom balance. However, that does not solve the problem completely because it introduces two new unknowns: y_6, the hydrogen concentration; and z, the hydrogen to carbon ratio of the exhaust hydrocarbons.

Spindt (1965) has found by experiment that it is satisfactory to assume

$$\frac{y_2}{y_1}\frac{y_5}{y_6} = 3.5 \tag{5.18}$$

Substitution of these relationships into the hydrogen balance gives, after much manipulation

$$y_2^0 = \frac{\dfrac{1}{2}\dfrac{\beta}{\alpha}\,(y_1^0 + y_5^0)}{1 + \dfrac{y_5^0}{3.5\,y_1^0}} \tag{5.19}$$

In experiments the measured fuel-air ratios and those determined from exhaust gas analysis agree to within $\pm 2\%$. For greater accuracy, Lynch et al. (1997) recommend that the NO concentration be measured and also included in the analysis.

5.5 RESIDUAL FRACTION

The residual fraction can be determined directly by use of a sampling valve to withdraw gases from the compression stroke for analysis with the same instruments already described. The mole fraction of carbon dioxide in those gases is

$$y_{CO_2} = y_r\, y''_{CO_2} \tag{5.20}$$

where y_r is the residual mole fraction and y''_{CO_2} is the carbon dioxide mole fraction in the exhaust gases. The residual mass fraction is, by Equations 5.20 and 3.53

$$f = \left[1 + \frac{m'}{m''}\left(\frac{y''_{CO_2}}{y_{CO_2}} - 1\right)\right]^{-1} \tag{5.21}$$

The molecular weights of the residual gas M″ and fuel-air mixture M′ are known because their compositions are known and Equation 3.14 can be applied.

A typical sampling valve is shown in Figure 5-13. The seat (2) has threads that are screwed into a receiving hole that provides access to the combustion chamber. In this application, the valve is mounted in the cylinder head, and when it is opened, gases in the cylinder are withdrawn. The poppet value (1) is opened by a trigger signal corresponding to a given crank angle and is open for 1 to 2 ms. It is electromagnetically opened by passing 2 to 5 amps of current through a coil (19). A spring (10) closes the valve when the current stops. A capacitance type of valve motion detector (24, 25) is incorporated into the valve, and the sampled gases flow out the port (4). By opening the valve at the same angle in successive cycles, it is possible to have a steady flow of gas for delivery to the exhaust gas analyzers.

It is also possible, though certainly more difficult, to determine the residual fraction by measuring the temperature at some angle during the compression stroke and applying the equation of state

$$PV = mRT \tag{5.22}$$

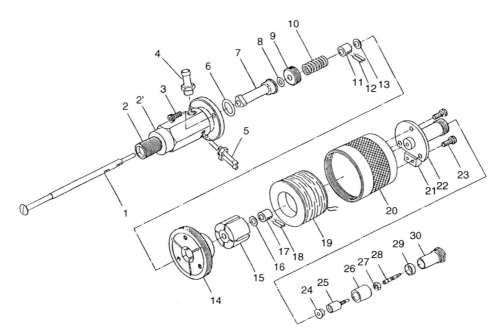

Figure 5-13 Typical sampling value. (Courtesy of Tsukasa Sokker, Ltd.)

5.6   PRESSURE-VOLUME MEASUREMENT
AND COMBUSTION ANALYSIS

A number of methods have been used to measure pressure as a function of cylinder volume. We will restrict our attention to piezoelectric transducers, since they are the method preferred by most engine laboratories.

The piezoelectric effect is the generation of an electric charge on a solid by a change in pressure. Consider that a crystalline solid is made up of positive and negative charges distributed over a space in a lattice structure. If the distribution of charges is nonsymmetrical, stressing the crystal will distort the lattice and displace positive charges relative to negative charges. A surface that was electrically neutral may become positive or negative. Substances such as table salt (NaCl) have a symmetrical distribution of charges and therefore stresses do not lead to piezoelectricity.

There are two primary piezoelectric effects: (1) the transversal effect in which charges on the x-planes of the crystal result from forces acting upon the y-plane, and (2) the longitudinal effect in which charges on the x-planes of the crystal result from forces acting upon the x-plane. In Figure 5.14(a), an example of the transversal effect, the quartz is cut as a cylinder with two 180° or three 120° sectors. The potential difference between the outer and inner curved surfaces of the cylinder is a measure of the gas pressure. In Figure 5.14(b), an example of the longitudinal effect, the quartz is cut into a number of wafers electrically connected in parallel. The potential difference is measured between the plane surfaces.

Piezoelectric transducers can be obtained with internal coolant passages and with a temperature compensator. Note that a rise in temperature will cause the housing to expand and thereby relieve the precompressed crystals from load. Piezoelectric transducers can also be obtained with flame shields to reduce flame impingement errors. Such errors are also reduced by coating the diaphragm with a silicone rubber to act as a heat shield.

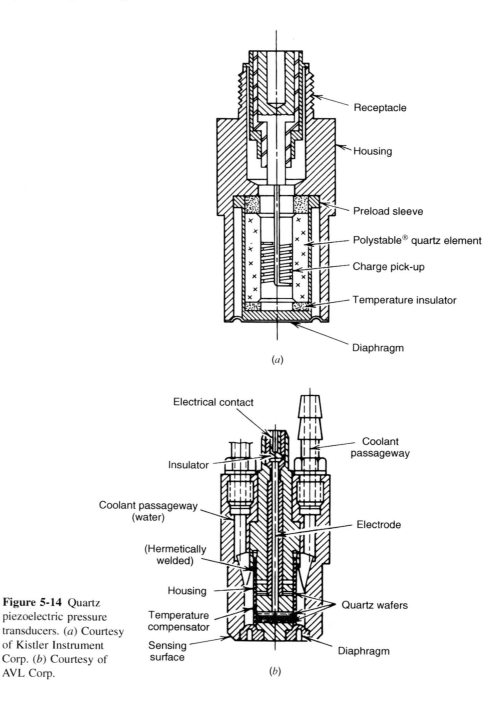

(a)

(b)

Figure 5-14 Quartz piezoelectric pressure transducers. (a) Courtesy of Kistler Instrument Corp. (b) Courtesy of AVL Corp.

Quantitative use of piezoelectric transducers is nontrivial. Care and methodical procedure is required. The reader who is using these transducers is advised to also consult SAE papers by Brown (1967), Lancaster, Krieger, and Lienesch (1975), and Randolph (1990). A typical system for measuring cylinder pressure as a function of cylinder volume is shown in Figure 5-15. A crank angle encoder is used to establish the top dead center position and the phasing of cylinder pressure to crank angle.

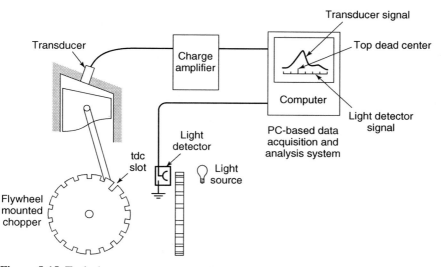

Figure 5-15 Typical pressure measurement system.

PC-based combustion analysis hardware and software are available to acquire and analyze the pressure data. The hardware consists of high speed A/D data acquisition systems and dedicated digital signal processors. The software performs statistical and thermodynamic analysis of the pressure data in real time. To study the effect of cycle to cycle variation, the analysis can be performed on an individual cycle and also on an ensemble average of many cycles. Measurements of cylinder pressure can be used to determine not only the location of peak pressure, but also the instantaneous heat release, burn fraction, and gas temperature. Foster (1985) reviews various models used to analyze the pressure data.

Representative pressure versus crank angle data is plotted in Figure 5-16. Using the known slider-crank geometry, the pressure data can be plotted as a function of cylinder volume, see Figure 5-17, and in the form $\log P$ versus $\log V$, see Figure 5-18. The nonreacting portions of the compression and expansion strokes are modeled as polytropic processes where $PV^n = $ constant. The polytropic exponent n can be found from the slope of the curve on the $\log P$ versus $\log V$ plot. The intake and exhaust pumping loop characteristics and the pressure sensor fluctuations are much more evident when the pressure data is plotted on logarithmic coordinates.

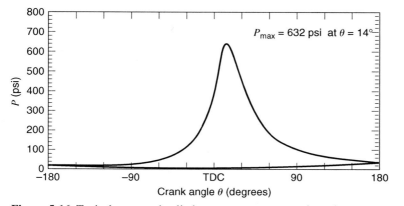

Figure 5-16 Typical measured cylinder pressure versus crank angle.

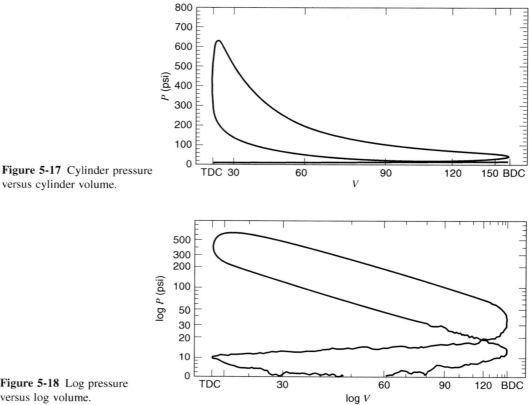

Figure 5-17 Cylinder pressure versus cylinder volume.

Figure 5-18 Log pressure versus log volume.

The instantaneous heat release, see Figure 5-19, is deduced from the cylinder pressure measurements through the use of the differential energy equation, Equation 2.30, rearranged in this chapter as Equation 5.23, where dQ_{wall} is the heat transfer to the wall.

$$\frac{dQ}{d\theta} = \frac{1}{\gamma - 1} V \frac{dP}{d\theta} + \frac{\gamma}{\gamma - 1} P \frac{dV}{d\theta} + \frac{dQ_{wall}}{d\theta} \tag{5.23}$$

The integral of the instantaneous heat release provides the burn fraction curve, shown in Figure 5-20. The average gas temperature can be computed from the measured pressure

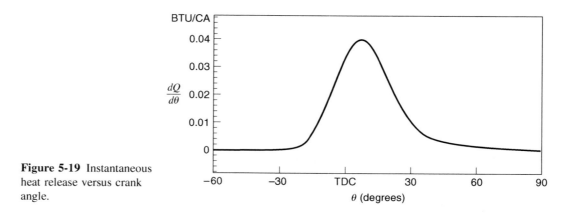

Figure 5-19 Instantaneous heat release versus crank angle.

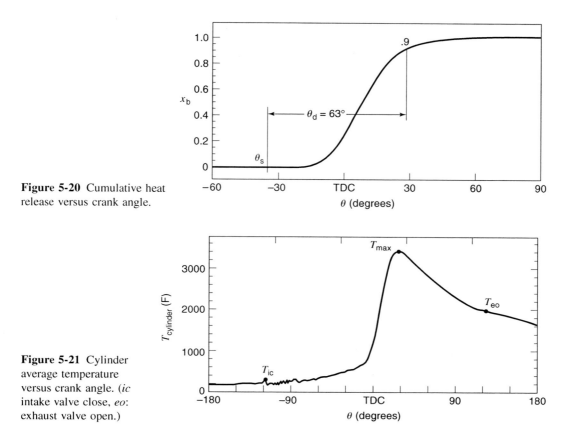

Figure 5-20 Cumulative heat release versus crank angle.

Figure 5-21 Cylinder average temperature versus crank angle. (*ic* intake valve close, *eo*: exhaust valve open.)

and the ideal gas equation, Equation 5.22. The temperature shown in Figure 5-21 is an average of the burned and unburned gas temperatures. The mixture mass m is evaluated from the conditions at a convenient reference, such as intake valve closing.

5.7   VEHICLE EMISSIONS TESTING

For emissions testing of engines in vehicles, a chassis dynamometer is used. The chassis dynamometer is used to put vehicles through a driving cycle, with the advantage of not having to instrument a moving vehicle. Chassis dynamometers were first developed for locomotives, and more recently for road vehicles. The U.S. Environmental Protection Agency requires chassis dynamometer testing of many classes of vehicular engines for emissions purposes. The word "homologation" is used to describe this certification process.

The chassis dynamometer is composed of a series of rollers, flywheels, and dynamometers, as shown in Figure 5-22. The vehicle to be tested is driven onto the top of the chassis dynamometer and its drive tires rotate between two rollers that are mechanically connected to flywheels, and electric dynamometers. The rolling inertia of the vehicle is simulated with rotating flywheels and electronic inertia. A cooling fan is needed to prevent the vehicle from overheating.

The United States, the European Community, and Japan have developed their own driving cycles that simulate a variety of driving conditions for various classes of vehicles. The United States driving cycle for passenger cars and light duty trucks is the Federal Test Procedure (FTP). The test procedure for heavy duty (gross vehicle weight > 8500 lbs.)

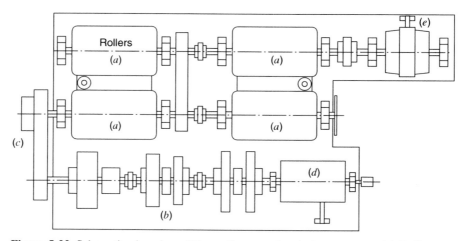

Figure 5-22 Schematic plan view of four roller type chassis dynamometer. (*a*) Rollers, (*b*) flywheel set, (*c*) clutchable chain drive, (*d*) d.c. machine, and (*e*) eddy current brake. (Plint and Martyr, 1999.)

highway engines is the EPA Transient Test Procedure and EPA Smoke Test Procedure. For locomotives, the test procedure is the Federal Locomotive Steady-State Test. Two United States driving schedules are shown in Figure 5-23.

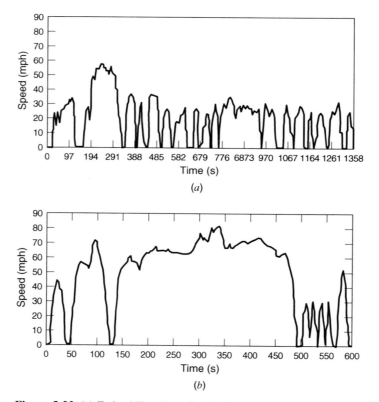

Figure 5-23 (a) Federal Test Procedure LA4 driving schedule. (b) US06 driving schedule. (Kwan et al., 1997.) Reprinted with permission. ©1997 Society of Automotive Engineers, Inc.

5.8 ENGINE SENSORS AND ACTUATORS IN VEHICLES

In automobile and truck engines, there is a need to provide real time information about the engine state to an engine control system, and inversely, the control system needs to be able to change the operating state of the engine. For example, to meet emission requirements, the fuel-air ratio needs to be measured and controlled to remain within a very narrow range. Various sensors and actuators are used for this purpose.

The measured parameters in an automobile engine include

- Exhaust gas oxygen concentration
- Crank angle
- Engine speed
- Throttle position
- Manifold and ambient pressure
- Inlet air and coolant temperature
- Intake air flow
- Knock

The operation of sensors that measure the above parameters is described below. In development for use in production vehicles are sensors for additional engine parameters. These include an optical combustion sensor to detect the peak combustion pressure, i.e., peak torque. Similarly, a crankshaft torque sensor is also under development. These two sensors can be used to maintain the engine at maximum brake torque conditions. In the emissions area a vehicle NO_x sensor is being developed to allow an engine to be operated with closed loop control of the NO_x levels.

Exhaust Gas Oxygen Concentration

The exhaust gas oxygen concentration is measured with an oxygen sensor. The oxygen sensor is used to control the air-fuel ratio, since the operation of a three-way catalyst requires that the air-fuel ratio be maintained within about 1% of stoichiometric. The sensor is constructed out of a thimble-shaped solid zirconium oxide (ZrO_3) electrolyte stabilized with yttrium oxide (Y_2O_3). The development of the zirconia sensor is detailed in Hamann et al. (1977). The interior and exterior surfaces of the electrolyte are coated with porous platinum to form interior and exterior electrodes. Electrochemical reactions on the electrodes produce negatively charged oxygen ions, which then produce a voltage across the electrolyte. The voltage produced depends on the oxygen ion flow rate, which in turn is proportional to the oxygen partial pressure at the electrodes, as indicated by Equation 5.24. The symbol F is the Faraday constant equal to 9.649×10^7 C/kmol. The oxygen partial pressure for a lean ($\phi = 0.8$) mixture is about 0.04 bar.

$$V_0 = \frac{RT}{4F} \ln \left(\frac{P_{O_2 \text{ atm}}}{P_{O_2 \text{ exhaust}}} \right) \qquad (5.24)$$

The voltage output is highly nonlinear at stoichiometric conditions, with a large change in the voltage between rich and lean conditions. If the exhaust mixture is rich, with a lack of O_2, oxygen ions will flow from the air side electrode across the electrolyte to the exhaust side. If the mixture is lean, with excess oxygen, oxygen ions also form at the exhaust gas electrode, and migration of oxygen ions across the electrode drops. For rich conditions, a voltage of about 800 mV is formed, and for lean conditions about 50 mV is produced,

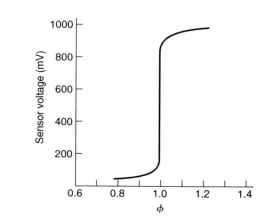

Figure 5-24 Oxygen sensor voltage versus equivalence ratio.

as indicated in Figure 5-24. A control set point voltage of about 0.5 V is used to maintain stoichiometric conditions. If the sensor voltage is below the set point voltage, the exhaust is considered by the control system to be lean, and vice versa.

The behavior of the oxygen sensor is temperature dependent, as the electrolyte needs to be above 280°C for proper operation. A heating electrode is sometimes embedded in the sensor to rapidly bring it up to operating temperature.

Crank Angle

The crankshaft position can be determined from measurements made in a number of locations on the mechanical drive train, for example, on the crankshaft, camshaft, or distributor shaft. Noncontacting methods are used, which are usually electrical, but optical methods have also been devised. A Hall effect sensor is commonly used on the camshaft or distributor shaft. The Hall effect, discovered by Hall in 1879, is due to electromagnetic forces acting on electrons in metals and semiconductors. If a current is passed through a semiconductor that is placed near a magnet, a voltage is developed across the semiconductor perpendicular to the direction of current flow $\vec{V}$ and perpendicular to the direction of magnetic flux $\vec{B}$. The voltage results from the Lorentz force $(\vec{B} \times \vec{V})$ acting on the electrons in the semiconductor. The voltage is proportional to the magnetic flux, so that if the magnetic flux is changed, the voltage will change.

There are a number of Hall effect sensor configurations. With a shielded field sensor, tabs are placed on a rotating disc, mounted on the distributor shaft, and the Hall sensor and magnet are placed on opposite sides of the tab. Each time the tab passes between the magnet and the Hall sensor, the magnetic reluctance of the tab will decrease the magnetic flux intensity at the sensor, which causes a corresponding decrease in the sensor Hall voltage. The voltage is independent of engine speed, so the Hall effect sensor can be used even if the engine is not running.

Engine Speed

The engine speed can be determined using the same sensing techniques used to measure the crankshaft position. The engine speed is found by measuring the frequency at which a tab passes by the position sensor. Many engines use a notch in the flywheel and a magnetic reluctance sensor. The sensor is an electromagnet whose induced voltage varies with the change in the magnetic flux. As the notch in the flywheel passes by the sensor, the induced voltage will first decrease, then increase. If the engine is not running, there will be no change in the magnetic flux, and the magnetic reluctance sensor will not produce any voltage.

Throttle Position

The throttle position is measured by a variable resistor or potentiometer attached to the axis of the sensor butterfly throttle valve. As the throttle rotates, the internal resistance of the sensor is changed proportional to the throttle angle change.

Manifold and Ambient Pressure

The manifold absolute pressure is used by the engine control system as an indication of engine load. Higher manifold pressures correspond to higher loads since the throttle is opened with increasing load. The manifold air pressure is measured by the displacement of a diaphragm which is deflected by manifold pressure. There are several types of diaphragm sensors. A common one is the silicon diffused strain gauge sensor. This sensor is a thin square silicon diaphragm with sensing resistors at each edge. One side of the diaphragm is sealed under vacuum, and the other side is exposed to the manifold pressure. The resistors are piezoresistive, so that their resistivity is proportional to the strain of the diaphragm. A Wheatstone bridge circuit is used to convert the resistance change to a voltage signal. The pressure fluctuations from the finite opening and closing of the intake valves are filtered out with a small diameter vacuum hose, so that the time-average pressure is measured.

Inlet Air and Coolant Temperature

The inlet air temperature and coolant temperature are measured with thermistors. The thermistors are mounted in a housing placed in the fluid stream. The coolant temperature is used to indicate engine warm-up and overheating states.

Intake Air Flow

The inlet air mass flow rate is measured by a constant temperature hot film anemometer. The principle of operation is very elegant. A small resistive wire or film placed in the air flow is heated by an electric current. The current required to maintain the film at a constant temperature above the ambient is proportional to the mass flow rate of the air. The sensor is placed at one leg of a Wheatstone bridge circuit so that the proper current is sent through the sensor to maintain a constant sensor temperature.

Knock

The onset of knock is detected by a knock detector. The methods used to detect knock include piezoelectric and magnetostrictive techniques. If a knock signal is sent to the engine control unit, the timing will be retarded until the knock ceases.

Actuators

The engine actuators control the exhaust gas recirculation (EGR), the fuel metering, and the ignition timing. The exhaust gas recirculation actuator is a vacuum operated spring

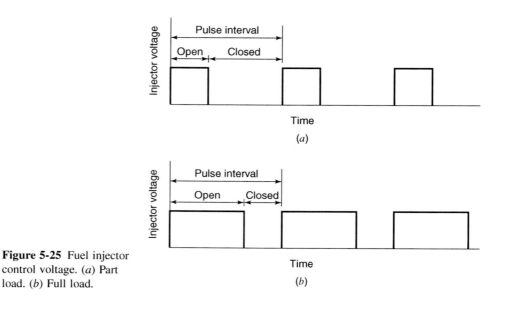

Figure 5-25 Fuel injector control voltage. (*a*) Part load. (*b*) Full load.

loaded diaphragm valve. The EGR valve is normally closed. When EGR is needed, the engine control system energizes a solenoid valve to supply vacuum to the diaphragm valve. The diaphragm valve opens to allow exhaust gases to flow directly into the intake manifold.

The fuel metering of a fuel injector is performed by a solenoid operated plunger attached to a needle valve. The plunger is normally closed, so that when the solenoid is not energized, no fuel can flow through the fuel injector. When the engine control unit energizes the solenoid, the valve is lifted and allows fuel to flow through the injector nozzle into the intake port. The fuel pressure is regulated, so the amount of fuel injected is proportional to the time that the valve is open. The fuel injector control voltage is pulse width modulated, as shown in Figure 5-25. The width of the solenoid voltage pulse depends on the engine load, and the frequency is proportional to engine speed. The pulse width and frequency are determined by the engine control system, as discussed next.

5.9   ENGINE CONTROL SYSTEMS

The function of an engine control system is not only to maintain a stable operating condition, but also to maximize the thermal efficiency and minimize the toxic emissions from the engine. These functions are carried out by a computerized engine control system, shown schematically for a laboratory test engine in Figure 5-26.

Engine control systems used in vehicles are composed of sensors, microprocessors, and actuators. The sensors measure temperatures, pressures, flow rates, and concentrations at various locations throughout the engine. The sensor information is provided to the engine control microprocessors to characterize the instantaneous thermodynamic state of the engine. Using various control algorithms, the engine control computer will use actuators to change the operating state of the engine as needed. For a spark ignition engine, the major control variables are the spark timing, valve timing, exhaust gas recirculation, and fuel injector flow. For a compression ignition engine, the major control variable is the fuel injector flow.

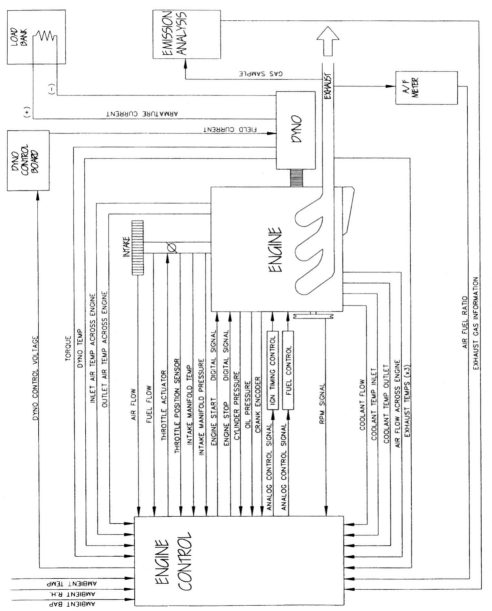

Figure 5-26 Engine control system. (Kirkpatrick and Willson, 1998.)

The engine control systems operate with both open and closed loop control. Some of the critical sensors, such as the oxygen sensor, operate properly only when an engine has warmed up. When an engine is cold, it operates on an open loop control without input from the oxygen sensor. The fuel flow rate is computed from the air mass flow rate and the requirement that the engine maintain a stoichiometric air-fuel ratio.

When the engine has warmed up, it switches to closed loop, and uses the oxygen sensor data to compute the required fuel flow rate. If the oxygen sensor indicates a rich mixture ($\phi > 1$), the pulse width of the fuel injector actuator signal is reduced to decrease the equivalence ratio. After a time lag, primarily the time required for the leaner fuel-air mixture to flow from the injector to the oxygen sensor located in the exhaust system, the oxygen sensor will indicate a lean mixture ($\phi < 1$), In response, the controller will increase the injection pulse width to enrich the mixture. With this type of closed loop control, known as a limit cycle, the air-fuel mixture continually oscillates about stoichiometric conditions, with a time average value of $\phi = 1$.

There are two types of control systems used on engines: memory based systems and adaptive systems. Memory systems store the optimum values of control variables such as spark timing and fuel injector pulse width for a range of engine operating conditions in a table. The optimum values include both efficiency and emissions considerations. For a given engine load (i.e., manifold air pressure) and engine speed, the engine control computer will "look up" the optimum timing and then change the spark timing to that optimum value. Engine maps obtained from tests described earlier in this chapter are used to locate the optimum values. Memory based control systems have the disadvantage of not accounting for part to part variation in engine components, the effect of deposits, and fuel property changes. Also, the optima determined from mapping measurements on a test engine are not exactly the same from engine to engine.

Adaptive systems determine an operating point from real time measurement of engine variables and subsequent correction of the look-up tables. The subsystems that have used adaptive control are the exhaust gas recirculation, evaporative emissions, idle air control, and air-fuel ratio control. Various calibrations need to be periodically reset due to wear, aging, and replacement of components, such as fuel injectors or sensors.

Subsystems that alternate between open and closed loop control, such as the air-fuel ratio, will perform periodic adaptive corrections. If a system is in closed loop control, it can compare the closed loop values with the open loop values. If there is a significant change, the open loop table values are corrected. The corrections to the look-up tables are also obtained by driving the vehicle through specified driving cycles, typically stop-and-go traffic with intermittent idle periods. During the driving cycle, the control system applies very small perturbations to the parameters, such as the ignition timing, and measures the response in other parameters, such as the fuel flow. Optimum values are then stored in the various tables such as the ignition-timing table. The type of memory used for adaptive memory reprogramming is a nonvolatile electrically erasable programmable read only memory (EEPROM).

The U.S. Environmental Protection Agency has mandated the use of on-board diagnostics (OBD) on passenger cars and light and heavy duty trucks built after 1994. These regulations are designed to detect emissions-related malfunctions. The diagnostic system uses various methods to communicate diagnostic information to operators. If a fault is detected, such as a faulty sensor, a fault code corresponding to the fault is stored. One type of diagnostic system flashes the "check engine" light using a variation of the Morse code signaling system. Alternatively, a computer or some similar type of digital analysis tool can be connected to the communications port of the diagnostic system.

If a sensor fails, the engine control system is able to maintain engine operation. It substitutes a fixed value for the sensor input, sends out a fault code, and continues to monitor the incorrect sensor input. If the sensor returns to normal limits, the engine control system will then return to processing the sensor data.

Some automotive engine control systems disable a number of fuel injectors if a cylinder head sensor indicates that the engine is overheating, perhaps from a loss of coolant. Varying and alternating the number of disabled fuel injectors controls the engine temperature. When a fuel injector is disabled, its cylinder works as a convective heat exchanger since air flow into and out of the cylinder continues to take place, and no combustion is occurring in the cylinder. One consequence of this strategy is that the engine will produce proportionally less power with disabled fuel injectors. Fuel injector disabling is also used if an engine or vehicle over-speed condition is detected. Once the speed is reduced, the engine returns to normal operating mode.

The sensor information stored by the engine control unit can also be sent via telemetry or wireless Internet to a host computer. Racing teams use this technique to debug and fine-tune high performance race cars. Many trucking firms are tracking the performance of their truck engines in this manner. In the near future, wireless communication technologies will be used by engine and vehicle manufacturers not only for routine engine diagnostics but also to collect information about long-term engine performance and reliability.

5.10   EFFECT OF AMBIENT PRESSURE AND TEMPERATURE

Engines are designed to operate over large ranges of ambient temperature, pressure, and humidity. Engine tests are corrected for the effects of ambient pressure and temperature using the ideal gas compressible flow equation, discussed in Chapter 7.

$$\dot{m}_a = A_{eff} \frac{P_0}{(RT_o)^{1/2}} \left\{ \frac{2\gamma}{\gamma - 1} \left[\left(\frac{P}{P_o} \right)^{2/\gamma} - \left(\frac{P}{P_o} \right)^{\frac{\gamma+1}{\gamma}} \right] \right\}^{1/2} \qquad (5.25)$$

If P/P_o is assumed to remain constant, then the mass flow rate through the engine varies as

$$\dot{m}_a \sim P_o/(T_o)^{1/2} \qquad (5.26)$$

For a constant fuel-air ratio, this implies that

$$\frac{\text{bmep}_m}{\text{bmep}_o} = \frac{P_m}{P_o} \left(\frac{T_o}{T_m} \right)^{1/2} \qquad (5.27)$$

where the subscript m denotes the measured conditions and the subscript o denotes standard atmosphere conditions at sea level.

For example, as an aircraft flies from sea level to an elevation of 6000 m, the density of the standard atmosphere decreases by 50% from 1.22 kg/m^3 to 0.66 kg/m^3. The bmep performance of a number of aircraft engines with altitude is shown in Figure 5-27. As a result of the density decrease, there is a 60% decrease in bmep at 6000 m relative to sea level. Equation 5.27 correlates the experimental data very well, as shown in Figure 5-27.

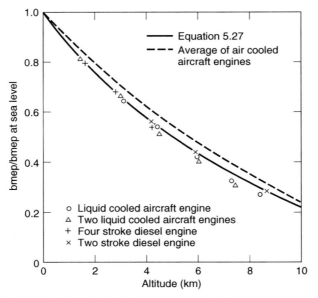

Figure 5-27 Effect of altitude on unthrottled engine performance at constant fuel-air ratio and coolant temperature.

5.11   REFERENCES

BROWN, W. L. (1967), "Methods for Evaluating Requirements and Errors in Cylinder Pressure Measurement," SAE paper 670008.

CHEUNG, H. and J. HEYWOOD (1993), "Evaluation of a One-zone Burn Rate Analysis Procedure Using Production SI Engine Pressure Data," SAE paper 932749.

FOSTER, D. (1985), "An Overview of Zero-dimensional Thermodynamic Models for IC Engine Data Analysis," SAE paper 852070.

GATOWSKI, J., E. BALLES, K. CHUN, F. NELSON, J. EKCHIAN, and J. HEYWOOD (1984), "Heat Release Analysis of Engine Pressure Data," SAE paper 841359.

HAMANN, E., H. MANGER, and L. STEINKE (1977), "Lambda-sensor with Y_2O_3-stabilized ZrO_2-ceramic for Application in Automotive Emission Control Systems," SAE paper 770401.

KASTNER, L. J. (1947), "An Investigation of the Airbox Method of Measuring the Air Consumption of Internal Combustion engines," *Proc. Inst. Mech. Eng,* Vol. 157, pp. 387–404.

KIRKPATRICK, A., and B. WILLSON (1998), "Computation and Experimentation on the Web with Application to Internal Combustion Engines," *ASEE Journal of Engineering Education,* Vol. 87, No. 5, pp. 529–537.

KWAN, S., D. PARKER, and K. NOLAN (1997), "Effectiveness of Engine Calibration Techniques to Reduce Off-Cycle Emissions," SAE paper 971602.

LANCASTER, D. R., R. B. KRIEGER, and J. H. LIENESCH (1975), "Measurement and Analysis of Engine Pressure Data," SAE paper 750026.

LYNCH, D. and W. SMITH (1997), "Comparison of AFR Calculation Methods Using Gas Analysis and Mass Flow Measurement," SAE paper 971013.

PLINT, M. and A. MARTYR (1999), *Engine Testing: Theory and Practice,* SAE, Warrendale, Pennsylvania.

RANDOLPH, A. (1990), "Methods of Processing Cylinder-Pressure Transducer Signals to Maximize Data Accuracy," SAE paper 900170.

RIBBENS, W. (1993), *Understanding Automotive Electronics,* Prentice Hall, Indianapolis, Indiana.

SPINDT, R. S. (1965), "Air-Fuel Ratios from Exhaust Gas Analysis," SAE paper 650507.

STONE, C. R. (1989), "Air Flow Measurement in Internal Combustion Engines," SAE paper 890242.

WANG, S. and R. FLAGAN (1990), "Scanning Electrical Mobility Spectrometer," *Aerosol Sci. and Tech.,* Vol. 13, pp. 257–261.

5.12   HOMEWORK

5.1   With reference to Section 5.4, the exhaust composition of a test engine is as follows:

$$CO_2 = 11.5\% \qquad NO = 310 \text{ ppm}$$
$$H_2O = 7.11\% \qquad NO_2 = 20 \text{ ppm}$$
$$N_2 = 77.99\% \qquad CH_4 = 350 \text{ ppm}$$
$$O_2 = 3.19\% \qquad C_3H_8 = 225 \text{ ppm}$$
$$CO = 0.06\% \qquad C_7H_{17} = 475 \text{ ppm}$$
$$H_2 = 0.01\%$$

Find the following:

(a) Wet concentration in ppm of HC and NO_x as would be indicated by heated flame ionization and chemiluminescence detectors, respectively. (Assume the FID responds to all carbon atoms equally.)

(b) Dry concentrations of CO_2, O_2 in percent, and CO in ppm.

(c) Fuel-air equivalence ratio if the hydrogen to carbon ratio of the fuel is 1.3.

5.2   Explain how Equation 5.22 could be used to measure the residual fraction.

5.3   A diesel engine operated on $C_{14}H_{27}$ produced exhaust gas of the following dry composition:

$$CO_2 = 6.22\% \qquad N_2 = 81.51\%$$
$$O_2 = 12.20\% \qquad NO_x = 400 \text{ ppm}$$
$$CO = 0.024\% \qquad HC = 200 \text{ ppm C}$$

(a) Explain how the method of hydrocarbon measurement can yield a situation wherein the sum of the exhaust constituents adds up to slightly greater than 100%.

(b) At what equivalence ratio was the engine operated? How would the answer differ if one neglected the carbon monoxide and hydrocarbons?

5.4   Manufacturers of laminar air-flow meters typically provide a calibration curve of the following form:

$$\dot{V}_{stp} = c_1 \Delta P + c_2 \Delta P^2$$

where

V_{stp} = volumetric flow rate at standard temperature and pressure

ΔP = pressure drop across the meter

(a) Use dimensional analysis to show how the constants c_1 and c_2 would change for measurements made at conditions other than standard temperature and pressure (STP).

(b) How can one determine the mass flow rate in (a) rather than the volumetric flow rate?

5.5   Assuming one-dimensional, isentropic steady flow of an ideal gas with constant specific heats, derive an expression for the constant C of the critical flow nozzle in Figure 5-7. The calibration constant depends on the nozzle throat area A, the gas constant R, and the ratio of specific heats γ. You may assume the upstream area is large enough that measured P_1, T_1 are stagnation properties.

5.6   Replot Figure 4-22 as ln P versus ln V. Estimate the polytropic exponent $n \equiv -d \ln V/d \ln P$ in the middle of both expansion and compression. How should these exponents relate to the specific heat ratio?

Chapter 6

Friction

6.1   INTRODUCTION

In this chapter we will examine the frictional processes in internal combustion engines. The friction forces in engines are a consequence of hydrodynamic stresses in oil films and metal to metal contact. Since frictional losses are a significant fraction of the power produced in an internal combustion engine, minimization of friction has been a major consideration in engine design and operation. Engines are lubricated to reduce friction and prevent engine failure. The friction energy is eventually removed as waste heat by the engine cooling system.

The frictional processes in an internal combustion engine can be categorized into three main components: (1) the mechanical friction, (2) the pumping work, and (3) the accessory work. The mechanical friction includes the friction of internal moving parts such as the crankshaft, piston, rings, and valve train. The pumping work is the net work done during the intake and exhaust strokes. The accessory work is the work required for operation of accessories such as the oil pump, fuel pump, alternator, and a fan.

We will use scaling arguments to develop relations for the dependence of the various modes of friction work on overall engine parameters such as the bore, stroke, and engine speed, then construct an overall engine friction model. The coefficients for the scaling relations are obtained from experimental data and implicitly include lubrication oil properties such as viscosity.

6.2   FRICTION MEAN EFFECTIVE PRESSURE

The engine friction reduces the indicated work of an engine, and is the difference between the indicated work and the brake work. It is useful to normalize the engine friction into a mean effective pressure (mep) form, as shown in Equation 6.1, so that the performance of engines of different displacement can be directly compared.

$$\text{fmep} = \text{imep} - \text{bmep} \qquad (6.1)$$

The indicated mean effective pressure (imep) is the net work per unit displacement volume done by the gas during compression and expansion, and the brake mean effective pressure (bmep) is the external shaft work per unit displacement volume done by the engine. With this definition, the pumping losses during the intake and exhaust strokes are considered to be part of the overall engine friction. It is customary to also include the work to run auxiliary components with the mechanical friction when the friction mean effective pressure is defined as in Equation 6.1. Accordingly, the friction mean effective pressure (fmep) is sum of the mechanical friction (mfmep), pumping (pmep), and accessory (amep) mean effective pressure.

$$\text{fmep} = \text{mfmep} + \text{pmep} + \text{amep} \qquad (6.2)$$

If a supercharger is connected to the engine crankshaft then

$$\text{fmep} = \text{imep} - \text{bmep} - \text{cmep} \qquad (6.3)$$

The term cmep is the work per unit volume to power the supercharger compressor.

The dependence of the mechanical friction mean effective pressure on the friction force, F_f, and other engine parameters depends on the friction regime and the lubrication surface geometry, such as sliding or rotating surfaces. The friction force, F_f, is the product of the friction coefficient f and the normal force F_n, and the friction power P_f is the product of the friction force F_f and a characteristic velocity U:

$$F_f = f F_n \tag{6.4}$$

$$P_f = F_f U \tag{6.5}$$

Therefore the mechanical friction mean effective pressure scales as

$$\text{mfmep} \sim \frac{P_f}{N V_d} \sim \frac{F_f U}{N V_d} \sim \frac{F_f U}{n_c N b^2 s} \tag{6.6}$$

where N is the engine speed, V_d is the displacement volume, b is the cylinder bore, s is the piston stroke, and n_c is the number of cylinders.

6.3   MEASUREMENTS OF THE FRICTION MEAN EFFECTIVE PRESSURE

There are a number of measurement methods used to determine the friction mean effective pressure. The most direct method is to use Equation 6.1. The indicated mean effective pressure is computed from cylinder pressure measurements during compression and expansion, and the brake mean effective pressure is determined from dynamometer measurements. Such measurements are discussed in Chapter 5.

A more commonly used method that does not require the measurement of cylinder pressure is motoring the engine without combustion. This method measures the motoring mean effective pressure (mmep), defined as the work per unit displacement volume required to rotate an engine operated without combustion. A direct current electric cradle type dynamometer is an appropriate apparatus for such a measurement. The dynamometer is mounted on bearings and is restrained from rotation by only a strut connected to a load beam. Whether the dynamometer is absorbing or providing power, a torque is applied to the dynamometer. The work done in rotating the crankshaft through one revolution is $2\pi\tau$. Hence for two- and four-stroke engines, respectively

$$\text{mep} = \frac{2\pi\tau}{V_d} \quad \text{(two stroke)}$$
$$\tag{6.7}$$
$$\text{mep} = \frac{4\pi\tau}{V_d} \quad \text{(four stroke)}$$

Equation 6.7 is valid whether or not the engine is firing. Although the sign of τ changes, we think of bmep and mmep as positive numbers. Practical application comes from the observation that under controlled conditions

$$\text{fmep} \cong \text{mmep} \tag{6.8}$$

Therefore, motoring tests are useful because it is much easier to measure mmep than fmep. Hence, most friction studies, for example, see Yagi et al. (1991), have been performed by application of Equation 6.8.

The motored friction and the fired friction are not the same since the thermodynamic state of the engine is different in each case. The motored friction can be more or less or greater than the fired friction, depending on the relative magnitudes of the various friction components. The piston and cylinder bore temperatures are lower for the motored state, which will increase the viscosity and thus the hydrodynamic friction of the lubricating film, but also increase the clearances which lowers the friction. Since the combustion pressure on the piston rings is much greater for the fired case, the ring friction for the fired case will be greater than the motored case. Similarly, the bearing loads will be greater for the fired case. The cooler exhaust gas for the motoring case will have a greater density, producing a larger motored pmep. Finally, temperature gradients within the engine are much less in the motored engine than in the fired engine. To obtain approximately the same heat loss, oil and coolant temperatures should be matched between the motored and fired engines.

Comparisons between motoring and firing friction are given in Figure 6-1 for a diesel engine, and Figure 6-2 for a spark ignition engine. The total fmep is of the order of 1 bar for both engines. The results illustrate the differences between a fired and a motored test. In both figures, the friction for motored and fired cases is determined by integrating the pressure-volume diagram to obtain the imep and subtracting the measured bmep as per Equation 6-1. Therefore the motored friction labeled in Figures 6-1 and 6-2 is the motored fmep, not the mmep, which is obtained from dynamometer measurements alone.

The diesel engine tested in Figure 6-1 is a six-cylinder, in-line engine. As the piston speed is increased from 4 to 12 m/s, the fmep increases nearly linearly, with no significant variation with load. The motored fmep is slightly greater than the fired fmep, a consequence of cooler wall temperatures and higher oil viscosity during motoring. Since the airflow in a diesel engine is unthrottled, the pumping friction is not a function of the load.

The spark ignition engine of Figure 6-2 is a four-cylinder, in-line engine with a 12:1 compression ratio. The mechanical friction (mfmep) and pumping (pmep) are plotted versus load for a constant mean piston speed of 6.1 m/s. As the throttle of the spark

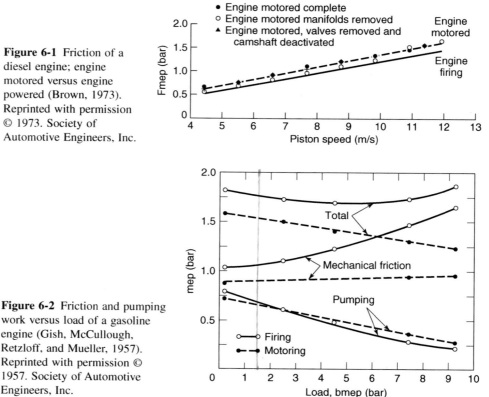

Figure 6-1 Friction of a diesel engine; engine motored versus engine powered (Brown, 1973). Reprinted with permission © 1973. Society of Automotive Engineers, Inc.

Figure 6-2 Friction and pumping work versus load of a gasoline engine (Gish, McCullough, Retzloff, and Mueller, 1957). Reprinted with permission © 1957. Society of Automotive Engineers, Inc.

ignition engine is opened to meet an increased load, the pmep decreases for both the fired and motored cases. However, the fired mechanical friction increases with load due to the increased gas pressure, while the motored mechanical friction is unchanged. Consequently, for this engine, the total fired friction is greater than the motored total friction. Note that Equation 6.8 is a rather crude approximation at high loads for this spark ignition engine.

The most important advantage of a motoring test is that the engine can be partially disassembled, which precludes firing, and motored to study the distribution of friction among the various parts. Figure 6-3 is typical of the results that can be obtained by this method. In this experiment by Brown (1973), a diesel engine was systematically disassembled, and the resulting fmep measured. To measure the pumping loss, the first items removed were the manifolds and turbocharger. Next the valves were removed and the camshafts disconnected. The resultant curve C is nearly identical to curve J, which is the mechanical friction, determined for a motored engine. The small difference is caused by a change in the gas load, which without valves is nonexistent. Curve K is the fmep of the fired engine and, as illustrated, there is a small difference between fired and motored

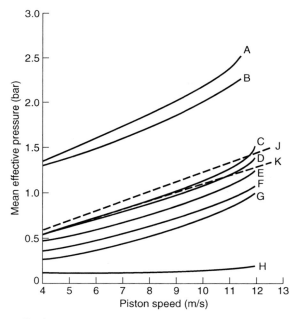

Figure 6-3 Motored friction (mmep) during disassembly of a diesel engine (Brown, 1973). Reprinted with permission © 1973. Society of Automotive Engineers, Inc.

Engine setup

A Complete engine
B Complete engine minus intake and exhaust manifolds
C Setup B minus all valves, camshaft and measured
    pumping loss
D Setup C minus water pump
E Setup D minus oil pump
F Setup E minus all top and intermediate piston rings
G Setup E minus all piston rings
H Crankshaft only
J Motored engine friction per imep meter measurements
    on complete engine
K Idle and load engine friction per imep meter measurements
    on complete engine

friction. It was concluded that the difference between curve A and curve C is due to pumping losses, and a nonzero imep during motoring. As disassembly proceeds, it becomes clear that pistons and rings are responsible for about one half to three fourths of the friction. The last curve, H, shows the friction due to spinning only the crankshaft.

6.4   FRICTION COEFFICIENT

The friction coefficient, defined in Equation 6.9, is the ratio of the shear or tangential stress to the normal stress or pressure acting on a surface. The friction coefficient is a function of the type of lubrication between two surfaces. Generally speaking, there are three friction regimes: hydrodynamic, mixed, and boundary friction.

The three friction regimes are shown on a Stribeck diagram in Figure 6-4. The Stribeck diagram plots the friction coefficient as a function of a dimensionless Stribeck variable $\mu N/P$, where μ is the lubricant dynamic viscosity, N is the relative rotational speed between surfaces, and P is a normal stress.

In hydrodynamic friction, the surfaces are completely separated by a liquid film. The liquid is pressurized to keep the surfaces separated. This is a preferred mode of operation since mechanical wear is minimized. The shear stress is entirely due to the lubricant viscosity. Therefore, the friction coefficient f is

$$f = \frac{F_f}{F_n} = \frac{\tau_f}{P} = \frac{\mu}{P}\frac{du}{dy} \tag{6.9}$$

Crankshafts, connecting rods, and piston rings are designed to operate in the hydrodynamic regime as much as possible.

As the lubricant pressure is increased or the speed decreased, the oil film thins out to the point where its thickness is comparable to the size of the surface irregularities. This is the mixed lubrication regime. The liquid film no longer separates the surfaces, and intermittent metal to metal contact occurs, causing an increase in the friction coefficient. The friction coefficient is a combination of hydrodynamic and metal to metal contact friction.

With further increase in load or decrease in speed, the boundary layer regime is reached. Boundary lubrication occurs at either end of the piston stroke when the piston velocity approaches zero, in relatively slow moving valve train components, and during engine start-up and shut down. In boundary lubrication, oil film patches separate the sliding surfaces where

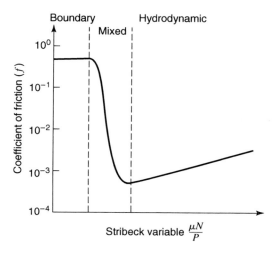

Figure 6-4 Stribeck diagram showing regimes of friction and orders of magnitude of the coefficient of friction.

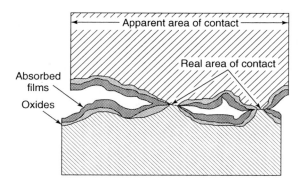

Figure 6-5 Metal to metal contact in boundary lubrication (Rosenberg, 1982). Reprinted with permission © 1982. Society of Automotive Engineers, Inc.

the thickness is just a few molecular diameters of the lubricant as shown in Figure 6-5. The force required to cause tangential motion in boundary lubrication is approximately the area of contact times the shear strength of the adsorbed oil layer, σ_o. It is important to know how the lubricant can be adsorbed by the surfaces, how rough the surfaces are, and whether or not the surface molecules themselves are prone to adhering to one another. The real area of contact depends primarily on the applied load, the yield strength, and the asperities of the softer material. The yield stress, σ_m, of the softer material balances the applied load, so that as the load increases, there is a proportional increase in the area of contact.

The coefficient of friction in the boundary layer regime is

$$f = \frac{\sigma_o}{\sigma_m} \tag{6.10}$$

which is independent of the engine design and operating parameters such as engine speed. The friction depends on the properties of the lubricant except the viscosity, and the properties of the sliding surfaces, such as the roughness, plasticity, shear strength, and hardness. Minimum friction is obtained by use of a low shear strength oil and hard surfaces. Additives are added to lubricating oils which preferentially adsorb to bearing surfaces and lower the coefficient of friction. The properties of lubricating oils are discussed in Chapter 10.

In the following sections, Sections 6.5 through 6.9, friction correlations are developed from scaling analysis, and correlated with experimental data for 2 L four-cylinder spark ignition automotive class engines.

6.5   JOURNAL BEARINGS

The journal bearings used in an internal combustion engine include the main crankshaft bearings, connecting rod bearings, and accessory bearings and seals. A crankshaft with main and connecting rod journal bearing shafts is shown in Figure 6-6. Note the lubrication ports on the shafts and the different main and connecting rod bearing diameters. A journal bearing is shown schematically in Figure 6-7. The journal bearing is cylindrical, with as smooth a finish as possible. The journal bearing operates in the hydrodynamic friction regime during normal operation. During startup and shutdown, the friction regimes are mixed and boundary friction due to the low bearing speed. The difference in the diameters of the inner shaft and the outer bearing create a thin annulus through which the lubricant flows. At rest the shaft sits in contact with the bottom of the bearing. As the shaft begins to rotate, its center-line shifts eccentrically in the bearing maintaining metal to metal contact with the bearing. As the engine speed increases, the bearing will "aquaplane" and enter the hydrodynamic

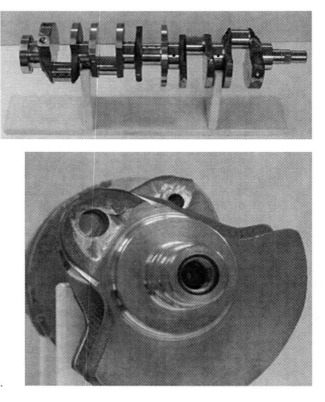

Figure 6-6 Engine crankshafts
(Courtesy Norton Manufacturing.).

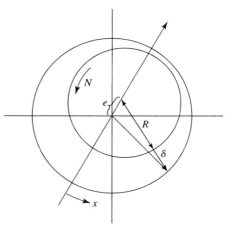

Figure 6-7 Two-dimensional journal bearing
geometry.

regime. The shaft pumps the oil around the annulus. If the bearing is not sealed, oil will leak out at the ends, so oil is pumped at relatively low pressures through internal passages to the bearing annulus.

The loads on the bearings vary significantly with crank angle, the connecting rod geometry, and the combustion gas pressure, as shown in Figure 6-8. To meet the bearing loads, crankshaft main bearings for automotive spark ignition engines are typically sized to be about 65% of the cylinder bore, and connecting rod bearings are sized to be about 55% of the cylinder bore. Bearing lengths are sized at 35 to 40% of the cylinder bore.

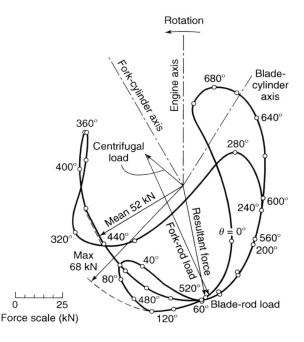

Figure 6-8 Bearing load diagram. Polar diagram showing the magnitude of the resultant force on the crankpin of the Allison V-1710 engine and its direction with respect to the engine axis. Engine speed = 3000 rpm, imep = 16.5 bar (Burwell, 1949).

Journal bearings are relatively soft to allow foreign particles to be embedded without damaging the journal, and have a low melting point to reduce the risk of seizure. Materials used in journal bearings in internal combustion engines are composed of babbit, which is an alloy of lead (Pb), tin (Sn), antimony (Sb), and copper (Cu) electroplated onto a steel substrate and various aluminum alloys. Typical babbit thicknesses vary from 10–100 μm, with the major component either lead or tin. A representative lead-based babbit is composed of approximately 89% Pb, 9% Sn, and 2% Cu.

A minimum oil film thickness is required to maintain hydrodynamic lubrication as operation in the mixed or boundary lubrication regime increases engine wear. This thickness is about 2μm. The oil film thickness can be determined by resistance and capacitance measurements (Tseregounis et al., 1998). A plot of oil film thickness is shown in Figure 6-9 for a six-cylinder engine. Note that the minimum film thickness of about 2μm occurs during the power stroke of the cylinder nearest the bearing.

In addition to the crankshaft and connecting rod journal bearings, there are journal bearings on the camshaft rotating at one half the engine speed. Also the piston pin on the connecting rod oscillates back and forth without completing a revolution.

To first order, the flow in a journal bearing can be modeled as Couette flow with dynamic viscosity μ in which the velocity gradient for a no-load condition in a bearing of diameter D_b, length L_b, and annular clearance c is given by

$$\frac{du}{dr} = \pi D_b N/c \tag{6.11}$$

The no-load friction force, F_f, in the bearing is therefore

$$F_f = A\mu \frac{du}{dr} = (\pi D_b L_b)\, \mu \pi D_b N/c \tag{6.12}$$

$$= \pi^2 \mu D_b^2 L_b N/c$$

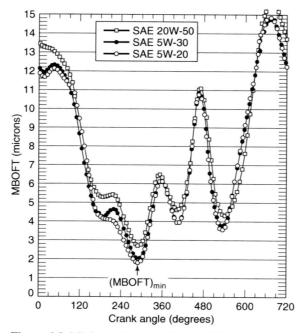

Figure 6-9 Minimum bearing oil film thickness
(MBOFT) (Tseregounis et al., 1998). Reprinted with
permission © 1998. Society of Automotive Engineers, Inc.

Equation 6.12 is known as Petroff's equation. The friction coefficient in the model is

$$f = \frac{F_f}{PDL} = \pi^2 \left(\frac{\mu N}{P}\right) \frac{D_b}{c} \tag{6.13}$$

In the hydrodynamic regime, the Petroff flow model predicts that the friction coefficient increases linearly with the Stribeck variable, and the slope is dependent on the ratio of the bearing diameter to clearance. The friction of real bearings approaches the Petroff value at high values of the Stribeck variable, $\mu N/P$. More rigorous analyses include the bearing load and eccentricity, but the results still retain the scaling of the Petroff equation.

The friction mean effective pressure of a journal bearing array with n_b bearings, such as the crankshaft main bearings or the connecting rod bearings, scales linearly with engine speed, assuming constant bearing clearance and oil viscosity, as shown in Equation 6.14:

$$\text{fmep}_{\text{bearings}} \sim \frac{F_f U}{n_c N b^2 s} \sim \frac{(\mu n_b D_b^2 L_b N/c)(ND_b)}{n_c N b^2 s} \tag{6.14}$$

$$\sim \frac{\mu n_b N D_b^3 L_b}{n_c b^2 s} = c_b \frac{n_b N D_b^3 L_b}{n_c b^2 s}$$

Patton et al. (1989) suggest a proportionality constant $c_b = 3.03 \times 10^{-4}$ (kPa-min/rev-mm) for spark ignition engines. The oil properties are included in the constant, c_b.

Bishop (1964) recommends a similar fmep equation, which also scales as ND_b^3, for both spark ignition and compression ignition engines. The crankshaft and connecting rod bearing friction are both included in the Bishop correlation:

$$\text{fmep}_{\text{bearings}} \sim 41.37 \, K \left(\frac{b}{s}\right)\left(\frac{N}{1000}\right) \tag{6.15}$$

where

$$K = \left(D_{mb}^2 \, L_{mb} + \frac{D_{cb}^2 L_{cb}}{m} + D_{as}^2 \, L_{as}\right)\frac{1}{b^3} \tag{6.16}$$

and D_{mb} is the main bearing diameter, L_{mb} is the total main bearing length per number of cylinders, D_{cb} is the connecting rod bearing diameter, L_{cb} is the rod bearing length, m is the number of pistons per rod bearing, D_{as} is the accessory or balancing shaft bearing diameter, and L_{as} is the total length of all accessory shaft bearings per number of cylinders. Typical values of K are 0.14 for spark ignition engines and 0.29 for diesel engines, as the larger journal bearings in diesel engines will have a greater fmep relative to spark ignition engines.

The crankshaft bearing seals operate in a boundary lubrication regime, since the seals directly contact the crankshaft surface. As the normal force, which is the seal lip load, is constant, the friction force will be constant, and the friction mean effective pressure of the crankshaft bearing seal will be independent of engine speed, and will scale as

$$\text{fmep}_{\text{seals}} \sim \frac{ND_b}{n_c N b^2 s} = c_s \frac{D_b}{n_c b^2 s} \tag{6.17}$$

Patton et al. (1989) suggest a proportionality constant $c_s = 1.22 \times 10^5$ kPa-mm^2.

6.6   PISTON AND RING FRICTION

A spark ignition engine piston head is shown in Figure 6-10. A piston and connecting rod are shown in Figure 6-11. A schematic of a diesel piston assembly is given in Figure 6-12. The friction of the piston and rings results from contact between the piston skirt

Figure 6-10 Piston head for a spark ignition engine. (Courtesy Mahle, Inc.)

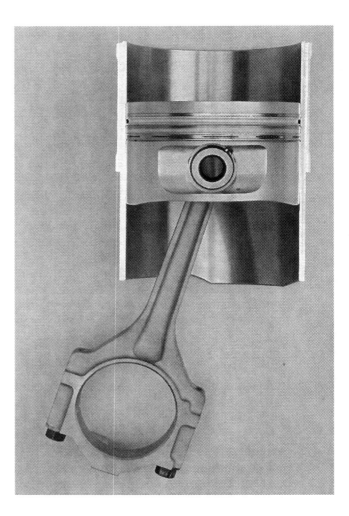

Figure 6-11 Piston and
connecting rod.
(Courtesy Mahle, Inc.)

and the ring pack with the cylinder bore. The cylinder bore is rougher than a journal bearing
bore since the cylinder bore must retain some oil during operation. The piston ring pack
has three main functions. The rings seal the combustion chamber, control the lubrication
oil flow, and transfer heat from the piston to the cylinder. In order to preserve a seal against
the cylinder bore, each ring has some amount of radial tension. Current ring pack designs
generally use three piston rings; two compression rings and an oil control ring.

Common types of piston rings are shown in Figure 6-13. Various cross sections are
available, such as rectangular, crown or barrel face, taper, and Dykes. The top compression ring can have a bevel on the upper inside edge of the ring can have a bevel on the
bottom inside edge to produce a "postive twist," and a seal on the bottom edge of the ring.
The second ring can have a bevel on the bottom inside edge to produce a "reverse twist,"
which will help scrape oil off the cylinder wall. The oil control ring typically has two narrow rails and an expander, which wipes off excess oil from the cylinder liner.

Ring materials include cast iron, ductile (nodular) iron, and stainless steel. A ring
pack will often have a ductile iron top ring, a cast iron second ring, and a stainless steel
oil ring. Coatings available for ring faces are molybdenum and chrome. The space between
the side of the ring and the groove is called the side clearance, and the space between the

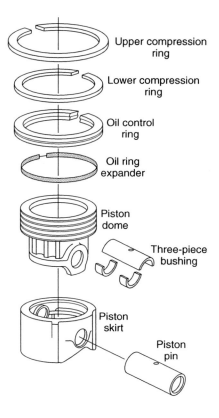

Figure 6-12 Piston assembly schematic for a
compression ignition engine (Merrion, 1994).
Reprinted with permission © 1994. Society of
Automotive Engineers, Inc.

back of the ring and the groove is termed, not unexpectedly, the back clearance. The back
clearance is minimized in high performance engines to increase the sealing effectiveness.
The side and back clearances allow the pressure in the piston groove to follow the cylinder
pressure, so as the cylinder pressure increases, the increasing groove pressure presses the
compression rings more firmly against the cylinder wall, increasing the sealing. Piston
rings are split, forming an end gap, so that they can be slipped onto the piston and also
accommodate thermal expansion of the piston ring. Engine power is sensitive to the size
of the end gap, so the end gap is minimized in high performance engines.

The piston skirt is designed to meet the side thrust forces originating from the rotation
of the connecting rod. The side thrust force on the piston skirt depends on the crank angle,
cylinder pressure, piston speed, acceleration, and connecting rod geometry. In addition,
the wrist pins are offset slightly to reduce the side thrust force. Offsetting the wrist pins
will also reduce piston noise. Figure 6-14 illustrates a force balance applied to the piston.

Results of piston force balance calculations for a particular engine are given in Figure 6-15. The computation requires data from the P-V diagram, the mass and moment of
inertia of the piston and connecting rod, and the engine speed. The largest side thrust forces
occur during the expansion stroke when the cylinder pressures are largest. The side thrust
force changes direction as the piston passes through top and bottom center since the connecting rod changes sides. As a consequence, the "left" side of the clockwise rotating engine is subjected to larger forces than the "right" side. In this case, the left side is referred
to as the major thrust side whereas the right side is called the minor thrust side. Since the
friction work is the product of the friction force and piston velocity, the friction work will
be the largest during the middle of the stroke where the piston velocity is greatest.

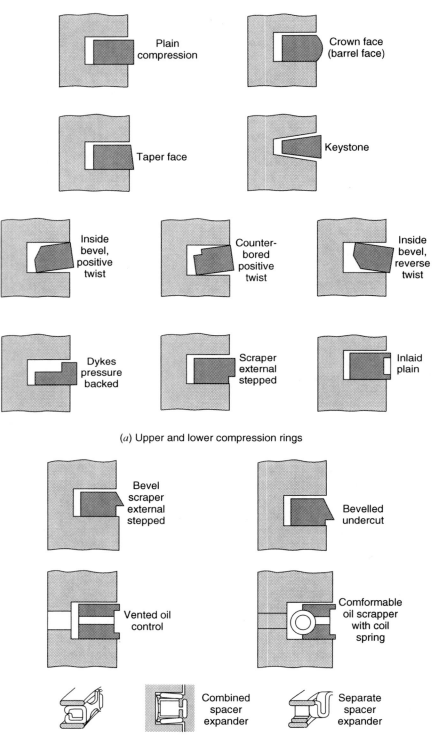

(a) Upper and lower compression rings

(b) Oil control rings

Figure 6-13 Common types of piston rings. (a) Upper and lower compression rings, (b) oil control rings.

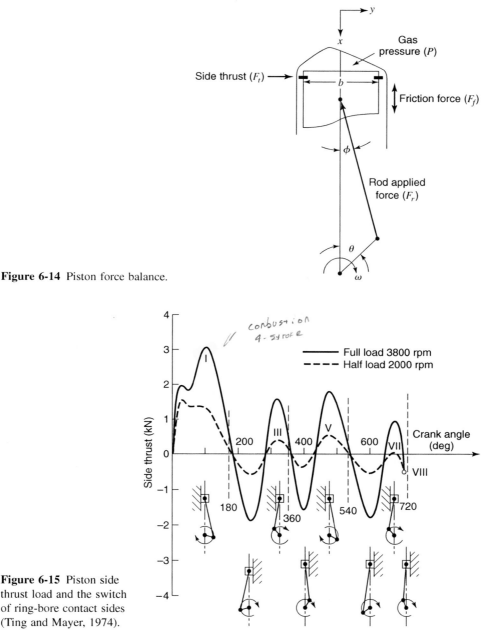

Figure 6-14 Piston force balance.

Figure 6-15 Piston side thrust load and the switch of ring-bore contact sides (Ting and Mayer, 1974).

The friction due to the piston-ring assembly has been directly measured in a friction research engine. Typical results are shown in Figure 6-16. The following features should be noted (Taylor, 1985):

- Friction forces are comparable on the compression and exhaust strokes.
- Friction forces occurring during expansion are about twice as large as those occurring during any other stroke.
- Friction forces tend to be high just after top and bottom dead center, which Taylor hypothesized was due to metallic contact between the rings and the cylinder wall.

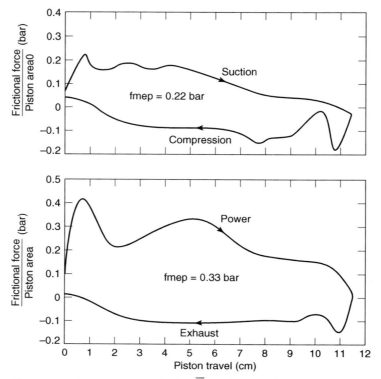

Figure 6-16 Piston and ring friction. $\overline{U}_p = 4.57$ m/s, bmep = 5.78 bar, $T_c = 356$ K, $T_{oil} = 356$ K (Leary and Jovellanos, 1944).

- The friction force is nonzero at the top and bottom dead centers, due to the type of friction research engine, which had a spring loaded cylinder head for measurement of the axial friction force.
- The fmep due to piston and ring friction is on the order of 30 kPa for this engine.

Support for Taylor's hypothesis that metallic contact occurs in the vicinity of top dead center comes from measurements of the oil film in running engines. Measurements of the oil film thickness have been performed using electrical resistance and capacitance techniques. The electrical resistance between a piston ring and the cylinder in a running engine is shown in Figure 6-17. The results show that resistance is low in the vicinity of top and bottom dead center, as one would expect, if there is metallic contact. The electrical resistance is higher at the middle of each stroke where piston speed is high and hydrodynamic lubrication is expected. Similar results have been reported by Arcoumanis et al. (1998) using a laser-induced fluorescence system. Therefore, it has generally been concluded that a boundary lubrication regime exists at the ends of the stroke where the piston speed is low, and hydrodynamic lubrication exists in the middle of each stroke where the piston speed is higher.

Figure 6-18 illustrates some concepts used for a theoretical analysis of ring friction. There is an oil layer separating the ring from the cylinder wall whose thickness, δ, is both time and spatially dependent. The oil pressure P_{oil} is also time and spatially dependent. Since the bore is much larger than the oil film thickness, a one-dimensional approximation can be used. In the situation shown, the coordinate system is defined so that the ring is stationary and the cylinder wall is moving with the instantaneous piston speed, U_p.

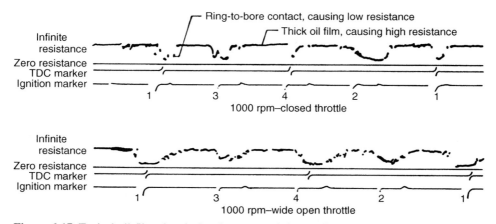

Figure 6-17 Typical oil film electrical resistance result in gasoline engine (McGeehan, 1978). Reprinted with permission © 1978. Society of Automotive Engineers, Inc.

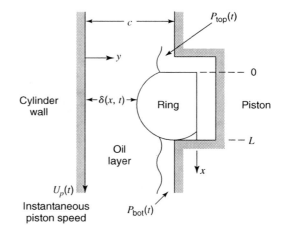

Figure 6-18 Essential features of a hydrodynamic analysis of ring friction.

For the case in which the oil film is much thinner than the ring width, that is

$$\delta(x, t) \ll L \tag{6.18}$$

the Navier-Stokes equations for the oil motion reduce to a Reynolds equation, which is

$$\frac{d}{dx}\left[\delta^3 \frac{dP_{oil}}{dx}\right] = 6\mu U \frac{d\delta}{dx} + 12\mu \frac{d\delta}{dt} \tag{6.19}$$

The boundary conditions to be applied to the Reynolds equation in the case illustrated by Figure 6-17 are

$$P_{oil}(0, t) = P_{top}(t) \tag{6.20}$$
$$P_{oil}(L, t) = P_{bot}(t) \tag{6.21}$$

where $P_{top}(t)$ and $P_{bot}(t)$ are periodic functions known either from direct measurements or from a combustion model complete with a blow-by model. Numerical solution of the Reynolds equation yields the oil film thickness $\delta(x, t)$ and the oil pressure $P_{oil}(x, t)$ such that the boundary conditions are satisfied.

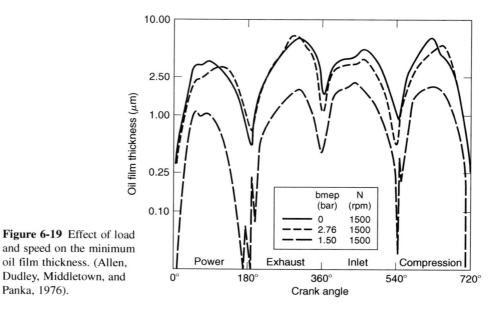

Figure 6-19 Effect of load and speed on the minimum oil film thickness. (Allen, Dudley, Middletown, and Panka, 1976).

Models that include both boundary and hydrodynamic lubrication are available for analysis of the piston ring pack. The models also account for piston tilt, blow-by and inter-ring pressures, ring and piston curvature, ring tension, and surface roughness. The results of the modeling predict the oil film thickness and pressure distribution for each ring.

Some calculated oil film thicknesses for a particular engine are shown in Figure 6-19. In the graph, the minimum oil thickness is plotted for different speeds and loads. The results are similar in shape for all conditions and are fairly insensitive to speed and load. Surface roughness in engines is such that for film thicknesses less than about a micron, metal to metal contact can be expected to occur. The results in Figure 6-19 show that metal to metal contact can be expected to occur at top and bottom dead center for all speeds and loads. The results also show that at high load, metal to metal contact can occur for most, if not all, of the power stroke.

Friction correlations for piston and ring friction have been developed which take both the boundary lubrication and the hydrodynamic friction regimes into account. The hydrodynamic friction component depends on the contact area. Setting L_s as the average piston skirt and ring contact length, and c as an average skirt clearance, the friction force, F_f, of the piston skirt scales as

$$F_f = A\mu \frac{du}{dy} \sim b\, L_s\, \mu\, \frac{\overline{U}_p}{c} \tag{6.22}$$

and for n_c pistons, the piston skirt fmep scales as

$$\text{fmep}_{\text{skirt}} \sim \frac{F_f u}{n_c N b^2 s} \sim \frac{(\mu n_c b L_s \overline{U}_p)\, \overline{U}_p}{c n_c N b^2 s}$$

$$\sim \mu \frac{L_s\, U_p}{bc} \tag{6.23}$$

It is reasonable to assume that the piston skirt length and the clearance scale directly with the bore, i.e., $L_s \sim b$, and $c \sim b$. The skirt length scaling is based on geometrical similarity, and the clearance scaling is based on thermal expansion considerations. Therefore, the piston skirt hydrodynamic fmep scales as

$$\text{fmep}_{\text{shirt}} \sim \mu \frac{\overline{U}_p}{b} = c_{ps} \frac{\overline{U}_p}{b} \tag{6.24}$$

Patton et al. (1989) suggest a proportionality constant $c_{ps} = 294$ kPa-mm-s/m for the piston hydrodynamic friction, again including the oil properties in the proportionality constant.

The friction force of the piston rings has two components, one resulting from the ring tension and the other component from the gas pressure loading. The component of piston friction due to ring tension in the mixed lubrication regime will have a friction coefficient inversely proportional to the engine speed. Patton et al. (1989) recommend a scaling for piston-ring friction which bridges the boundary and hydrodynamic lubrication regimes given by

$$F_f \sim fF_n \sim \left(1 + \frac{1000}{N}\right) \tag{6.25}$$

The piston ring fmep scaling is

$$\text{fmep}_{\text{rings}} \sim \frac{n_c\left(1 + \dfrac{1000}{N}\right)\overline{U}_p}{n_c\,Nb^2\,s} = c_{pr}\left(1 + \frac{1000}{N}\right)\frac{1}{b^2} \tag{6.26}$$

The proportionality constant recommended by Patton et al. (1989) is $c_{pr} = 4.06 \times 10^{+4}$ kPa-mm^2.

A correlation for the component of piston friction due to the gas pressure loading recommended by Bishop (1964) is

$$\text{fmep}_{\text{gas load}} = c_g \frac{P_i}{P_a}\left[0.088\,r + 0.182\,r^{(1.33 - K \cdot \overline{U}_p)}\right] \tag{6.27}$$

where P_i is the intake manifold pressure, P_a is the atmospheric pressure, r is the compression ratio, $c_g = 6.89$, and $K = 2.38 \times 10^{-2}$ s/m. The correlation includes the effect of compression ratio, and a decrease in the friction coefficient in the mixed lubrication regime.

The relative magnitudes of the three piston friction terms, skirt, ring tension, and gas pressure, are shown in Figure 6-20 for the 86-mm bore and stroke engine specified in

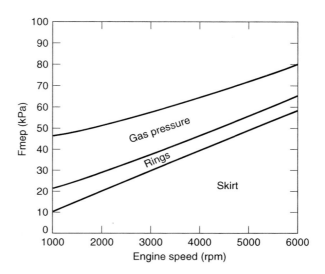

Figure 6-20 Components of piston friction.

Table 6-1   Representative Engine Parameters for Example 6.1

Bore (mm)	86
Stroke (mm)	86
Number of Cylinders	4
Compression Ratio	9
Atmospheric Pressure (kPa)	101
Intake Pressure (kPa)	101
Exhaust Pressure (kPa)	103
Intake Valves/Cylinder	2
Exhaust Valves/Cylinder	2
Intake Valve Diameter (mm)	35
Exhaust Valve Diameter (mm)	31
Maximum Valve Lift (mm)	11
Number of Crankshaft Bearings	5
Crankshaft Bearing Dia. (mm)	56
Crankshaft Bearing Length (mm)	21
Number of Connecting Rod Bearings	4
Connecting Rod Bearing Dia. (mm)	48
Connecting Rod Bearing Length (mm)	42
Number of Camshafts	1

Table 6-1. The skirt and rod bearing fmeps increase linearly with engine speed, while the piston ring fmep decreases with engine speed. At low speeds, most of the friction is due to the piston rings, and at higher speeds, the majority of the friction is from the piston skirt.

6.7 VALVE TRAIN FRICTION

The valve train friction results from the camshaft, cam follower, and valve components. Common valve train designs are listed in Table 6-2. The designs listed in Table 6-2 include overhead cam (OHC), and cam-in-block (CIB) with push rods. Either flat (*ff*) or roller (*rf*) cam followers are used. A schematic of these valve train designs is given in Figure 6-21. Figure 6-22 is a photograph of a camshaft for a type V push rod V8 engine showing the shape and orientation of the camshaft lobes. A type V rocker arm with a roller follower is shown in Figure 6-23 and rocker arms mounted on a cylinder head are shown in Figure 6-24. Figure 6-25 is a photograph of engine poppet valves.

The valve train frictional losses are due to the following; hydrodynamic friction in the camshaft bearing, mixed lubrication in the flat followers, rolling contact friction in the roller followers, and both mixed and hydrodynamic friction due to the oscillating motion of the lifters and valves.

Table 6-2   Valve train types

Type I	OHC	Direct acting/flat or roller follower
Type II	OHC	End pivot rocker/flat or roller follower
Type III	OHC	Center pivot rocker/flat or roller follower
Type V	CIB	Rocker arm/flat or roller follower

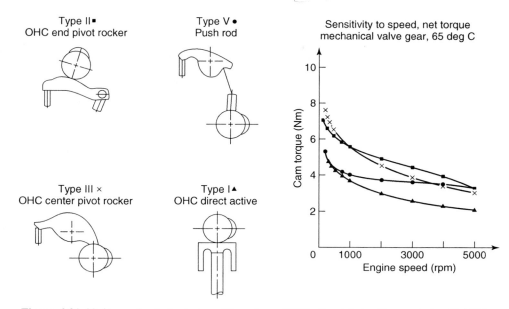

Type II■
OHC end pivot rocker

Type V ●
Push rod

Type III ×
OHC center pivot rocker

Type I▲
OHC direct active

Sensitivity to speed, net torque
mechanical valve gear, 65 deg C

Figure 6-21 Various valve train designs (Rosenberg, 1982). Reprinted with permission © 1982. Society of Automotive Engineers, Inc.

Figure 6-22 Portion of a V8 engine camshaft. (Courtesy COMP Cams.)

Figure 6-23 Type V push rod rocker arm with roller follower. (Courtesy Jesel Valvetrain Components.)

The hydrodynamic friction in the camshaft is similar to that of the main and connecting rod bearings. If the camshaft bearing diameter and length are assumed to be constant and not a function of engine size, the fmep scales as

$$\text{femp}_{\text{cam}} \sim \frac{N\, n_{cs}}{n_c\, b^2 s} = c_c \frac{N\, n_{cs}}{n_c\, b^2 s} \tag{6.28}$$

Figure 6-24 Rocker arms on cylinder head. (Courtesy Jesel Valvetrain Components.)

Figure 6-25 Engine poppet valves. (Courtesy Wesco Valve & Manufacturing Company, a wholly owned subsidiary of Safety Seal Piston Ring Company.)

where n_{cs} is the number of camshaft bearings, assumed equal to the product of the number of camshafts and the number of main bearings. Patton et al. (1989) suggest a value of $c_c = 2.44 \times 10^2$ kPa-mm³-min/rev as the proportionality constant, plus an additional value of 4.12 kPa to account for the camshaft seals.

For scaling purposes, the normal force in the valve mechanism components such as followers, rocker arms, valve lifters, and valves is assumed to be proportional to the product of the effective valve train mass and acceleration. The valve train effective mass is proportional to valve area, which is in turn proportional to the cylinder cross sectional area. Therefore the normal force scales as the square of the cylinder bore.

A flat follower (ff) is assumed to operate in the mixed lubrication regime, and can be scaled with a friction coefficient inversely proportional to engine speed.

$$F_f \sim fF_n \sim \left(1 + \frac{1000}{N}\right)b^2 \tag{6.29}$$

The fmep scales as

$$\text{fmep}_{ff} \sim \left(1 + \frac{1000}{N}\right)\frac{n_v}{n_c s} = c_{ff}\left(1 + \frac{1000}{N}\right)\frac{n_v}{n_c s} \tag{6.30}$$

where n_v is the total number of valves, and c_{ff} is the flat follower coefficient.

A roller follower (rf) operates in the rolling contact friction regime and is scaled with a friction coefficient proportional to engine speed. Patton et al. (1989) recommend the following fmep scaling,

$$F_f \sim fF_n \sim Nb^2 \tag{6.31}$$

$$\text{fmep}_{rf} \sim \frac{n_v N}{n_c s} = c_{rf} \frac{n_v N}{n_c s} \tag{6.32}$$

where c_{rf} is the roller follower coefficient.

The oscillating hydrodynamic (*oh*) friction in the lifters and valve guides scales with square root of the Stribeck variable, in which the relative velocity is proportional to the maximum valve lift L_v and the engine speed N. The fmep scaling is

$$F_f \sim fF_n \sim (L_v N F_n)^{1/2} \sim (L_v Nb^2)^{1/2} \tag{6.33}$$

$$\text{fmep}_{oh} \sim \frac{n_v L_v^{3/2} N^{1/2}}{n_c bs} = c_{oh} \frac{n_v L_v^{3/2} N^{1/2}}{n_c bs} \tag{6.34}$$

where c_{oh} is the oscillating hydrodynamic coefficient.

The oscillating mixed (*om*) lubrication in the valve stem and tip can be scaled with a friction coefficient proportional to valve lift, and inversely proportional to engine speed.

$$F_f \sim fF_n \sim \left(1 + \frac{1000}{N}\right) b^2 \tag{6.35}$$

$$\text{fmep}_{om} \sim \left(1 + \frac{1000}{N}\right) \frac{n_v L_v}{n_c s} = c_{om} \left(1 + \frac{1000}{N}\right) \frac{n_v L_v}{n_c s} \tag{6.36}$$

where c_{om} is the oscillating mixed coefficient.

The coefficients for the valve train friction terms are given in Table 6-3 for the four types of valve train mechanisms. In Table 6-3, the coefficients for the flat follower, the roller follower, oscillating hydrodynamic, and oscillating mixed friction have the units of kPa-mm, kPa-mm-min/rev, kPa-(mm-min/rev)$^{1/2}$, and kPa, respectively.

Bishop (1964) has recommended a simpler overall expression for valve train friction

$$\text{fmep}_{\text{valve train}} = 393 \left(30 - \frac{4N}{1000}\right) \frac{n_{iv} D_{iv}^{7/4}}{b^2 s} \tag{6.37}$$

Table 6-3   Coefficients for valve train friction terms (Patton et al., 1989)

Configuration	Type	Flat Follower c_{ff} (kPa-mm)	Roller Follower c_{rf} (kPa-mm-min/rev)	Oscillating Hydrodynamic c_{oh} (kPa-mm-min/rev)$^{1/2}$	Oscillating Mixed c_{om} (kPa)
Single overhead cam (SOHC)	Type I	200	0.0076	0.5	10.7
Double overhead cam (DOHC)	Type I	133	0.0050	0.5	10.7
Single overhead cam (SOHC)	Type II	600	0.0227	0.2	42.8
Single overhead cam (SOHC)	Type III	400	0.0151	0.5	21.4
Cam in block (CIB)	Type IV	400	0.0151	0.5	32.1

where D_{iv} is the intake valve head diameter and D_{iv}, b, and s have units of mm. For current SOHC valve train systems, Lee et al. (1999) recommend a value of 246 for the leading coefficient in Equation 6.37, an indication that current technology has reduced the valve train friction. The Bishop correlation takes into account the decrease of the friction coefficient in the mixed lubrication regime with engine speed, and is inversely proportional to the piston stroke.

6.8   PUMPING MEAN EFFECTIVE PRESSURE

The pumping mean effective pressure is the sum of the pressure drops across flow restrictions during the intake and exhaust strokes. It is a measure of the work required to move the fuel-air mixture into and out of an engine. The flow restrictions are located in the intake system, the intake valves, the exhaust valves, and the exhaust system. The intake system elements include the air filter, intake manifold, and throttle valve. The exhaust system flow restrictions include the exhaust manifold, catalytic converter, muffler, and tail pipe.

The pressure drop in the intake manifold ΔP_{im} is

$$\Delta P_{im} = P_a - P_i \tag{6.38}$$

where P_i is the intake manifold pressure and P_a is the atmospheric pressure. The pressure drop across the inlet valves scales with the density, mass flow rate, and the open valve area

$$\Delta P_{iv} \sim \rho U^2 \sim \frac{1}{\rho}\left(\frac{\dot{m}}{n_{iv} A_{iv}}\right)^2 \tag{6.39}$$

where n_{iv} is the number of intake valves per cylinder. The mass flow rate scales with the volumetric efficiency, which in turn is proportional to the intake/atmospheric pressure ratio

$$\dot{m} \sim V_d N \sim e_v b^2 s N \sim \frac{P_i}{P_a} b^2 \overline{U}_p \tag{6.40}$$

Neglecting the density change across the valves, the inlet valve pressure drop therefore scales as the square of the mean piston speed

$$\Delta P_{iv} = c_v \left(\frac{P_i}{P_a} \frac{\overline{U}_p b^2}{n_{iv} D_{iv}^2}\right)^2 \tag{6.41}$$

where D_{iv} is the intake valve diameter. Similarily, the exhaust valve pressure drop ΔP_{ev} scaling is

$$\Delta P_{ev} = c_v \left(\frac{P_i}{P_a} \frac{\overline{U}_p b^2}{n_{ev} D_{ev}^2}\right)^2 \tag{6.42}$$

where D_{ev} is the exhaust valve diameter, and n_{ev} is the number of exhaust valves per engine. The constant of proportionality determined by Millington and Hartles (1968) for small high-speed diesel engines is $c_v = 4.12 \times 10^{-3}$ kPa-s^2/m^2.

The exhaust system pressure drop ΔP_{es} also scales with the square of the mass flow rate

$$\Delta P_{es} \sim \frac{\dot{m}^2}{\rho_{es}} \sim \left(\frac{P_i}{P_a} \overline{U}_p\right)^2 \tag{6.43}$$

$$= k_{es}\left(\frac{P_i}{P_a} \overline{U}_p\right)^2$$

If the exhaust system pressure drop is assumed to be 40 kPa at a piston speed of 15 m/s at wide open throttle conditions, then the proportionality constant c_{es} equals 0.178 kPa s^2/m^2 (Patton et al., 1989).

The total pmep is

$$\text{pmep} = \Delta P_{im} + \Delta P_{iv} + \Delta P_{ev} + \Delta P_{es} \qquad (6.44)$$

6.9   ACCESSORY FRICTION

The accessory mean effective pressure (amep) is the sum of the remaining crankcase friction terms other than the journal bearing, piston and rings, and valve train losses. It includes the oil pump (Figure 6-26), water pump (Figure 6-27), and noncharging alternator friction. If these terms are assumed to be proportional to engine displacement, then by reference to Equation 6.6, the mep of the accessories is a function of engine speed only.

Patton et al. (1989) suggest an accessory mean effective pressure of the form

$$\text{amep} = a_1 + a_2\left(\frac{N}{1000}\right) + a_3\left(\frac{N}{1000}\right)^2 \qquad (6.45)$$

where $a_1 = 6.23$ kPa

$a_2 = 5.22$ kPa-min/rev

$a_3 = -0.179$ kPa-min^2/rev^2

Figure 6-26 Automotive engine oil pump.
(Courtesy Melling Engine Parts.)

Figure 6-27 Automotive water pump. (Courtesy
Airtex Products.)

The accessory mean effective pressure from Bishop (1964) is

$$\text{amep} = 2.69 \left(\frac{N}{1000} \right)^{3/2} \tag{6.46}$$

For the engine specified in Table 6-1, with $N = 3000$ rpm, the accessory mean effective pressure from the Patton correlation is 20.2 kPa, and the Bishop accessory mean effective pressure is 14.0 kPa. Even though the Patton and Bishop correlations have different engine speed scaling, they both predict that the slope of the amep curve will decrease with increasing engine speed.

6.10   OVERALL ENGINE FRICTION MEAN EFFECTIVE PRESSURE

The preceding component analyses can be combined to form an overall engine fmep model. The component equations are summarized in Table 6-4, and have been used to develop a fmep model that is included in the text web page. It should be noted that the coefficients used in the fmep model are likely to depend on oil properties, such as the viscosity, that have not been explicitly included in the analysis.

EXAMPLE 6.1   *Engine friction mean effective pressure*

What are the crankshaft, piston, valve train, pumping, and accessory fmeps for the four-cylinder engine specified in Table 6-1, if it is operated at wide open throttle (WOT) at speeds varying from 1000 to 6000 rpm?

SOLUTION   The web page contains an applet entitled *Friction Mean Effective Pressure*, which can be used to compute engine component fmep. The input parameters for a SOHC four-cylinder in-line engine with two intake and two exhaust valves per cylinder are given in Table 6-1. Note that a wide open throttle condition is one in which the intake pressure is the same as the atmospheric pressure. The input parameters are entered into the applet as shown in Figure 6-28. The calculation results are also shown in Figure 6-28.

Table 6-4   Fmep component equations

Component	Subcomponent	Equation
Crankshaft	Main bearings	(6.14)
	Seals	(6.17)
	Connecting rod bearings	(6.15)
Piston	Skirt	(6.24)
	Rings	(6.26)
	Gas pressure	(6.27)
Valve train	Camshaft bearings	(6.28)
	Flat or roller follower	Table 6-3
	Oscillating hydrodynamic	Table 6-3
	Oscillating mixed	Table 6-3
Pumping	Intake manifold	(6.38)
	Inlet valves	(6.39)
	Exhaust valves	(6.42)
	Exhaust system	(6.43)
Accessory		(6.45), (6.46)

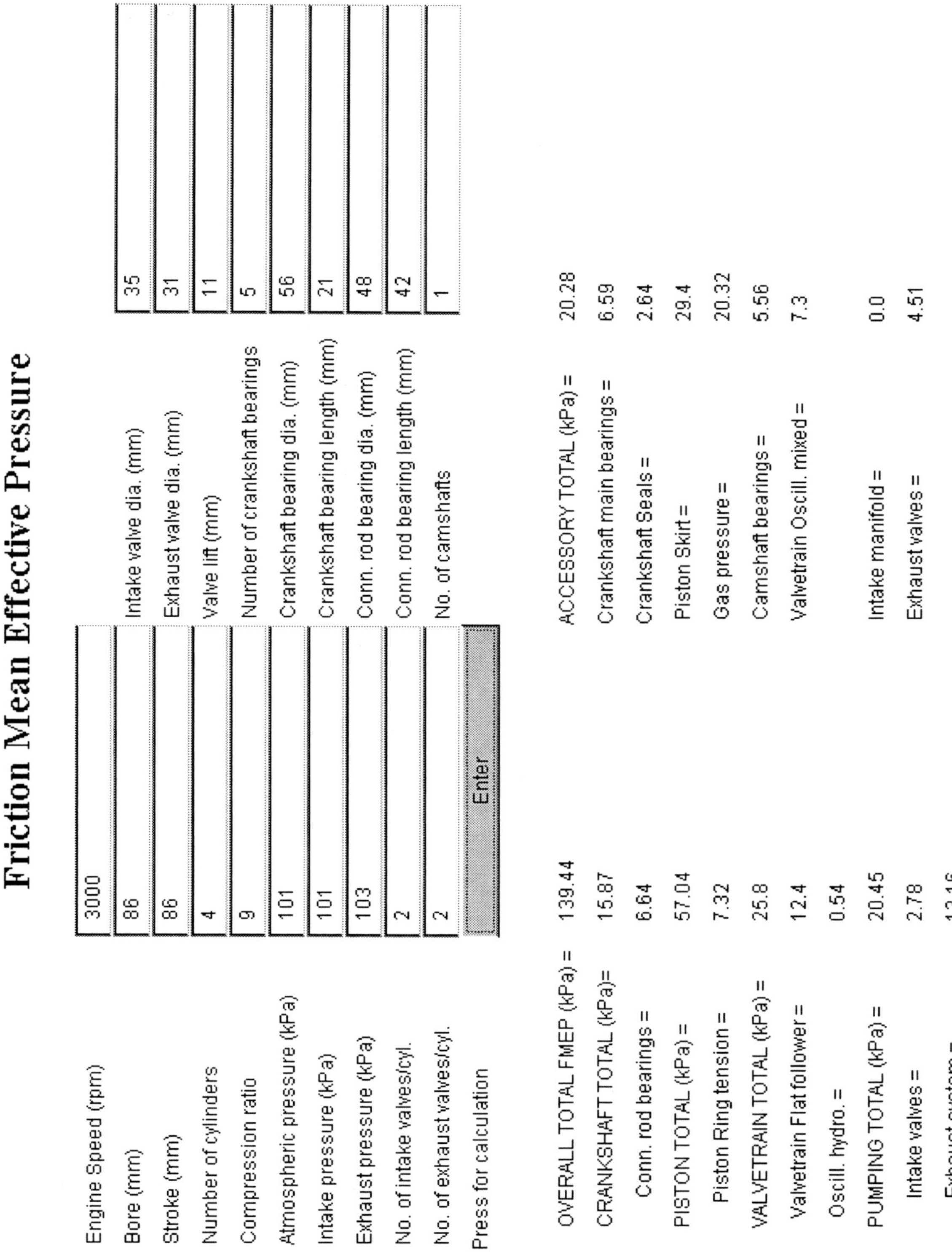

Friction Mean Effective Pressure

Engine Speed (rpm)	3000	Intake valve dia. (mm)	35
Bore (mm)	86	Exhaust valve dia. (mm)	31
Stroke (mm)	86	Valve lift (mm)	11
Number of cylinders	4	Number of crankshaft bearings	5
Compression ratio	9	Crankshaft bearing dia. (mm)	56
Atmospheric pressure (kPa)	101	Crankshaft bearing length (mm)	21
Intake pressure (kPa)	101	Conn. rod bearing dia. (mm)	48
Exhaust pressure (kPa)	103	Conn. rod bearing length (mm)	42
No. of intake valves/cyl.	2	No. of camshafts	1
No. of exhaust valves/cyl.	2		
Press for calculation	Enter		

OVERALL TOTAL FMEP (kPa) =	139.44	ACCESSORY TOTAL (kPa) =	20.28
CRANKSHAFT TOTAL (kPa)=	15.87	Crankshaft main bearings =	6.59
Conn. rod bearings =	6.64	Crankshaft Seals =	2.64
PISTON TOTAL (kPa) =	57.04	Piston Skirt =	29.4
Piston Ring tension =	7.32	Gas pressure =	20.32
VALVETRAIN TOTAL (kPa) =	25.8	Camshaft bearings =	5.56
Valvetrain Flat follower =	12.4	Valvetrain Oscill. mixed =	7.3
Oscill. hydro. =	0.54		
PUMPING TOTAL (kPa) =	20.45	Intake manifold =	0.0
Intake valves =	2.78	Exhaust valves =	4.51
Exhaust system =	13.16		

Figure 6-28 Friction mean effective pressure applet.

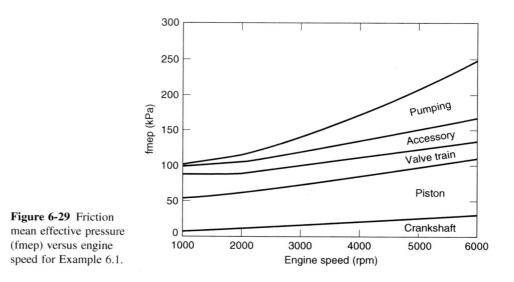

Figure 6-29 Friction mean effective pressure (fmep) versus engine speed for Example 6.1.

The fmep components are plotted as a function of engine speed in Figure 6-29. The friction from the piston rings and skirt is the largest component. As the engine speed increases from 1000 to 6000 rpm, the total fmep increases nonlinearly from about 100 kPa to 260 kPa, with the pumping friction contributing the largest increase.

A quadratic correlation of the total fmep as a function of engine speed for the engine of Table 6.1 is

$$\text{fmep}_{\text{total}} = a + b\left(\frac{N}{1000}\right) + c\left(\frac{N}{1000}\right)^2 \qquad (6.47)$$

The curvefit coefficients are: $a = 94.8$, $b = 2.3$, and $c = 4.0$.

The relative contributions of the component fmeps: crankshaft, piston, valve train, accessory, and pumping mean effective pressures for the engine of Table 6-1 are plotted in Figure 6-30. For wide open throttle conditions at low engine speeds, the overall engine fmep

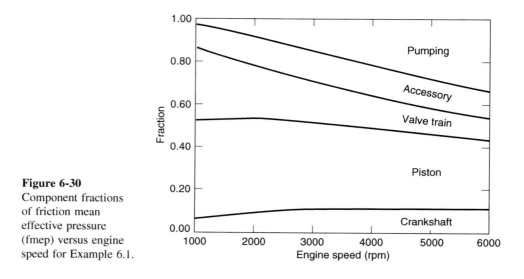

Figure 6-30 Component fractions of friction mean effective pressure (fmep) versus engine speed for Example 6.1.

is primarily due to piston and valve train friction. As the engine speed increases, the pmep fraction increases to about 35%, and the valve train fmep fraction decreases from 35% to 10%. The effect of throttling on fmep is examined in homework Problem 6-10.

6.11   REFERENCES

ALLEN, D. G., B. R. DUDLEY, J. MIDDLETOWN, and D. A. PANKA (1976), "Prediction of Piston Ring-Cylinder Bore Oil Film Thickness in Two Particular Engines and Correlation with Experimental Evidences," *Piston Ring Scuffing,* Mechanical Engineering Pub. Ltd., London, p. 107.

ARCOUMANIS, C., M. DUSZYNSKI, H. LINDENKAMP, and H. PRESTON (1998), "Measurements of Lubricant Film Thickness in the Cylinder of a Firing Diesel Engine Using LIF," SAE paper 982435.

BISHOP, I. N. (1964), "Effect of Design Variables on Friction and Economy," SAE paper 640807.

BROWN, W. L. (1973), "The Caterpiller imep Meter and Engine Friction," SAE paper 730150.

BURWELL S., (1949), "The Calculated Performance of Dynamically Loaded Sleeve Bearings," *J. Appl. Mech.*, 16(4), p. 358.

CHOI, J., B. MIN, and D. OH (1995), "A Study on the Friction Characteristics of Engine Bearing and Cam/Tappet Contacts From the Measurement of Temperature and Oil Film Thickness," SAE paper 952472.

GISH, R., S. MCCULLOUGH, J. RETZLOFF, and H. MUELLER (1957), "Determination of True Engine Friction," SAE paper 117.

LEARY, W. A. and J. U. JOVELLANOS (1944), "A Study of Piston and Piston-Ring Friction," NACA ARR-4J06.

LEE, S., B. A. SHANNON, A. MIKULEC, and G. VRSEK (1999), "Applications of Friction Algorithms for Rapid Engine Concept Assessments," SAE paper 1999-01-0558.

McGEEHAN, J. A. (1978), "A Literature Review of the Effects of Piston and Ring Friction and Lubricating Oil Viscosity on Fuel Economy," SAE paper 780673.

MERRION, D. (1994), "Diesel Engine Design for the 1990s," SAE SP-1011.

MILLINGTON B. and E. HARTLES (1968), "Frictional Losses in Diesel Engines," SAE paper 680590.

PATTON, K. J., R. G. NITSCHKE, and J. B. HEYWOOD (1989), "Development and Evaluation of a Friction Model for Spark Ignition Engines," SAE paper 890836.

ROSENBERG, R. C. (1982), "General Friction Considerations for Engine Design," SAE paper 821576.

STANLEY, R., D. TARAZA, N. HENEIN, and W. BRYZIK (1999), "A Simplified Friction Model of the Piston Ring Assembly," SAE paper 1999-01-0974.

TAYLOR C. (1985), *The Internal Combustion Engine in Theory and Practice,* MIT Press, Cambridge, Massachusetts.

TING, L. L. and J. E. MAYER, Jr. (1974), "Piston Ring Lubrication and Cylinder Bore Analysis, Part I–Theory and Part II–Theory Verification," *J. Lubr. Tech.*, p. 96.

TING, L. L. (1993), "Development of a Reciprocating Test Rig for Tribological Studies of Piston Engine Moving Components–Part 1: Rig Design and Piston Ring Friction Coefficients Measuring Method," SAE paper 930685.

TSEREGOUNIS, S., M. VIOLA, and R. PARANJPE (1998), "Determination of Bearing Oil Film Thickness (BOFT) for Various Engine Oils in an Automotive Gasoline Engine Using Capacitance Measurements and Analytical Predictions," SAE paper 982661.

YAGI, S., K. FUJIWARA, N. KUROKI, and Y. MAEDA (1991), "Estimate of Total Engine Loss and Engine Output in Four Stroke S.I. Engines," SAE paper 910347.

6.12   HOMEWORK

6.1 Oil companies are advertising "slippery oils," i.e., non-Newtonian, that if used in your automobile will slightly reduce the fuel consumption. The implication is that when the slippery oil is compared to conventional oil of equal viscosity, the slippery oil will have a reduced friction coefficient. How is this possible?

refer to
eqn. 6.9

6.2 Racing mechanics will often modify an engine by increasing the bearing clearances above the manufacturer specifications. This can increase the power of an engine. With reference to Equation 6-10, discuss why the power is increased, and the tradeoffs that should be considered in choosing a clearance.

refer to
Fig. 6.16

$F_f = f \cdot F_N$
$F_f = A \cdot \mu \frac{du}{dy}$

6.3 How is the fmep computed from the data in Figure ~~6-9~~? 6-16

6.4 It has been observed that the oil film thickness on the cylinder wall tends to be greater at lower loads. Relate this observation to the fmep versus load curves in Figure 6-1.

6.5 How do the piston and ring friction depend on the mass of the piston?

6.6 Explain the zero crossings in Figure 6-12. Why is there not a zero crossing during the expansion stroke in the range $0 < \theta < 180$?

6.7 If one cylinder of a multicylinder spark ignition engine is motored by disconnecting the spark plug, it has been observed that the motored cylinder pressure increases with load. Discuss why this is to be expected.

Use Ex. 6.1
Use applet
to finish

6.8 For the piston specifications given in Table 6-1, at what engine speed will the piston skirt friction be equal to the piston gas and ring friction?

6.9 What are the pumping, accessory, valve train, piston, and crankshaft fmeps for a six-cylinder engine at 3000 rpm and WOT with 0.1-m bore and stroke, and compression ratio of 11? The number of crankshaft bearings is seven. Assume the other engine specifications are as given in Table 6-1.

6.10 Compute the fmep components (pumping, accessory, valve train, piston, and crankshaft) for the engine specified in Table 6-1, but throttled to 50 kPa inlet pressure. Plot the results versus engine speed from 1000 to 6000 rpm.

6.11 Compare the Patton valve train fmep with the Bishop valve train fmep for the engine specified in Table 6-1 at N = 2500 rpm.

Chapter 7

Air, Fuel, and Exhaust Flow

7.1 INTRODUCTION

In this chapter we examine the flow of the air, fuel, and exhaust through internal combustion engines. So far we have restricted our attention to the ideal four-stroke model that assumes constant pressure intake and exhaust flows. The ideal four-stroke model is qualitatively useful but quantitatively lacking, since no account is made of such important factors affecting the intake and exhaust flow and pressure as unsteady and compressible flow phenomena, wall friction, heat transfer, area changes, branches, and bends. The airflow in two-stroke engines is also examined in this chapter. A model is presented for two-stroke engine scavenging, including parameters such as the delivery ratio and the residual fraction. Finally, the basic principles of fuel-injection systems and carburetors are laid out, providing an introduction to the various means employed to deliver fuel to the combustion chamber.

7.2 VALVE FLOW

Valve Flow and Discharge Coefficients

The most significant airflow restriction in an internal combustion engine is the flow through the intake and exhaust valves. Typically the minimum cross sectional area in the intake and exhaust system occurs at the valve. In accounting for the pressure drop across the intake and exhaust valves considerable success has been realized by modeling the gas flow through the valves as one-dimensional quasi-steady compressible flow.

The actual mass flow rate, $\dot{m}$, and the isentropic mass flow rate, $\dot{m}_{is}$, through a valve are given by

$$\dot{m} = C_f \dot{m}_{is} \tag{7.1}$$

$$\dot{m}_{is} = \rho_v A_v U_{is} \tag{7.2}$$

where C_f is a valve flow coefficient, U_{is} is the reference isentropic velocity, and A_v and ρ_v are the cross sectional area and fluid density at the valve, respectively. The isentropic velocity U_{is} depends on the pressure ratio and is calculated from the isentropic relation for a flow in a converging nozzle emptying into a large volume:

$$U_{is} = \left[2 \frac{\gamma}{\gamma - 1} \frac{P_o}{\rho_o} \left(1 - \left(\frac{P_v}{P_o} \right)^{(\gamma - 1)/\gamma} \right) \right]^{1/2} \tag{7.3}$$

where

P_o = upsteam total or stagnation pressure

P_v = valve static pressure

ρ_o = upstream total or stagnation density

The isentropic equation relating the valve pressure and density to the upstream stagnation pressure and density is

$$\rho_v = \rho_o \left(\frac{P_v}{P_o}\right)^{1/\gamma} \tag{7.4}$$

and the ideal gas equation at stagnation conditions is

$$P_o = \rho_o R T_o \tag{7.5}$$

The stagnation sound speed c_o is

$$c_o = (\gamma R T_o)^{1/2} \tag{7.6}$$

Substituting Equations 7.2–7.6 into Equation 7.1 we obtain

$$\dot{m} = \rho_o C_f A_v c_o \left[\frac{2}{\gamma - 1} \left(\left(\frac{P_v}{P_o}\right)^{2/\gamma} - \left(\frac{P_v}{P_o}\right)^{(\gamma + 1)/\gamma} \right) \right]^{1/2} \tag{7.7}$$

For intake flow into the cylinder, the stagnation conditions refer to conditions in the intake port upstream of the valve. For exhaust flow out of the cylinder, the stagnation conditions refer to conditions in the cylinder.

Choked flow occurs at a valve throat if the ratio of the upstream pressure to downstream pressure exceeds a critical value. The critical pressure ratio is

$$\left(\frac{P_{up}}{P_{down}}\right)_{cr} = \left(\frac{\gamma + 1}{2}\right)^{\gamma/(\gamma - 1)} \tag{7.8}$$

For $\gamma = 1.35$ the critical pressure ratio is 1.86. If the flow is choked, the pressure at the throat is

$$\frac{P_v}{P_o} = \left(\frac{2}{\gamma + 1}\right)^{\gamma/(\gamma - 1)} \tag{7.9}$$

Note that for choked flow, the valve static pressure P_v depends only on the upstream stagnation pressure P_o and is independent of the downstream pressure. The choked flow rate $\dot{m}_{cr}$ is

$$\dot{m}_{cr} = \rho_o C_f A_v c_o \left(\frac{2}{\gamma + 1}\right)^{(\gamma + 1)/2(\gamma - 1)} \tag{7.10}$$

For nonchoked flow into the cylinder, it may generally be assumed that the throat pressure is equal to the cylinder pressure. If the kinetic energy in the cylinder is relatively negligible, one need not distinguish between static and stagnation cylinder pressure. However, for exhaust flow from the cylinder in nonchoked situations, one equates the throat pressure to the exhaust port static pressure and this may differ significantly from the exhaust port stagnation pressure.

Equation 7.2 assumes isentropic flow from an upstream reservoir to a minimum valve throat area, A_v. In the idealized model of a poppet valve shown in Figure 7-1, two minimum area A_v possibilities are evident; the valve curtain area $A_1 = \pi d l$ or the port area $A_2 = \pi d^2/4$. For low lift the minimum area is the valve curtain area, and for larger lifts the minimum area is the valve seat area. In this idealized model, the geometric effects of the valve stem and valve seat angle are neglected. These considerations are addressed in Homework Problem 7.7. There is little reason to open a valve much beyond $l/d \sim 1/4$, since the flow area at such lifts would be limited by the port size. For intake ports, the maximum l/d is about 0.4, accounting for the flow coefficient of the port.

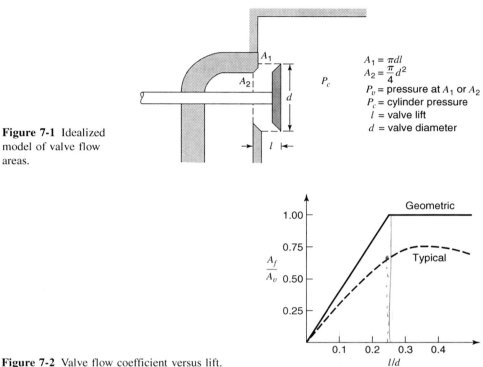

Figure 7-1 Idealized model of valve flow areas.

$A_1 = \pi dl$
$A_2 = \dfrac{\pi}{4}d^2$
P_v = pressure at A_1 or A_2
P_c = cylinder pressure
l = valve lift
d = valve diameter

Figure 7-2 Valve flow coefficient versus lift.

As shown in Figure 7-2, the flow coefficient is the ratio of an effective flow area, A_f, to a representative area, A_v. The representative valve area can be defined using either the valve curtain area or the seat area. If the valve seat area is chosen the flow coefficient is denoted as C_f, as indicated by Equation 7.11a, and if the valve curtain area is used, the flow coefficient is called a discharge coefficient and is denoted as C_d, as indicated by Equation 7.11b:

$$A_f = C_f A_v = C_f \frac{\pi}{4} d^2 \qquad (7.11a)$$

$$A_f = C_d A_v = C_d \pi dl \qquad (7.11b)$$

Flow coefficients are measured using steady flow benches like that illustrated in Figure 7-3. The flow rate and pressure drop across the valve are measured for a number of different valve lifts and pressure ratios. Equation 7.7 is then solved for the flow coefficient for a particular choice of representative valve area. Considerations for use in engine simulation are given in Blair and Drouin (1996).

Some experimental results obtained using flow benches are given in Figure 7-4. Figure 7-4 is a typical plot of C_f versus lift. The flow coefficient C_f increases monotonically from zero with lift since the effective flow area through the valve increases with lift, while the representative valve area $\pi d^2 / 4$ is a constant. The maximum value of C_f is seen to be about 0.6.

The discharge coefficient C_d is plotted versus Reynolds number in Figure 7-5 and versus lift in Figure 7-6. The dependence of the discharge coefficient C_d on Reynolds number can be understood in terms of the flow patterns shown in Figure 7-6. At low lifts the inlet jet is attached to both the valve and the seat, and thus affected by viscous shear.

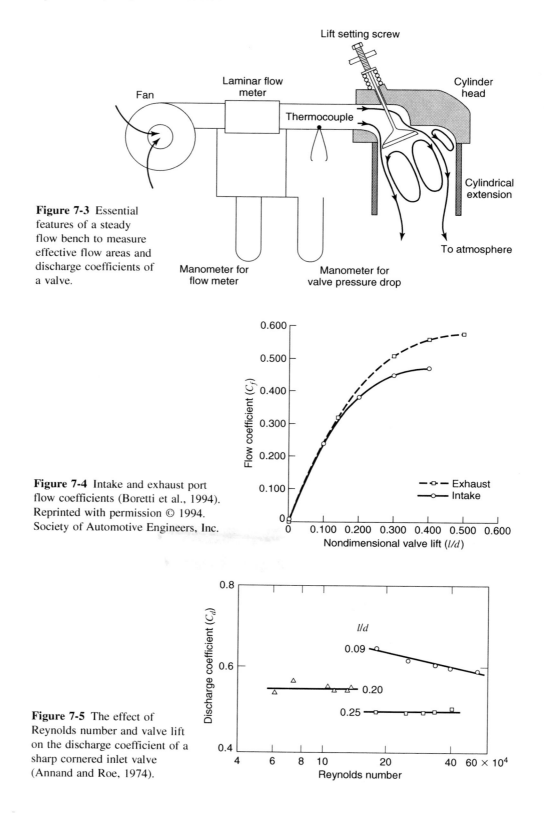

Figure 7-3 Essential features of a steady flow bench to measure effective flow areas and discharge coefficients of a valve.

Figure 7-4 Intake and exhaust port flow coefficients (Boretti et al., 1994). Reprinted with permission © 1994. Society of Automotive Engineers, Inc.

Figure 7-5 The effect of Reynolds number and valve lift on the discharge coefficient of a sharp cornered inlet valve (Annand and Roe, 1974).

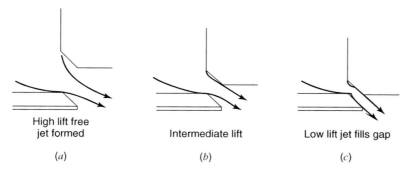

High lift free
jet formed

Intermediate lift

Low lift jet fills gap

(*a*)   (*b*)   (*c*)

Figure 7-6 Flow patterns through a sharp-cornered inlet valve (Annand and Roe, 1974).

If the jet is attached, then the discharge coefficient decreases slightly with increasing Reynolds number since the viscous effects in the jet decrease. At high lifts, the fluid inertia prevents the flow from turning along the valve seat, so the flow breaks away, forming a free jet. The flow area of a free jet is more or less independent of viscosity, thus the flow coefficient at high lifts is independent of Reynolds number.

The discharge coefficient is not as strong a function of lift as is the flow coefficient since they have different reference areas. The discharge coefficient, C_d, decreases slightly with lift since the jet fills less of the reference curtain area as it transforms from an attached jet to a separated free jet.

Discharge coefficient results for exhaust valves are shown in Figure 7-7. The exhaust flow patterns drawn in Figure 7-8 are basically unchanged as the exhaust valve opens, so the flow coefficient is a weak function of the exhaust valve lift. Separation of the exhaust jet from the valve seat at high lift will cause the flow coefficient to decrease slightly at high lifts. It should be noted that the flow bench pressure drops are of the order of 5 kPa, whereas the actual pressure drops across exhaust valves are about two orders of magnitude larger, since the cylinder pressure at the exhaust valve opening is of the order of 500 kPa.

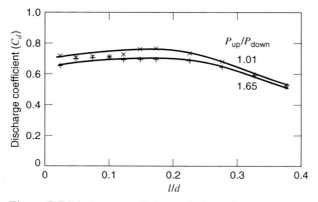

Figure 7-7 Discharge coefficient variation with lift/diameter ratio for a sharp-cornered poppet exhaust valve with 45° seat angle (Annand and Roe, 1974).

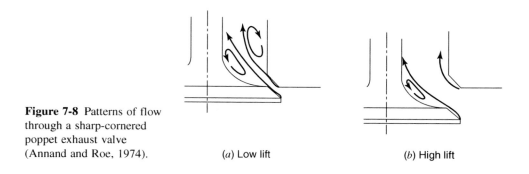

Figure 7-8 Patterns of flow through a sharp-cornered poppet exhaust valve (Annand and Roe, 1974).

(a) Low lift   (b) High lift

EXAMPLE 7-1

What is the maximum flow rate through an exhaust valve, if the valve throat area is 2.7×10^{-3} m^2, the valve C_f is 0.6 and the cylinder pressure and temperature are 500 kPa and 1000 K? Assume exhaust system pressure is 105 kPa, $\gamma = 1.35$, and $R = 287$ J/kg K.

SOLUTION   First compute the pressure ratio and compare it to the critical pressure ratio to determine if the flow is choked

$$\frac{P_{up}}{P_{down}} = \frac{500}{105} = 4.76$$

$$\left(\frac{P_{up}}{P_{down}}\right)_{cr} = \left(\frac{\gamma + 1}{2}\right)^{\gamma/(\gamma - 1)} = 1.86$$

Therefore the flow is choked, and $P_o/P_v = 1.86$.

Second, use the choked flow equation (Equation 7.10) to compute the flow rate:

$$\dot{m} = \rho_o \, C_f A_v \, c_o \left(\frac{2}{\gamma + 1}\right)^{(\gamma + 1)/2(\gamma - 1)}$$

$$\rho_o = P_o/RT_o = \frac{500 \times 10^3}{287 \cdot 1000} = 1.74 \text{ kg/m}^3$$

$$\dot{m} = (1.74)(0.6)(2.7 \times 10^{-3})(1.35 \cdot 287 \cdot 1000)^{1/2}\left(\frac{2}{2.35}\right)^{3.36}$$

$$= 1.02 \text{ kg/s}$$

Valve Performance

An arrangement found convenient for determining the effect of valves on engine volumetric efficiency is shown in Figure 7-9. The intake and exhaust pipe runners are short and the plenums are large (on the order of 100 times the engine displacement). With large plenums, intake and exhaust pressure fluctuations are reduced in magnitude and the pressure drops in the intake and exhaust system are primarily at the intake and exhaust valves. The pressure in the intake and exhaust ports may be assumed to be equal to the pressure in the intake and exhaust plenums, respectively. The mass inducted during the valve open

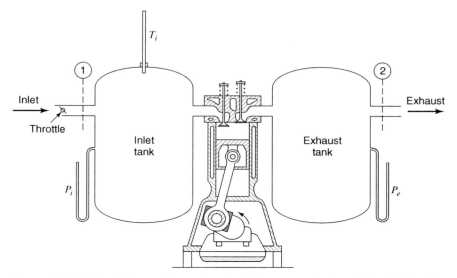

Figure 7-9 Single-cylinder engine equipped with large surge tanks and short inlet and exhaust pipes (Taylor, 1985). Copyright © 1985 MIT Press.

period is

$$m_i = \frac{1}{\omega} \int_{\theta_{io}}^{\theta_{ic}} \dot{m} \, d\theta = \frac{1}{\omega} \int_{\theta_{io}}^{\theta_{ic}} A_f \rho c \left[\frac{2}{\gamma - 1} \left(\left(\frac{P_v}{P_o} \right)^{2/\gamma} - \left(\frac{P_v}{P_o} \right)^{(\gamma + 1)/\gamma} \right) \right]^{1/2} d\theta \quad (7.12)$$

where θ_{io} is the crank angle at which the intake valve opens and θ_{ic} is the angle at which it closes. The terms A_f, ρ, c, P_v/P_o, and γ depend on whether flow is into the cylinder or out of the cylinder. Let us normalize Equation 7.12 by the average effective intake flow area, $\overline{A}_f$,

$$\overline{A}_f = \frac{1}{\theta_{ic} - \theta_{io}} \int_{\theta_{io}}^{\theta_{ic}} A_f \, d\theta = \overline{C}_f A_v \qquad (7.13)$$

the intake plenum density, ρ_i, and sound speed, c_i. The term $\overline{C}_f$ is the average flow coefficient. We then have for the volumetric efficiency

$$e_v = \frac{m_i}{\rho_i V_d} = \frac{\overline{A}_f c_i}{\omega V_d} \int_{\theta_{io}}^{\theta_{ic}} \frac{A_f}{\overline{A}_f} \frac{\rho}{\rho_i} \frac{c}{c_i} \left[\frac{2}{\gamma - 1} \left(\left(\frac{P_v}{P_o} \right)^{2/\gamma} - \left(\frac{P_v}{P_o} \right)^{(\gamma + 1)/\gamma} \right) \right]^{1/2} d\theta \quad (7.14)$$

Note that in the absence of reverse flow during induction $\rho/\rho_i = c/c_i = 1.0$; the terms are included as shown to cover the more general case in which reverse flow occurs. A more rigorous analysis would include the effect of engine speed.

Let us consider the limiting case in which the flow is always choked and into the cylinder. The pressure ratio given by Equation 7.9 is independent of crank angle, so

$$e_v = \left(\frac{2}{\gamma + 1} \right)^{(\gamma + 1)/2(\gamma - 1)} \frac{\overline{A}_f c_i}{\omega V_d} (\theta_{ic} - \theta_{io}) \qquad (7.15)$$

Introducing $\gamma = 1.4$ and the inlet Mach index Z defined by Taylor (1985) as

$$Z = \frac{\frac{\pi}{4} b^2}{\overline{A}_f} \frac{\overline{U}_p}{c_i} \qquad (7.16)$$

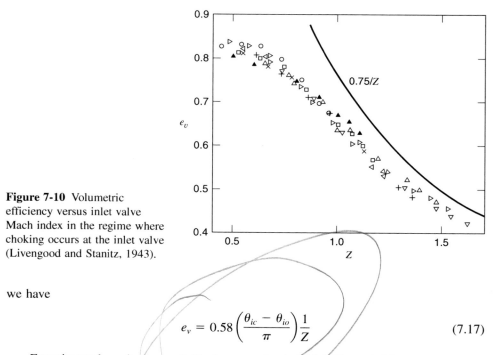

Figure 7-10 Volumetric efficiency versus inlet valve Mach index in the regime where choking occurs at the inlet valve (Livengood and Stanitz, 1943).

we have

$$e_v = 0.58 \left(\frac{\theta_{ic} - \theta_{io}}{\pi} \right) \frac{1}{Z} \tag{7.17}$$

Experimental results are available for an engine in which $(\theta_{ic} - \theta_{io})/\pi = 1.3$ and thus $e_v = 0.75/Z$. They are given in Figure 7-10. Equation 7.15 is an upper bound for the volumetric efficiency valid for large Z. It should be noted that the Mach index is not a parameter that characterizes an actual gas speed; rather, it characterizes what the average gas speed through the inlet valve would have to be to realize complete filling of the cylinder at that particular piston speed. The Mach number for that average inlet gas speed would be $Z/0.58$ for $\gamma = 1.4$.

The results in Figure 7-7 show that for good volumetric efficiency one should keep the Mach index down to less than about $Z = 0.6$. Based on the analyses that led to Equation 7.17, we can interpret this to mean that the average gas speed through the inlet valve should be less than the sonic velocity, so that the intake flow is not choked. Hence, inlet valves can be sized on the basis of the maximum piston speed for which the engine is designed; choose $Z = 0.6$ at this speed and it follows that the average effective area $\overline{A}_i$ of the intake $(f = i)$ valves is

$$\overline{A}_i \geq 1.3 \, b^2 \frac{\overline{U}_p}{c_i} \qquad \text{(intake)} \tag{7.18}$$

Likewise for efficient expulsion of the exhaust gas, the average effective area $\overline{A}_e$ of the exhaust $(f = e)$ valves should be such that their Mach index is less than about 0.6, in which case, relative to intake conditions

$$\frac{\overline{A}_e}{\overline{A}_i} \approx \frac{c_i}{c_e} = \left(\frac{T_i}{T_e} \right)^{1/2} \tag{7.19}$$

As shown in Equation 7.19, a smaller exhaust valve diameter and lift $(l \sim d/4)$ can be used because the speed of sound is higher in the exhaust gases than in the inlet gas flow. Current practice dictates that the exhaust to intake valve area ratio is on the order of 70 to 80%.

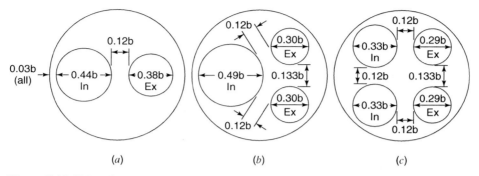

Figure 7-11 Valve diameter ratios for a flat cylinder head (b: bore, In: intake, Ex: exhaust).

In many situations it turns out that the intake valves are sized as large as possible while being consistent with Equation 7.19. This is because there is only so much room available for valves and it may not be possible to satisfy Equation 7.18, thereby compromising the maximum speed of the engine. The use of multiple valves increases the valve area per unit piston area, and hence the speed at which the engine power becomes flow limited. Heads are often wedge-shaped or domed to increase the valve area to piston area, so that intake valve areas to piston area ratios of up to 0.5 can be obtained. Typical valve diameter ratios for two, three, and four valves per cylinder are given for a flat cylinder head in Figure 7-11.

EXAMPLE 7-2

What is the intake valve area A_i and the ratio of the intake valve area to piston area required for a Mach Index of 0.6 for an engine with a maximum speed of 8000 rpm, bore and stroke of 0.1 m, and inlet air temperature of 330 K? Assume $\gamma = 1.4$, $R = 287$ J/kg K, and an average flow coefficient $\overline{C}_f = 0.35$.

SOLUTION

$$\overline{A}_i = 1.3\, b^2\, \frac{\overline{U}_p}{c_i}$$

$$c_i = (\gamma R T_o)^{1/2} = (1.4 \cdot 287 \cdot 330)^{1/2} = 364 \text{ m/s}$$

$$\overline{U}_p = 2 \cdot 5 \cdot N = 2 \cdot 0.1 \cdot 8000/60 = 26 \text{ m/s}$$

$$\overline{A}_i = (1.3)(0.1)^2\, 26/364 = 9.3 \times 10^{-4} \text{ m}^2$$

$$A_v = \overline{A}_i/\overline{C}_f = 2.65 \times 10^{-3} \text{ m}^2$$

$$\frac{A_v}{A_p} = \frac{A_v}{\frac{\pi}{4} b^2} = \frac{2.65 \times 10^{-3}}{\frac{\pi}{4}(0.1)^2} = 0.34$$

In a four-stroke engine, the pumping work is defined as the work required to push out the exhaust gas and to pull in the fresh charge. It is evaluated from a pressure-volume diagram from bottom center at the start of the exhaust stroke to bottom center at the end of the induction stroke. The ideal model, valid for engines operated at low Mach indices, predicts that the pmep is the difference between the exhaust pressure and the inlet pressure. At higher speeds, the pressure difference across the valves at closing reduces or increases the pumping depending on whether or not the engine is supercharged. Data given by Taylor (1985) for an engine with short intake and exhaust pipes are correlated by the following expression.

$$\text{pmep} = (P_e - P_i) - (1.4\,P_e - 2.6\,P_i)\,Z^{1.5} \qquad (7.20)$$

As $Z \rightarrow 0$, the pmep goes to the ideal case of $P_e - P_i$.

Valve Timing

Intake and exhaust valve lifts are plotted as a function of crank angle in Figures 1-13 and 7-12. In order to ensure that a valve is fully open during a stroke for good volumetric efficiency, the valves are open for longer than 180 degrees. The exhaust valve will open before bottom dead center and close after top dead center. Likewise, the intake valve will open before top dead center and close after bottom dead center.

There is a valve overlap period at top dead center where the exhaust and intake valves are both open. This creates a number of flow effects. With a spark ignition engine at part throttle, there will be back flow of the exhaust into the inlet manifold since the exhaust pressure is greater than the throttled intake pressure. This will reduce the part load performance since the volume available to the intake charge is less, reducing the volumetric efficiency. Rough idle can also result due to unstable combustion. On the other hand, since this dilution will reduce the peak combustion temperatures, the NO_x pollution levels will also be reduced.

At wide open throttle, with both valves open, there will be some shortcircuiting of the inlet charge directly to the exhaust, since in this case, the intake pressure is greater than the exhaust pressure. This will reduce the full load performance, since a fraction of the fuel is not burning in the cylinder.

Typical valve timing angles for a conventional and a high performance automatic spark engine are given in Table 7-1. The high performance engine operates at much higher piston speeds at wide open throttle, with power and volumetric efficiency as the important factors, whereas the conventional engine operates at lower rpm, with idle and part load performance of importance. Therefore, the high performance intake valve opens about 25° before the conventional intake valve, and closes about 30° after the conventional intake valve. As the engine design speed increases, to maintain a maximum valve opening during the intake and exhaust strokes, the intake and exhaust valves are open for a longer duration, from about 230° to about 285°. Early opening of the exhaust valve will reduce the expansion ratio, but will also reduce the exhaust stroke pumping work.

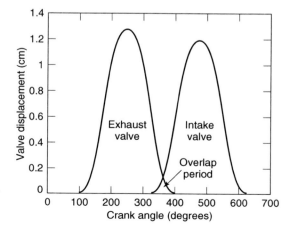

Figure 7-12 Exhaust and intake valve profiles (Olsen et al., 1998). Reprinted with permission © 1998. Society of Automotive Engineers, Inc.

Table 7-1   Valve Timing Angles

		Open	Close	Duration
Intake	Conventional	5° before tdc	45° after bdc	230°
	High performance	30° before tdc	75° after bdc	285°
Exhaust	Conventional	45° before bdc	10° after tdc	235°
	High performance	70° before bdc	35° after tdc	285°

To minimize engine size and produce a given torque versus speed curve (with torque proportional to the volumetric efficiency at fixed thermal efficiency), it is clearly desirable to be able to vary valve timing with engine speed. Variable valve timing (VVT) is a technique that can address the problem of obtaining optimal engine performance over a range of throttle and engine speed. Variable valve timing allows the intake and exhaust valves to open and close at varying angles, depending on the speed and load conditions. At idle, with a nearly closed throttle, the intake and exhaust valve overlap is minimized to reduce exhaust back flow. At low speed, the intake valves are closed earlier to increase volumetric efficiency and torque. At high speed, with an open throttle, the intake valves are closed later to increase volumetric efficiency and power.

There have been a number of VVT mechanisms built and commercialized. As one might expect the mechanics of a VVT device are complex. Hydraulic mechanisms, dual lob camshafts with followers have been developed, and electromagnetic and electrohydraulic actuators which replace the camshaft can also be used.

Effect of Valve Timing on Volumetric Efficiency and Residual Fraction

The first law of thermodynamics applied to an open system doing boundary work is

$$\Delta E = -\int PdV + \int (\dot{m}_{in}h_{in} - \dot{m}_{out}h_{out})\, dt + \int \dot{Q}\, dt \qquad (7.21)$$

For an ideal gas with constant specific heat, it can be shown that

$$\Delta E = c_v\, \Delta(mT) = \frac{1}{\gamma - 1} \Delta(PV) \qquad (7.22)$$

Let us assume that during overlap, exhaust gas flows into the intake manifold and later an equal amount flows out. The flow into the cylinder is then composed of residual exhaust gas until the total residuals pushed into the intake manifold return to the cylinder. It follows that for the intake process

$$\int_{io}^{ic} [(\dot{m}c_pT)_{in} - (\dot{m}c_pT)_{out}]dt = \int_{io}^{ec} [\ ] + \int_{ec}^{is} [\ ] + \int_{is}^{ic} [\ ] \qquad (7.23)$$

The integrand is the same in all integrals and is abbreviated on the right-hand side by brackets to save space. The time notation is:

io = intake valve opens

ic = intake valve closes

ec = exhaust valve closes

is = intake of fresh mixture starts

The first integral on the right-hand side is assumed to be zero since during overlap the enthalpy flow into the cylinder from the exhaust port is balanced by the enthalpy flow from the cylinder to the intake manifold. (We are neglecting the small decrease in temperature that occurs between the exhaust coming in and going out, because of heat loss while in the intake port or pipe.) The second integral is equal to the amount of enthalpy that flowed into the cylinder from the exhaust port during the overlap period since we assume that this gas returns prior to the start of induction. Hence,

$$P_{ic}V_{ic} - P_{io}V_{io} = (\gamma - 1)\left[-\int_{io}^{ic} P dV + \int_{io}^{ec} (\dot{m} c_p T)_{ov} dt \right.$$

$$\left. + c_p T_i \int_{is}^{ic} \dot{m}_{in} dt + \int_{io}^{ic} \dot{Q} dt \right] \tag{7.24}$$

Let us introduce the mass inducted

$$m_i = \int_{is}^{ic} \dot{m}_{in} dt \tag{7.25}$$

and the mass of exhaust that flows into the cylinder from the exhaust system during overlap

$$m_{ov} = \int_{io}^{ec} \dot{m}_{ov} dt \tag{7.26}$$

We then have for the volumetric efficiency

$$e_v = \frac{m_i}{\rho_i V_d} = \frac{1}{\gamma} \frac{P_{ic}V_{ic} - P_{io}V_{io}}{P_i V_d} + \frac{\gamma - 1}{\gamma} \int_{io}^{ic} \frac{P dV}{P_i V_d} - \frac{\gamma - 1}{\gamma} \frac{Q}{P_i V_d} - \frac{T_{ov}}{T_i} \frac{m_{ov}}{\rho_i V_d} \tag{7.27}$$

Both heat transfer to the gas and gas exchange during overlap decrease the volumetric efficiency.

Now let us consider the limiting case in which the piston speed is small, $U_p \to 0$. In this case there should be no pressure drop across the valves at closure, therefore

$$U_p \to 0 \Rightarrow P_{io} = P_e \quad \text{and} \quad P_{ic} = P_i \tag{7.28}$$

Equation 7.27 then becomes

$$e_v = \frac{V_{ic} - V_{io}}{V_d} - \frac{1}{\gamma}\left(\frac{P_e}{P_i} - 1\right)\frac{V_{io}}{V_d} - \frac{\gamma - 1}{\gamma}\frac{Q}{P_i V_d} - \frac{T_{ov}}{T_i}\frac{m_{ov}}{\rho_i V_d} \tag{7.29}$$

For engines with a short stroke to rod ratio, the cylinder volume is given by

$$\frac{V}{V_o} = 1 + \frac{r - 1}{2}(1 - \cos\theta) \tag{7.30}$$

where θ is the crank angle measured from top dead center. Finally, we can write

$$e_v = \frac{\cos\theta_{io} - \cos\theta_{ic}}{2} - \frac{P_e/P_i - 1}{\gamma(r - 1)}\left[1 + \frac{r - 1}{2}(1 - \cos\theta_{io})\right]$$

$$- \frac{\gamma - 1}{\gamma}\frac{Q}{P_i V_d} - \frac{T_{ov}}{T_i}\frac{m_{ov}}{\rho_i V_d} \tag{7.31}$$

Data are available for three geometrically similar engines in which the valve overlap is small and the cylinders are not much warmer than the inlet air; the overlap and heat loss

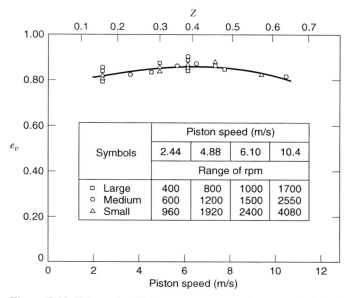

Figure 7-13 Volumetric efficiency versus mean piston speed of the MIT geometrically similar engines under similar operating conditions: $r = 5.74$, $\phi = 1.10$, $P_i = 0.95$ bar, $P_e = 1.08$ bar, $T_i = 339$ K, $T_c = 356$ K, optimum spark advance (Taylor, 1985). Copyright © 1985 MIT Press.

terms are negligible in this case. Figure 7-13 presents the volumetric efficiency of those engines as a function of piston speed. Using the specified valve timings, compression ratio, exhaust to intake pressure ratio, and a specific heat ratio of $\gamma = 1.4$, Equation 7.31 predicts that $e_v \rightarrow 0.78$ as $U_p \rightarrow 0$. The prediction is seen to be quite good and similar agreement would be realized for different pressure ratios.

Our analysis shows that opening and closing valves at angles other than top and bottom center hurts the volumetric efficiency as the piston speed $\rightarrow 0$. Why then are valves opened earlier and closed later than the ideal case? In addition to the finite valve opening times discussed above, one needs to also consider that this analysis is only valid in the limit of zero piston speed.

For a finite piston speed there will be a pressure drop across the valves, the most important of which is at the intake valve at closing. In the limiting case, air is pushed out of the cylinder as the piston moves up during the time from bottom dead center to inlet valve closure. However, at a finite engine speed, the cylinder pressure at bottom center will be less than the inlet pressure because of the pressure drop across the valve as the charge was entering. Hence, as the piston begins the compression stroke, mixture can continue to flow into the cylinder until the pressure rises because of the filling and the upward moving piston. The flow will reverse itself when the two pressures are equal, and then flow back into the intake system until valve closure.

The volumetric efficiency increases with piston speed until a point is reached where the flow reversal starts at intake valve closure. For speeds beyond that point, volumetric efficiency will drop because the valve will close during a time in which mixture is still flowing in the engine. The trend of volumetric efficiency with speed discussed here is shown very clearly in Figure 7-13.

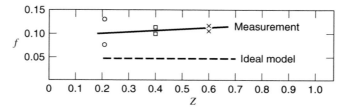

Figure 7-14 Comparison of residual gas fraction for fuel–air cycles with the ideal four-stroke inlet process with measured values (Livengood, Taylor, and Wu, 1958).

Now consider the exhaust process. At any instant, the energy equation can be written as

$$-\dot{Q}_l - P\frac{dV}{dt} = c_v\left(m\frac{dT}{dt} + \frac{dm}{dt}T\right) - \frac{dm}{dt}c_p T \qquad (7.32)$$

Combined with the equation of state and integrated, we have

$$f = \frac{m_{io}}{m_{eo}} = \left(\frac{P_i}{P_{eo}}\right)^{1/\gamma}\frac{V_{io}}{V_{eo}}\exp\left(\int_{eo}^{io}\frac{\dot{Q}_l}{PV}dt\right) \qquad (7.33)$$

Integration is carried only to the time when the intake valve opens because during overlap it is assumed that the gases are pushed into the intake manifold only to return later. Notice that heat loss increases the residual fraction and is important here because the exhaust gases are considerably hotter than the cylinder walls. This is the main reason for the discrepancy between fuel-air cycle calculations with ideal intake and exhaust and the experiments noted in Figure 7-14.

7.3   INTAKE AND EXHAUST FLOW

In engines, the configuration of the inlet and exhaust flow networks employed plays an important role in determining the volumetric efficiency and residual fraction. Intake manifolds (see Figure 7-15) consisting of plenums and pipes are usually required to deliver the inlet air charge from some preparation device such as an air cleaner or compressor, and exhaust manifolds are used to duct the exhaust gases to a point of expulsion, often far removed from the engine. In multicylinder engines, manifolds are used so cylinders can share the same compressor, muffler, and catalytic converter.

The flow in the inlet and exhaust manifolds is unsteady due to the periodic piston and valve motion. The opening and closing of the intake and exhaust valves or ports create finite amplitude compression and rarefaction pressure waves that propagate at sonic velocity through the intake and exhaust airflow. Measured and computed pressure and frequency profiles in an intake manifold are shown in Figure 7-16a in an engine operating at a low speed (~ 3000 rpm), and in Figure 7-16b at a higher speed (~ 6000 rpm). Note that as the engine speed increases, the frequency and amplitude of the pressure waves increase.

Inlet and exhaust manifolds are sized or "tuned" to use the pressure waves to optimize the volumetric efficiency at a chosen engine speed. A tuned intake manifold will have a locally higher pressure when the intake valve is open, increasing the charge density. A tuned exhaust manifold will have a locally lower pressure when the exhaust valve is open, increasing the exhaust outflow.

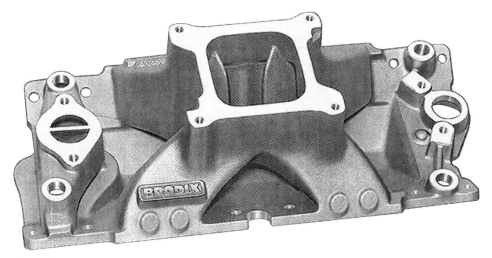

Figure 7-15 Automotive engine intake manifold photograph. (Courtesy Brodix, Inc.)

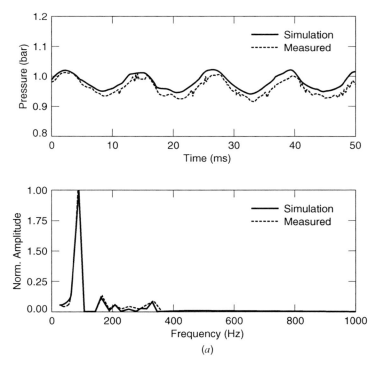

(a)

Figure 7-16a Intake manifold pressure profile and frequency at low speed (approx. 3000 rpm) (Silvestri, Morel, and Costello, 1994). Reprinted with permission © 1994. Society of Automotive Engineers, Inc.

The behavior of the pressure waves is modeled using the equations of compressible gas dynamics. The governing equations for the intake and exhaust flow are the unsteady mass, momentum, and energy conservation equations, which for one dimension in vector

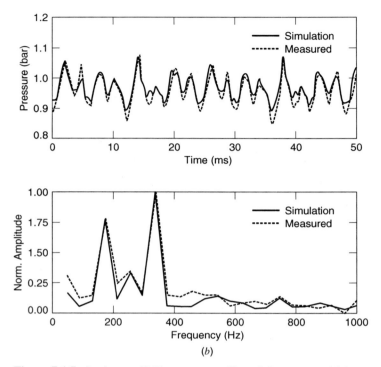

Figure 7-16b Intake manifold pressure profile and frequency at high speed (approx. 6000 rpm) (Silvestri, Morel, and Costello, 1994).

form are:

$$\frac{\partial \vec{f}}{\partial t} + \frac{\partial \vec{F}}{\partial x} = \vec{T} \tag{7.34}$$

$$\vec{f} = \begin{bmatrix} \rho A \\ \rho A U \\ \rho A E \end{bmatrix} \tag{7.35}$$

$$\vec{F} = \begin{bmatrix} \rho A U \\ A(\rho U^2 + P) \\ \rho A U H \end{bmatrix} \tag{7.36}$$

$$\vec{T} = \begin{bmatrix} 0 \\ \tau_\omega \sqrt{4\pi A} + P\dfrac{\delta A}{\delta x} \\ q_\omega \sqrt{4\pi A} \end{bmatrix} \tag{7.37}$$

where ρ is the fluid density, A is the cross sectional area of the duct, U is the fluid velocity, P is the pressure, E is the total specific energy, H is the total specific enthalpy, τ_w is the wall shear stress, and q_w is the wall heat flux. Physical effects such as wall friction, heat transfer, area changes, branches, and bends that occur in actual manifolds are accounted for in Equations 7.34 through 7.37.

Computational fluid dynamics codes are presently used for the design of intake and exhaust plenums, runners, and ports, and to assess the effect of design changes on the

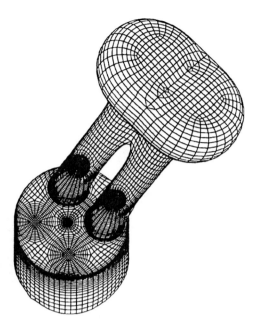

Figure 7-17 Geometric mesh of intake port (Boretti et al., 1994). Reprinted with permission © 1994. Society of Automotive Engineers, Inc.

flow patterns and the volumetric efficiency. There are a number of gas dynamics programs that numerically solve the compressible conservation equations to predict intake and exhaust system flow over an engine cycle. An extensively used 1-D gas dynamics program is WAVE (1999).

The specific boundary condition geometry of the intake and exhaust ports is required for solution of Equations 7.34 through 7.37. A representative three-dimensional intake port geometry divided into computational mesh elements is shown in Figure 7-17. The modeling includes the intake manifold, the intake ports, valves, and cylinder volume. The computed 1-D pressure profiles for an intake geometry similar to Figure 7-17 are compared to measurements in Figures 7-16a and 7-16b. Note the agreement with respect to amplitude, frequency, and harmonic content. The CFD grid for a three cylinder engine intake manifold is shown in Figure 7-18a, and the computed velocities with the middle cylinder open are shown in Figure 7-18b.

A CFD grid for an exhaust manifold is shown in Figure 7-19a. The assembly in Figure 7-19a includes four exhaust pipes, a collector, and a catalytic converter. The one- and

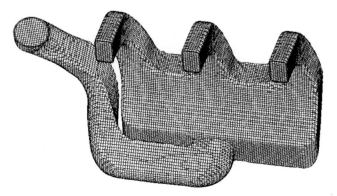

Figure 7-18a CFD grid for intake manifold flow (Courtesy Adapco).

V6 Manifold (samm Mesh)

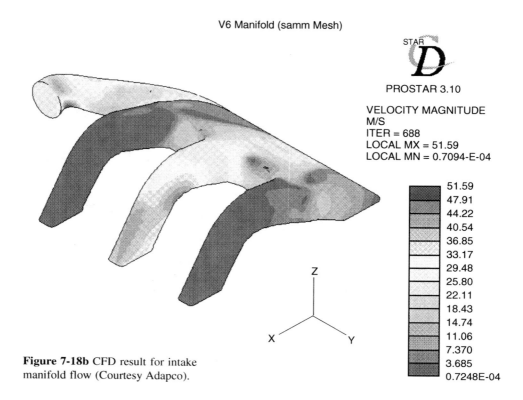

STAR
PROSTAR 3.10

VELOCITY MAGNITUDE
M/S
ITER = 688
LOCAL MX = 51.59
LOCAL MN = 0.7094-E-04

	51.59
	47.91
	44.22
	40.54
	36.85
	33.17
	29.48
	25.80
	22.11
	18.43
	14.74
	11.06
	7.370
	3.685
	0.7248E-04

Figure 7-18b CFD result for intake manifold flow (Courtesy Adapco).

three-dimensional pressure profiles in the exhaust port of Figure 7-19a are compared versus crank angle in Figure 7-19b. The change in amplitude is significant, as the peak pressure is predicted to be about 1.6 bar, corresponding to choked flow conditions, and the minimum pressure is predicted to be about 0.5 bar. The CFD grid for the exhaust manifold shown in Figure 7-20a is for a three cylinder engine with a closely-coupled catalytic converter. The computed velocities with the middle cylinder exhaust valve open are shown in Figure 7-20b.

The volumetric efficiency of the four cylinder engine of Figures 7-17 and 7-19a is plotted as a function of engine speed in Figure 7-21. The compressible flow creates three

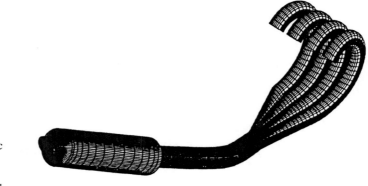

Figure 7-19a Dynamic three-dimensional computations: exhaust manifold computational domain (Boretti et al., 1994).

(a)

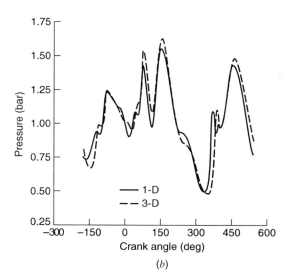

Figure 7-19b Exhaust manifold pressure
variation: comparison of 1-D and 3-D
computations (Boretti et al., 1994).

(b)

local maxima in volumetric efficiency, with the largest volumetric efficiency located at
about 14,000 rpm for this particular engine.

The effect of inlet runner length on the volumetric efficiency of a racing engine is
plotted as a function of engine speed in Figure 7-22. As the engine speed goes to zero,
the volumetric efficiency goes to about 0.8, and with no intake pipe or runner, the volumetric efficiency follows a shallow curve with a maximum at about 0.9, consistent with
the analysis of Section 7.2. The runner length for maximum volumetric efficiency is
inversely proportional to the engine speed.

The sensitivity of the volumetric efficiency to runner length and engine speed has
been a challenge to engine design engineers. By choosing an intake runner of a given
length, the volumetric efficiency can be increased for a particular engine speed, but it
drops off more sharply at other speeds. A fixed length vertical intake runner is desirable
in engines with throttle body injectors or carburetors to minimize wall wetting and mald-
istribution of the fuel air mixture. With port fuel injection it is possible to use a variable
runner length, and production engines with variable intake runner length are now becoming
common in vehicles.

Acoustic analytical models of inlet and exhaust flow have also been developed. The
acoustic analyses assume that valve opening and closing produces infinitesimal pressure

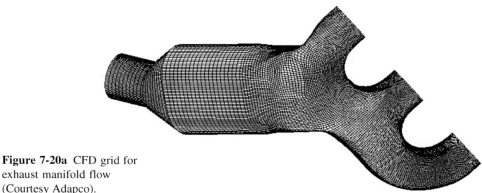

Figure 7-20a CFD grid for
exhaust manifold flow
(Courtesy Adapco).

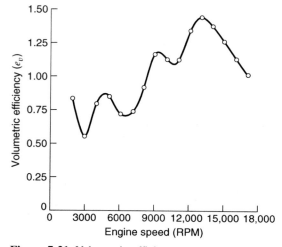

RIGHT-HAND EXHAUST MANIFOLD AND CATYLIST
Case 2: Runner 2 open, Velocity = 4 m/s, other runners closed.

4.000	
3.778	
3.556	
3.333	PROSTAR 2.1
3.111	19 Jul 92
2.889	VELOCITY MAGNITUDE
2.667	
2.444	M/SEC
2.222	LOCASL MX = 5.044
2.000	LOCAL MN = .2621E-05
1.778	
1.556	
1.333	
1.111	
.8889	
.6667	
.4444	
.2222	
−.2980E-07	

Figure 7-20b CFD results for exhaust manifold flow (Courtesy Adapco).

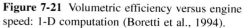

Figure 7-21 Volumetric efficiency versus engine speed: 1-D computation (Boretti et al., 1994).

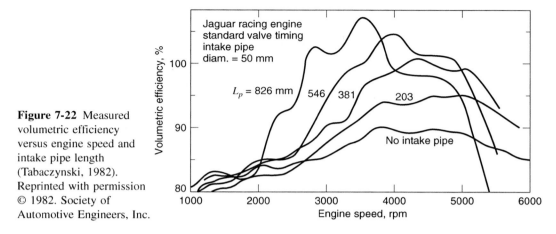

Figure 7-22 Measured
volumetric efficiency
versus engine speed and
intake pipe length
(Tabaczynski, 1982).
Reprinted with permission
© 1982. Society of
Automotive Engineers, Inc.

waves, so the coefficients of the acoustic equations are empirically determined from experiment, as discussed by Winterbone and Pearson (1999). A representative acoustic equation relating engine rpm, N_t, to a tuned intake runner length, L_t, is

$$N_t = 7.5 \, c_o/L_t \tag{7.38}$$

where c_o is the speed of sound and L_t is the tuned intake runner length.

7.4   FLUID FLOW IN THE CYLINDER

Measurement Techniques

Using lasers, it has become possible to measure local and instantaneous velocity, temperature, and some species concentrations within the cylinder without insertion of intru-
sive probes. Research engines and test rigs are built with optical access for the laser as
one of the primary design features.

Figure 7-23 shows an arrangement for measuring velocity using a laser Doppler
velocimetry (LDV) measurement technique. The arrangement shown is a steady flow test rig.
The beam from an argon ion laser is split into two beams that are then focused to a small
volume within the flow. Small particles, about 0.5 μm in diameter, are deliberately added to

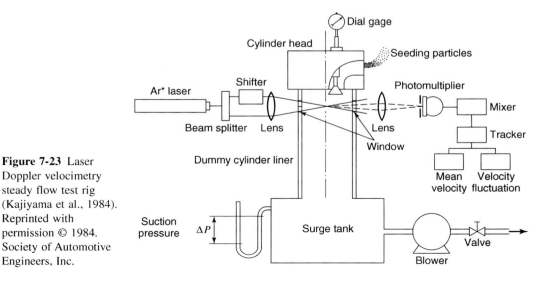

Figure 7-23 Laser
Doppler velocimetry
steady flow test rig
(Kajiyama et al., 1984).
Reprinted with
permission © 1984.
Society of Automotive
Engineers, Inc.

the flow to track the gas speed. As these particles pass through the probe volume made by the intersecting laser beams, they scatter radiation in all directions. The Doppler effect shifts the frequency of the scattered light. The frequency shift is proportional to the particle velocity. The electronics of the LDV system filter and process the signal to detect the frequency shifts. Both the mean and turbulent velocity components are measured. By moving the probe volume, the velocity can be measured at different points within the cylinder.

Particle image velocimetry (PIV) systems measure velocity by determining particle displacement over time using a double pulsed laser technique. A pulsed laser light sheet illuminates a plane in the flow, and the positions of particles in that plane are recorded by a video camera. A fraction of a second later, another laser pulse illuminates the same plane, creating a second image. Images on the two planes are analyzed using cross-correlation techniques to compute the velocity field. Additional information about PIV and other laser based measurement techniques is given in Adrian (1991).

Swirl, Squish, and Tumble

Three parameters that are used to characterize the large-scale fluid motion in the cylinder are swirl, squish, and tumble. This type of mixing is called large-scale mixing since the characteristic length of the fluid motion is on the order of the combustion chamber diameter, whereas the small scale vortices due to turbulence are much smaller. Tumble is the vortex motion induced by the inlet valve. Swirl refers to a rotational flow within the cylinder about its axis. Squish is a radial flow occuring at the end of compression in which the compressed gases flow into a cup located within the piston or cylinder head. A CFD grid for a four valve SI engine is given in Figure 7-24a, and a close up cutaway of the port and valve region is shown in Figure 7-24b. The CFD flow results at 120° atdc during the intake stroke are shown in Figure 7-24c. The tumble downstream of the intake valve is indicated by the vortex motion.

TRANSIENT IN-CYLINDER ANALYSIS OF A 4-VALVE SI ENGINE
120 Degrees ATDC During the Intake Stroke

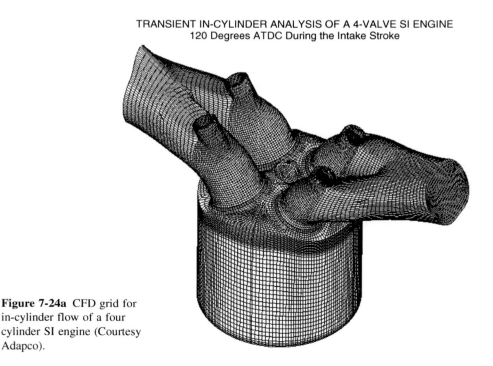

Figure 7-24a CFD grid for in-cylinder flow of a four cylinder SI engine (Courtesy Adapco).

TRASNSIENT IN-CYLINDER ANALYSIS OF A 4-VALVE SI ENGINE
120 Degrees ATDC During the Intake Stroke

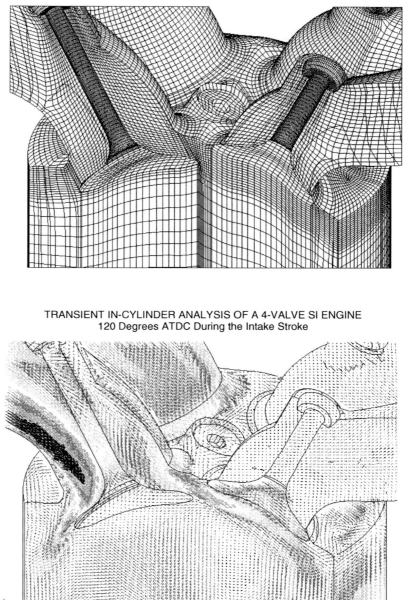

Figure 7-24b Close up of CFD grid (Courtesy Adapco).

TRANSIENT IN-CYLINDER ANALYSIS OF A 4-VALVE SI ENGINE
120 Degrees ATDC During the Intake Stroke

Figure 7-24c CFD flow results (Courtesy Adapco).

Swirl is one of the principal means to ensure rapid mixing between fuel and air in direct-injected diesel engines, and is used in gasoline engines to promote rapid combustion. The swirl level at the end of the compression process is dependent upon the swirl generated during the intake process and how much it is amplified during the compression process. In direct-injected diesel engines, as fuel is injected, the swirl convects it away from the fuel injector making fresh air available for the fuel about to be injected.

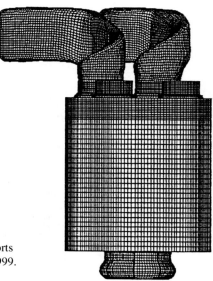

Figure 7-24d Computational model of helical inlet ports
(Bianchi et al., 1999). Reprinted with permission © 1999.
Society of Automotive Engineers, Inc.

The induction swirl is generated either by tangentially directing the flow into the
cylinder using directed ports or by preswirling the incoming flow by use of a helical port.
Helical ports are generally more compact than directed ports. They are capable of
producing more swirl than directed ports at low lifts, but are inferior at higher lifts. Either
design creates swirl at the expense of volumetric efficiency. In trying to optimize the port
design for both good swirl and volumetric efficiency, current high swirl ports are in part
both directed and helical. A numerical grid of a cylinder with helical intake ports is shown
in Figure 7-24d.

Some parameters to consider in the design of a port for swirl are shown in Figure 7-25.
These are the radius of the valve offset R_v and the orientation angle α. Development work,
like that for maximizing the discharge coefficient, is typically done on a steady-flow bench.
One way in which the swirl produced can be measured is shown in Figure 7-26. A honeycomb
structure of low mass, supported by a low-friction air bearing straightens the flow. The change
in angular momentum of the flow applies a torque to the honeycomb, which is measured by
recording the force required to restrain it. The swirl is proportional to that torque.

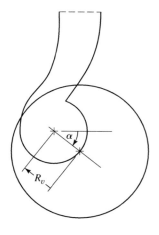

Figure 7-25 Sketch of an intake port showing the definition of
the swirl parameters R_v and α (Uzkan, Borgnakke, and Morel,
1983). Reprinted with permission © 1983. Society of
Automotive Engineers, Inc.

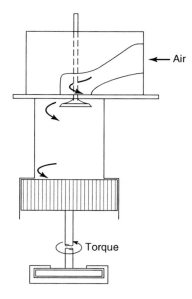

Figure 7-26 Steady-state flow and swirl system (Uzkan et al., 1983).

The efficiency of the port as a swirl producer is characterized by a swirl coefficient C_s defined as

$$C_s = \tau/(\dot{m}\, U\, b/2) \qquad\qquad (7.39)$$

where

τ = torque applied to honeycomb

$\dot{m}$ = mass flow rate

U = discharge velocity of gas

b = cylinder bore

The swirl coefficient is equal to one for the limiting case where the inlet flow enters tangentially at the cylinder wall. Some results obtained using hardware that allowed variation in the valve offset and port orientation are given in Figure 7-27. For this experiment the swirl coefficient varies in magnitude from 0 to 0.3, it increases with the valve lift or

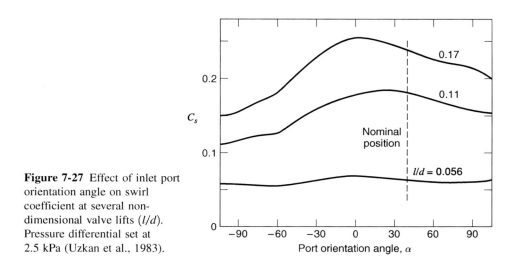

Figure 7-27 Effect of inlet port orientation angle on swirl coefficient at several non-dimensional valve lifts (l/d). Pressure differential set at 2.5 kPa (Uzkan et al., 1983).

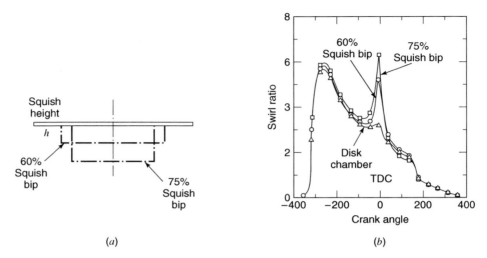

(a)   (b)

Figure 7-28 (*a*) Combustion chambers: $h = 1.27$ mm, 10:1 compression ratio. 75% squish bip means that the cross-sectional area of the "bowl in piston" is 25% of the piston area. (*b*) Computed swirl ratio (swirl rpm/engine rpm) versus crank angle: 1500 rpm, WOT, 16:1 A/F ratio, 0% EGR (Belaire et al., 1983). Reprinted with permission © 1983. Society of Automotive Engineers, Inc.

offset, and that the port orientation is important only at the larger lifts. Notice too that even at zero offset, the port is producing swirl because of the helical path upstream of the valve. Although the swirl coefficient characterizes the angular momentum of the flow by a single number, in fact, many different velocity distributions within the cylinder can yield the same angular momentum.

In operating engines a swirl ratio $R_s = \omega_s/2\pi N$ is used to characterize the swirl. The swirl ratio is defined as the ratio of the solid body rotational speed of the intake flow ω_s to the engine speed $2\pi N$. The solid body rotational speed is defined to have the same angular momentum as the actual flow. The swirl ratio in Figure 7-28 varies from zero to six times the engine speed. Using a moveable flap in an intake port, Kawashima et al. (1998) were able to vary the swirl ratio from 3.5 to as high as 10.

A bowl within the piston crown or cylinder head can be used to amplify swirl during the compression stroke. The swirl is proportional to the angular momentum, but it is also inversely proportional to the moment of inertia. At top center the moment of inertia goes through a minimum in a manner dependent upon the design of the piston bowl. As seen in Figure 7-28, near top dead center of compression the swirl increases and decreases in a rather short period. The deeper the bowl, at constant compression ratio, the greater is the change in the moment of inertia and the greater is the swirl amplification.

Incorporation of a bowl into the piston not only amplifies swirl, but also induces squish. The squish flow results from the cup geometry. This can be appreciated in terms of a rather simple argument based on Figure 7-29. The density within the cylinder at any time is more or less uniform (though time dependent) during the compression stroke. Thus, at any instant, the mass within any of the zones labeled (1), (2), and (3) is proportional to the volume in these zones at any time. During compression, zones (1) and (2) are getting smaller, whereas zone (3) remains fixed. Thus, during compression, mass must flow out of zones (1) and (2), into zone (3). The velocity of the gas crossing the control surface between zones (1) and (2) is called the squish velocity and zone (1) is called the squish zone.

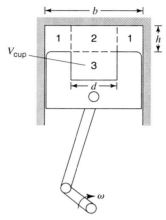

Figure 7-29 Schematic of piston squish.

Large-scale mixing and turbulence generation can also be achieved by high pressure fuel injection. Diesel engines designed without swirl are termed quiescent and the high pressure fuel injection is instead relied upon to mix the fuel and air. In this case, the fuel rushes into the cylinder dragging air with it, thus creating air motion to ensure that fresh air is available to the fuel about to be injected.

Design of the mixing process is a compromise of many conflicting demands. Increased swirl hinders volumetric efficiency and increases convective heat losses. Increased fuel injection induced air motion requires a higher injection pressure, and thus a more costly fuel injection system.

There are limits to the amount of swirl that can be used effectively to minimize demands on the fuel injection system. Herein lies one of the primary reasons for building divided-chamber, or, as they are often called, indirect-injection engines; less reliance on air motion induced by the fuel injection is required to effect large-scale mixing.

With a prechamber or swirl chamber, air is forced to flow into the chamber during the compression stroke establishing three-dimensional air motion and generating turbulence. A prechamber natural gas engine is shown in Figure 1-24. The pressure rise during combustion in the prechamber creates a flow out of the prechamber and back into the cylinder. The velocities of that backflow can be rather high, creating turbulence at the expense of an attendant head loss. Flow passages are often contained within the piston top to organize the back flow into the cylinder to create large-scale mixing of the combustion products and the cylinder air.

7.5   TURBULENT FLOW

Characterization of Turbulence

Turbulent flow can be envisioned as a mean fluid flow upon which are superimposed vortices of different sizes randomly dispersed in the flow. The vortices begin to appear above critical values, about 2300, of the mean flow Reynolds number.

$$Re = \overline{U}_p\, b/\nu \qquad (7.40)$$

For example, an engine with a bore of 0.1 m and mean piston speed of 10 m/s will have a cylinder Reynolds number of approximately 50,000, so the flow field in the cylinder will be turbulent. The turbulent vortices have finite lifetimes and appear to be born at random times. The axes of the vortices also assume random orientations. There are even vortices within vortices. It is not until the flow is analyzed statistically that any regularity in the flow field begins to appear.

Flows that are statistically steady are treated using time averaging. Flows that are statistically periodic, as in the case with intermittent internal combustion engines, are treated using ensemble averaging. The ensemble average velocity is given by

$$\overline{U}(x, \theta) = \frac{1}{n} \sum_{j=1}^{n} U(x, \theta, j) \tag{7.41}$$

where n is the number of cycles averaged and θ varies from 0 to 4π for a four-stroke engine and from 0 to 2π for a two-stroke engine. The left-hand side of Equation 7.41 is read as the ensemble average of the velocity at position x within the flow and at a time corresponding to the crank angle θ. The velocity summed on the right-hand side is the velocity at position x and angle θ for the jth cycle.

To define the instantaneous turbulence within a cycle, one writes

$$U(x, \theta, j) = \overline{U}(x, \theta) + u'(x, \theta, j) \tag{7.42}$$

where u' is the turbulent fluctuation, i.e., the difference between the ensemble average and the instantaneous velocity. To quantify the magnitude of the turbulent fluctuations, a root mean square is determined by ensemble averaging

$$u_t(x, \theta) = \left[\frac{1}{n} \sum_{j=1}^{n} u'^2 (x, \theta, j) \right]^{1/2} \tag{7.43}$$

Turbulence characteristics of flows in engine cylinders have been measured using both hot-wire anemometry and laser Doppler velocimetry Figure 7-30 shows results obtained for flows near top dead center by a number of investigators using motored engines. These results cover a range of engine configurations including engines with and without swirl.

One of the most important conclusions reached to date is that the magnitude of the fluctuating component increases with engine speed. Liou, Hall, Santavicca, and Bracco

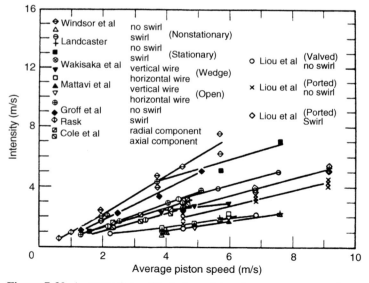

Figure 7-30 A comparison of turbulence intensity versus average piston speed measured by various researchers (Liou et al., 1984). Reprinted with permission © 1984. Society of Automotive Engineers, Inc.

(1984) conclude from these results that that the order of magnitude of the top center turbulent velocity is one half the mean piston speed

$$u_t \approx \frac{1}{2}\,\overline{U}_p \tag{7.44}$$

Of course, there are differences from engine to engine at the same piston speed. The differences are caused partly by differences in the engine design and partly because flow cannot be quantitatively characterized by a measurement of just one velocity component at just one point. In the same engine with and without swirl, it can be seen that the turbulent velocity is increased by swirling the flow.

To fully characterize a turbulent flow, one needs to also specify the size distribution of the random vortices that make up the turbulence. Researchers in this field deal with four different sizes:

1. the characteristic length L of the enclosure that represents the largest possible eddy size that confining geometry of the walls will allow, such as the cylinder bore or clearance height.
2. the integral scale l, that represents the largest turbulent vortex size, quantified as the distance where the correlation between the velocities at two points goes to zero.
3. the Taylor microscale λ, which is useful in estimating the mean strain rate of the turbulence.
4. the Kolmogorov microscale η, which is the smallest size viscous damping will allow.

For a cylindrical combustion chamber, one should expect near top center, that the characteristic length be roughly equal to the clearance height h; whereas near bottom center, the characteristic length should be roughly equal to the bore. With a cylindrical cup in the piston, near top center the characteristic length would be roughly the cup diameter.

Dimensional analysis of simple turbulent flows leads to the following relationships between the four scales.

$$l = C_l L \tag{7.45}$$

$$\frac{\lambda}{l} = \left(\frac{15}{C_\lambda}\right)^{1/2} Re_t^{-1/2} \tag{7.46}$$

$$\frac{\eta}{l} = (C_\eta)^{-1/4} Re_t^{-3/4} \tag{7.47}$$

where the constants C_l, C_λ, and C_η are numbers unique to the flow of interest and whose order of magnitude is unity (Reynolds, 1974). The turbulent Reynolds number Re_t is based on the integral scale and the turbulent velocity

$$Re_t = \frac{u_t l}{\nu} \tag{7.48}$$

Thus we see that if the integral scale can be determined, so can the other scales. As the turbulent Reynolds number increases, the smaller microscales decrease in size according to Equations 7.46 and 7.47. Since the turbulence in an engine increases with piston speed and the integral scale is independent of engine speed, we should expect that as engine speed goes up, the microscales of the turbulence will go down. In examining the microshadowgraphs of flames shown in Figure 7-31, one cannot help but be impressed with this prediction. Figure 7-31 clearly shows that the flame wrinkling increases as the engine speed increases.

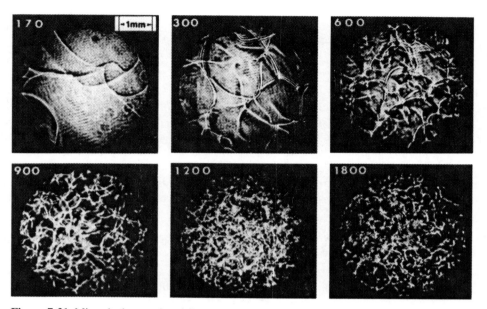

Figure 7-31 Microshadowgraphs of flame propagating toward the viewer showing increased wrinkling as engine speed increases. Engine rpm is indicated in the upper left corner of each photo (Smith, 1982).

Turbulence Models

Turbulent flow fields in engines have been modeled with a turbulent eddy viscosity v_t. There are a number of turbulent eddy viscosity models currently being used by engine modelers. The most widely used is the k-ε model. The k-ε model is a two equation model based on both a transport equation for turbulent kinetic energy k, and a transport equation for the dissipation of turbulent kinetic energy ε. The various forms of the k-ε model assume that the turbulent eddy viscosity depends on k and ε as shown by the formulation given by Equations 7.49 to 7.51:

$$v_t = C_\mu k^2/\varepsilon \tag{7.49}$$

$$\frac{Dk}{Dt} = \frac{\partial}{\partial x_j}\left(\frac{v_t}{\sigma_k}\frac{\partial k}{\partial x_j}\right) + v_t \frac{\partial \overline{U}_i}{\partial x_j}\left(\frac{\partial \overline{U}_i}{\partial x_j} + \frac{\partial \overline{U}_j}{\partial x_i}\right) - \varepsilon \tag{7.50}$$

$$\frac{D\varepsilon}{Dt} = \frac{\partial}{\partial x_j}\left(\frac{v_t}{\sigma_\varepsilon}\frac{\partial \varepsilon}{\partial x_j}\right) + C_1 v_t \frac{\varepsilon}{k}\frac{\partial \overline{U}_i}{\partial x_j}\left(\frac{\partial \overline{U}_i}{\partial x_j} + \frac{\partial \overline{U}_j}{\partial x_i}\right) - C_2 \frac{\varepsilon^2}{k} \tag{7.51}$$

The constants σ_k, σ_ε, C_μ, C_1, and C_2 are empirical constants which are flow field dependent. The k-ε model is based on a scalar eddy viscosity, so it does not take into account nonisotropic effects on the turbulence field such as streamline curvature resulting from cylinder swirl and tumble. It strictly is valid only for a fully developed turbulent flow field. It should be noted that modeling internal combustion engine turbulence with the k-ε model can lead to errors in prediction of the turbulence level and the corresponding reaction rate. Use of a compressible renormalized group (RNG) k-ε model has been recommended by Han and Reitz (1995) for in-cylinder mixing computations.

Equations 7.49 to 7.51 are combined with the continuity, momentum, and energy equations to form a complete system for numerical analysis. The k-ε model is available in CFD programs such as VECTIS (1999), STAR-CD (1999), FLUENT (1999), and KIVA (Amsden, 1997), and other turbulence models may be offered.

Large eddy simulation (LES) is a turbulence modeling procedure in which the large eddies are computed, and the smallest eddies are modeled. LES turbulence models are more computationally intensive than k-ε models. With LES modeling the time-dependent Navier Stokes equations are spatially filtered over the computational grid. The smallest eddies are more amenable to modeling as they have greater isotropic turbulence characteristics than the larger eddies. The grid cells can be much larger than the Kolmogorov length scale. LES models are available in the CFD computer programs mentioned previously. Galperin and Orszag (1993) give further details about various aspects of LES modeling.

Direct numerical simulation (DNS) is a complete time dependent solution of the Navier-Stokes and continuity equations. Since no turbulence model is used at any length scale, the grid must be small enough to resolve the smallest turbulent eddy whose size is of the order of the Kolmogorov length scale.

Turbulence modeling is an active research area. The validation process of turbulence models is on-going, and there are significant issues that need to be dealt with, for example, specifying the initial conditions throughout the flow and boundary conditions at the valves. In addition, the turbulence defined by Equation 7.43 does not recognize that a part of the fluctuation may be due to cycle-to-cycle variations in an organized flow that in any one cycle is different from the ensemble mean flow. Extensive lists of turbulence modeling references can be found in the book by Wilcox (1994).

EXAMPLE 7-3

An engine has a mean piston speed $\overline{U}_p$ of 5.0 m/s and a clearance volume height h of 10 mm. What is the characteristic length L, integral scale l, Taylor microscale λ, and Kolmogorov microscale η at the end of compression? Assume the fluid kinematic viscosity at the end of compression is 100×10^{-7} m²/s and $C_\eta = C_\lambda = 1$, $C_l = 0.2$.

SOLUTION $L = h$ since the flow is constrained by the clearance volume geometry

$$\therefore L = 10 \text{ mm}$$

$$l = C_l L = (0.2)(10) = 2 \text{ mm}$$

$$\frac{\lambda}{l} = \left(\frac{15}{C_\lambda}\right)^{1/2} Re_t^{-1/2}$$

$$Re_t = \frac{u_t l}{\nu}, \quad u_t = \frac{1}{2}\overline{U}_p = 2.5 \text{ m/s}$$

$$Re_t = \frac{(2.5)(2 \times 10^{-3})}{(100 \times 10^{-7})} = 500$$

$$\lambda = \left(\frac{15}{1}\right)^{1/2}(500)^{1/2}(2) = 0.30 \text{ mm}$$

$$\frac{\eta}{l} = (C_\eta)^{-1/4} Re^{-3/4}$$

$$\eta = (1)^{-1/4}(500)^{-3/4}(2) = 0.018 \text{ mm}$$

7.6   AIR FLOW IN TWO-STROKE ENGINES

Two-Stroke Configurations and Terminology

The two-stroke engine combines the intake and compression stroke and the expansion and exhaust stroke in order to produce power every downward stroke. Two-stroke engines can be either spark or compression ignition. A large number of different two-stroke engine configurations have been designed, each with different scavenging or air-charging characteristics, air flow paths, and valve arrangements. A crankcase-scavenged engine was discussed in Chapter 1. Air was inducted into the crankcase, subsequently compressed, and pumped into the cylinder. Another class of two-stroke engines is the separately scavenged engine in which a separate compressor, driven by the crank or perhaps an exhaust turbine, delivers the air.

Two-stroke engines are also classified on the basis of the air path during the course of scavenging. The configurations include cross, loop and through scavenging. Figure 7-32 illustrates various ways in which cross, loop, and through scavenging can be realized. Finally, there are many the different valve arrangements used to open and close ports, including piston control, poppet valves, rotary valves, or sleeve valves. Considering the large

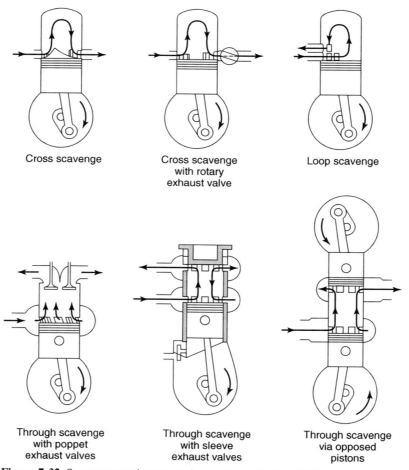

Figure 7-32 Some two-stroke scavenging methods (Taylor, 1985).
Copyright © 1985 MIT Press.

number of permutations possible, based on classification of the pumping method, the air
path, and the valving arrangement, it is clear why so many different types of two-stroke
engines exist.

The best air path is achieved via through scavenging in which air is admitted at one
end of the cylinder and exhaust gas is discharged at the other end. Ideally, this method
could result in perfect scavenging in which the incoming air displaces the exhaust gas
without mixing.

In cross scavenging, care must be taken to avoid what is referred to as short-circuiting.
Notice in Figure 7-32 that without the hump on the piston top the incoming air would
have a tendency to simply go in and out of the cylinder without displacing exhaust gas,
i.e., short-circuit the intake and exhaust process. Insertion of the hump forces the gas to
turn and mix with the exhaust gas, thus expelling a mixture of air and exhaust. Experimental
data suggests that the best scavenging that can be achieved via the cross-scavenging method
occurs when there is perfect mixing in which the fresh air introduced successively dilutes
the residual exhaust gas. If sufficient air is used, at the end of scavenging an acceptable
scavenging efficiency is then achieved.

Inspection of Figure 7-32 reveals that more often than not, two-stroke engines use
piston-controlled ports rather than cam-actuated valves to admit the fresh charge and expel the exhaust. Therefore, for two-stroke flow modeling, one must specify the effective
flow areas of the ports as functions of crank angle. A steady-flow apparatus for determining the effective flow area of piston-controlled ports is shown in Figure 7-33. Note
the similarity with the valve flow bench apparatus in Figure 7-2. Once again, solution of
Equations 7.7 and 7.11 yields the effective port area A_f from measurements of the mass
flow rate and the pressure ratio.

Some measured discharge coefficients, using the exposed geometric port area as
the reference area, for a piston-controlled inlet port are shown in Figure 7-34. Part (b) of
Figure 7-34 shows an important difference between results obtained for poppet valves and
those obtained for simple ports. The discharge coefficient increases with Mach number;
whereas, with poppet valves, it is nearly independent of Mach number. As the Reynolds
number is not constant in Figure 7-34b, the attribution of the observed effects to Mach
number tacitly assumes that there is no dependence upon Reynolds number.

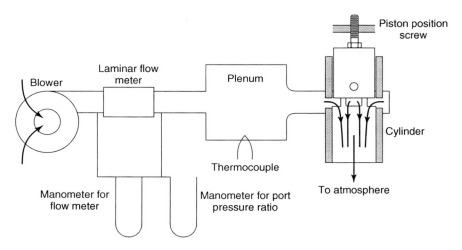

Figure 7-33 Essential features of a steady flow bench to measure effective flow areas
and discharge coefficients of a piston controlled port.

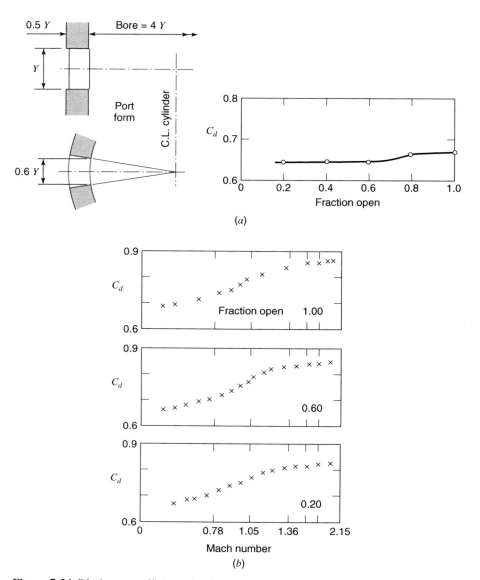

Figure 7-34 Discharge coefficient of a simple radial port as measured by Benson, see Annand and Roe (1974). (*a*) Variation with port opening at low Mach number. (*b*) Variation with Mach number based on the velocity and sound speed at the throat.

Crankcase and inlet pressures are plotted in Figure 7-35 for a loop scavenged two-stroke motorcycle engine with piston controlled induction. Corresponding cylinder and exhaust pressures are plotted in Figure 7-36. The crankcase pressure increases fairly linearly as the piston moves downward until the transfer port is uncovered (TO). Finite amplitude pressure waves occur in the intake and exhaust pipes. The pressure waves are a significant factor in the design of the intake and exhaust manifolds of two-stroke engines.

Table 7-2 gives performance terminology according to SAE recommended practice. This table as given applies to fuel-injected engines. For fuel-inducted engines, as with a throttle-body injector, the word mixture referring to the air-fuel mixture is to be substi-

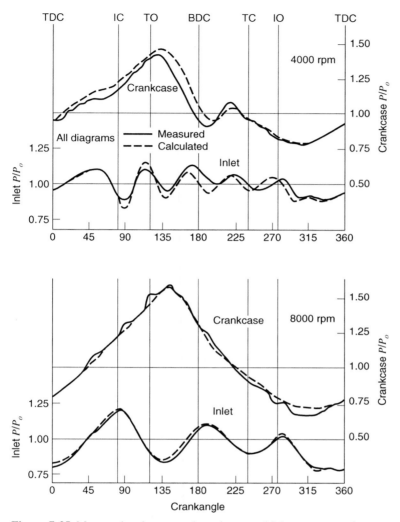

Figure 7-35 Measured and computed crankcase and inlet pressures of a two-stroke motorcycle engine. P_o is ambient pressure (Blair and Ashe, 1976). Reprinted with permission © 1976. Society of Automotive Engineers, Inc.

tuted for air, and mixture density at ambient pressure and temperature is to be substituted for ambient density. Scavenging efficiency, charging efficiency, and purity all are metrics of the success in clearing the cylinder of residual gases from the preceding cycle and as such can be mathematically related. By definition it follows that

$$e_c = D_r \, \Gamma \tag{7.52}$$

With excess air, the purity and scavenging efficiency differ because of "left over air" in the residual gas. It can be shown that (Schweitzer, 1949):

$$e_s = \mathscr{P} \quad \text{if} \quad \lambda \leq 1$$

$$e_s = \frac{1}{1 + \lambda\left(\dfrac{1}{\mathscr{P}} - 1\right)} \quad \text{if} \quad \lambda > 1 \tag{7.53}$$

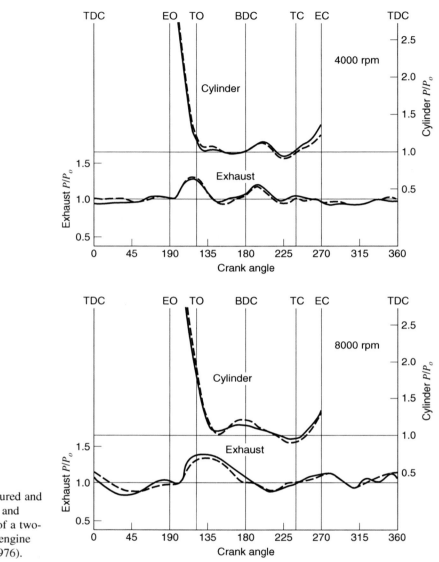

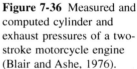

Figure 7-36 Measured and computed cylinder and exhaust pressures of a two-stroke motorcycle engine (Blair and Ashe, 1976).

The scavenging efficiency is less than or equal to the purity. However, as the difference is usually small, the two quantities are often confused. The residual mass fraction f required for thermodynamic analysis is

$$f = 1 - e_s \qquad (7.54)$$

Two-Stroke Scavenging Models

We will limit our analysis to the simplest of cases corresponding to short circuiting, perfect mixing, and perfect scavenging. Let us consider first the case of perfect scavenging. Recall that no mixing occurs and air simply displaces the exhaust gas to expel it. The trapping and scavenging efficiencies as functions of the delivery ratio are given in Figure 7-37.

Table 7-2   Two Stroke Performance Terminology

Delivery ratio, D_r

$$D_r = \frac{\text{mass of delivered air per cycle}}{\text{displaced volume} \times \text{ambient density}}$$

Trapping efficiency, Γ

$$\Gamma = \frac{\text{mass of delivered air retained}}{\text{mass of delivered air}}$$

Scavenging efficiency, e_s

$$e_s = \frac{\text{mass of delivered air retained}}{\text{mass of trapped cylinder charge}}$$

Purity, $\mathcal{P}$

$$\mathcal{P} = \frac{\text{mass of air in trapped cylinder charge}}{\text{mass of trapped cylinder charge}}$$

Relative charge, R_c

$$R_c = \frac{\text{mass of trapped cylinder charge}}{\text{displaced volume} \times \text{ambient density}}$$

Charging efficiency, e_c

$$e_c = \frac{\text{mass of delivered air retained}}{\text{displaced volume} \times \text{ambient density}}$$

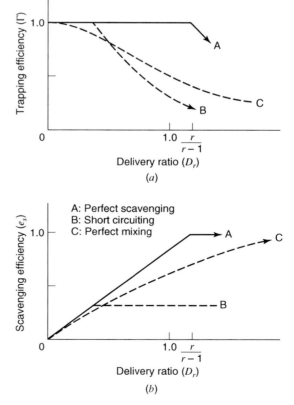

Figure 7-37 Scavenging characteristics of two-stroke engines according to three different idealizations. (*a*) Trapping efficiency versus delivery ratio. (*b*) Scavenging efficiency versus delivery ratio.

At a delivery ratio D_r given by

$$D_r = \frac{V_{bdc}}{V_d} = \frac{r}{r-1} \tag{7.55}$$

the cylinder volume at bottom center is filled with pure air ($\mathcal{P} = e_s = 1.0$), and if any more air is delivered, it is not retained. This occurs at a delivery ratio greater than one (see curve A in Figure 7-37) and dependent upon the compression ratio because the delivery ratio is defined in terms of the displacement volume rather than the maximum cylinder volume corresponding to bottom center.

In the case of short-circuiting, the air initially displaces all the gas within the path of the short circuit and then simply flows into and out of the cylinder along that path. Thus, initially, the scavenging efficiency e_s increases with delivery ratio as if scavenging were perfect. The scavenging efficiency then remains constant once the path has been displaced, see curve B in Figure 7-37b.

For perfect mixing, the first air to come in is instantaneously mixed with the exhaust and the first gas expelled is low in purity, being nearly all residual gas. For large delivery ratios, the gas being expelled is now rather pure and the trapping efficiency is thus low. The scavenging efficiency as a function of the delivery ratio can be expressed via an analysis based on the conservation of delivered air. Let m_a denote delivered air and m_a' denote delivered air retained. It follows that

$$\left(\frac{dm_a'}{dt}\right)_{cv} = \dot{m}_{a,in} - x\,\dot{m}_{out} \tag{7.56}$$

where the control volume cv is defined as the cylinder volume and x denotes the mass fraction of retained air to charge in the cylinder at any time.

Since

$$m_a' = x\,m \tag{7.57a}$$

it follows that

$$\frac{dm_a'}{dt} = x\frac{dm}{dt} + m\frac{dx}{dt} \tag{7.57b}$$

Assuming that the mass flow rates into and out of the cylinder are equal, hence $dm/dt = 0$, one obtains

$$\frac{1}{(1-x)}\frac{dx}{dt} = \frac{\dot{m}_{in}}{m} \tag{7.58}$$

Integration of Equation 7.58 over the scavenging event yields

$$e_s = 1 - \exp\left(-\int\frac{\dot{m}_{in}}{m}\,dt\right) = 1 - \exp\left(\frac{-D_r}{R_c}\right) \tag{7.59}$$

The trapping efficiency Γ is then

$$\Gamma = \frac{R_c}{D_r}\left[1 - \exp\left(\frac{-D_r}{R_c}\right)\right] \tag{7.60}$$

The perfect mixing curves (C) are drawn in Figure 7-37 accordingly.

The measured and predicted scavenging efficiencies using the perfect mixing model are compared in Figure 7-38 for a two-stroke motorcycle engine. The test engine is loop scavenged with piston controlled induction. The scavenging efficiencies are about 90%.

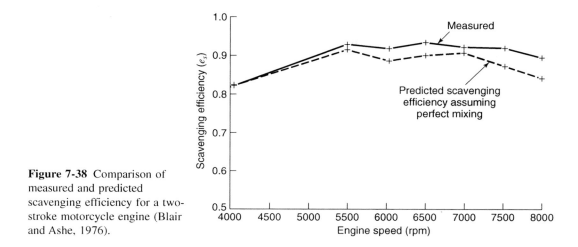

Figure 7-38 Comparison of measured and predicted scavenging efficiency for a two-stroke motorcycle engine (Blair and Ashe, 1976).

A review of scavenging modeling is given in Sher (1990). Additional information about air flow in two-stroke engines, including the unsteady compressible flow characteristics of two-stroke engines, is given in the books by Heywood and Sher (1999) and by Blair (1996).

7.7   SUPERCHARGERS AND TURBOCHARGERS

The power and efficiency of an internal combustion engine can be increased with the use of an air compression device such as a supercharger or turbocharger. Increasing the pressure and density of the inlet air will allow additional fuel to be injected into the cylinder, increasing the power produced by the engine. Spark ignition engines are knock limited, restricting the allowable compressor pressure increase. In many cases, the compression ratio of a spark ignition engine is reduced to mitigate knock when an air compressor is used. Superchargers and turbochargers are used extensively on a wide range of diesel engines, since they are not knock limited.

As shown in Figure 7-39, superchargers are classified as compressors that are mechanically driven off of the engine crankshaft. P. H. Roots first invented the supercharger in 1859, for use in the then-emerging steel industry. Superchargers have also been used in piston driven airplane engines since about 1910 to compensate for the decrease in air pressure and density with altitude, and to increase the flight ceiling. Since it is mechanically driven, the rotational speed of a supercharger is limited to about speeds of the order of 10,000 rpm. Superchargers are used in applications in which the increased density and pressure is desirable at all engine speeds.

The types of compressors used on internal combustion engines are primarily of two types: positive displacement and dynamic. With a positive displacement compressor, a volume of gas is trapped, and compressed by movement of a compressor boundary element. Three types of positive displacement compressors are the Roots, vane, and screw compressor, as shown in Figure 7-39. The efficiency of positive displacement compressors varies from about 50% for the Roots compressor to over 90% for the screw compressor. A dynamic compressor has a rotating element that adds tangential velocity to the flow which is converted to pressure in a diffuser. Two types of dynamic compressors and turbines are radial (centrifugal) and axial.

Turbochargers are defined as devices that couple a compressor with a turbine driven by the exhaust gases, so that the pressure increase is proportional to the engine speed. The

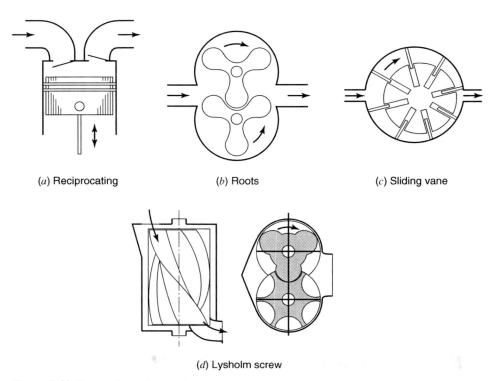

(*a*) Reciprocating   (*b*) Roots   (*c*) Sliding vane

(*d*) Lysholm screw

Figure 7-39 Types of positive displacement compressors.

turbocharger was first invented in 1906, and the applications have expanded from marine diesel engines, to vehicle diesel engines, and then to spark ignition engines. The potential increase in overall system efficiency with a turbocharger can be seen by inspection of Figure 7-40, in which a portion of the available work obtained from the blowdown of the exhaust gas can be used to compress the intake gas.

Turbochargers are usually composed of dynamic compressors and turbines, due to the high rotational speeds, of the order of 100,000 rpm, required for efficient operation at typical internal combustion engine flow rates and pressure ratios. A cross section of a turbocharger with a radial compressor and turbine is shown in Figures 7-41a and 7-41b. There are many turbocharger configurations. An intercooler heat exchanger is used with turbochargers and superchargers to cool the intake air and increase its density after the compression process has raised its temperature. A waste gate is used to control the exhaust gas flow rate to the turbine. The waste gate is a butterfly or poppet valve controlled by

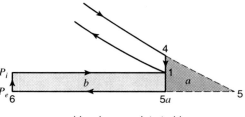

Figure 7-40 Comparison of IC engine cycle turbine and compressor work.

a: blowdown work to turbine
b: compression work

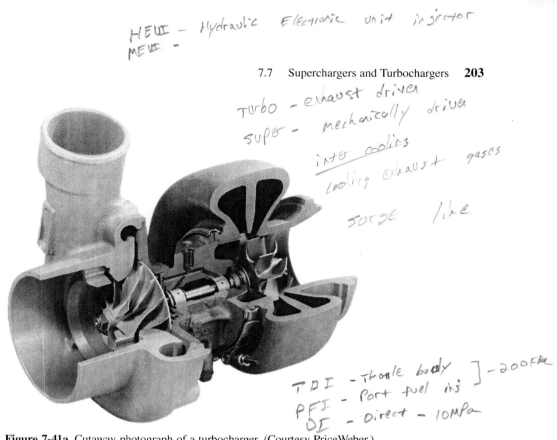

HEUI – Hydraulic Electronic unit injector
MEUE –

Turbo – exhaust driven
super – mechanically driven
inter cooling
cooling exhaust gases
surge line

TBI – Throttle body   ]–200kPa
PFI – Port fuel inj
DI – Direct – 10MPa

Figure 7-41a Cutaway photograph of a turbocharger. (Courtesy PriceWeber.)

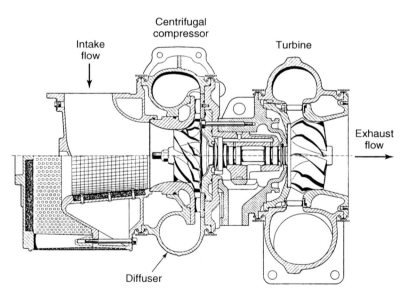

Figure 7-41b Turbocharger cross-section (Laustela et al., 1995).

the intake manifold pressure to prevent the turbocharger from compressing the intake air above a set knock or engine stress pressure limit. A turbine can also be mechanically connected to the engine drive shaft, a configuration called "compounding."

The adiabatic efficiency η_c of a compressor is defined as the ideal work required to compress the gas over the specified pressure ratio divided by the actual work required to compress the gas over the same pressure ratio. The pressure ratio of compressors used for

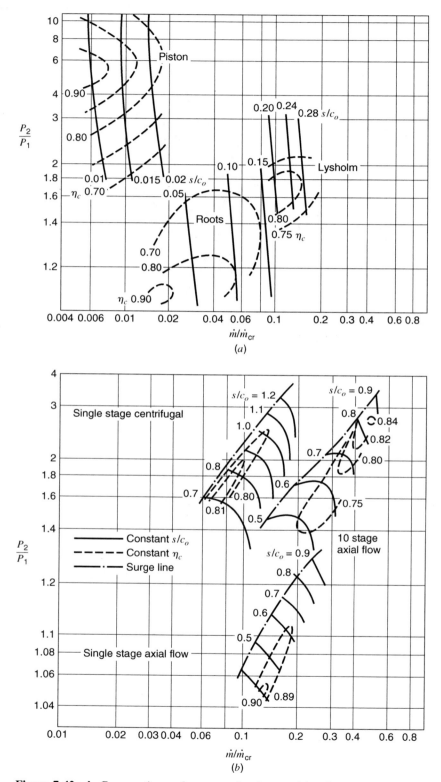

Figure 7-42a, b Comparative performance of various positive displacement and dynamic compressors (Taylor, 1985). Copyright © 1985 MIT Press.

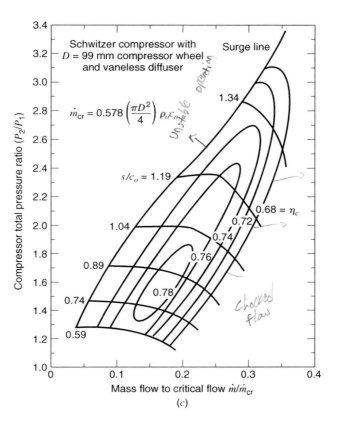

Figure 7-42c Centrifugal compressor map (Courtesy R. Hehman of Schwitzer).

internal combustion engines is generally small enough that the gas may be assumed to have constant specific heats. It follows then that the ideal work required per unit mass of gas is given by

$$w_{1-2} = c_p T_1 \left[\left(\frac{P_2}{P_1} \right)^{(\gamma-1)/\gamma} - 1 \right] \qquad (7.61)$$

In deriving Equation 7.61 it was tacitly assumed that the change in kinetic energy across the compressor was negligible compared to the change in enthalpy, an assumption usually valid in practice.

Experiments with compressors show that the adiabatic efficiency is dependent primarily upon a pressure ratio, rotational speed, and flow rate, expressed as dimensionless variables:

P_2/P_1	Outlet-inlet pressure ratio
s/c_o	Mach number based on rotor tip speed, $s = \omega D/2$
$\dot{m}/\dot{m}_{cr}$	Ratio of mass flow rate to the critical mass flow rate, Equation 7.10.

Experimental data for piston, Roots, Lysholm, screw, centrifugal, and axial compressors are given in Figure 7-42a, b, and c. The data shown are plotted on a compressor map, in which the adiabatic efficiency is plotted as a function of flow rate, pressure ratio, and the tip speed. It can be seen that the various compressors occupy different regions of the compressor map.

Dynamic compressors have surge and choking performance limits. The surge limit on the left side of the dynamic compressor map represents a boundary between stable and unstable operating points. For stable operation dynamic compressors operate to the right of the surge line. Surge is a self-sustaining flow oscillation. When the mass flow rate is reduced at constant pressure ratio, a point arises where somewhere within the internal boundary layers on the compressor blades a flow reversal occurs. If the flow rate is further reduced, then a complete reversal occurs which relieves the adverse pressure gradient. That relief means a flow reversal is no longer needed and the flow then begins to return to its initial condition. When the initial condition is reached, the process will repeat itself, creating surge.

On the right side of the dynamic compressor map is a zone where efficiencies fall rapidly with increasing mass flow rate. The gas speeds are quite high in this zone and the attendant fluid friction losses are increasing with the square of the gas speed. In this region there is also the choke limit which occurs at a slightly different value of $\dot{m}/\dot{m}_{cr}$ for each tip speed. Choking occurs when at some point within the compressor the flow reaches the speed of sound. It occurs at values of $\dot{m}/\dot{m}_{cr}$ less than 1 because $\dot{m}_{cr}$ is based on the compressor wheel diameter rather than on the cross section where choking is occurring. The value of $\dot{m}/\dot{m}_{cr}$ at choking varies with tip speed because the location within the compressor at which choking occurs depends on the structure of the internal boundary layers.

The general topic of turbomachinery is addressed in the book by Wilson (1984), and applications to internal combustion engines are treated in the book by Watson and Janota (1982). Turbomachinery issues relative to engines include:

- Coupling compressors and turbines and matching them to the needs of the reciprocating engine;
- Aftercooling of the compressed charge;
- Relating steady-flow bench tests to actual periodic flow conditions; and
- Transient response of the whole engine system.

There are alternative devices that compete with positive displacement and dynamic turbomachines. Such alternatives include shock wave compressors (Weber, 1995), an example of which is a device called the Comprex in which air is compressed by means of exhaust pressure waves and momentary direct contact between the exhaust stream and the fresh air (Gaschler, Eib, and Rhode, 1983).

7.8 FUEL INJECTORS

Spark Ignition Engines

A fuel injector is an electrically controlled valve that sprays fuel into the air flow. Spark ignition engines use fuel injectors to spray fuel into the air stream at the intake manifold (throttle body injection), inlet port (port fuel injection), or directly into the cylinder (direct injection). Figure 7-43 shows an example of a system using port fuel injection and Figure 7-44 shows the fuel injector in a direct injection engine. With port fuel injection, the fuel is sprayed into the port and onto the inlet valve to cool the valve and begin vaporization of the fuel. The amount of fuel required can be large enough so that the fuel injector will spray into the port even when the valve is closed. The fuel pressure depends on the location of the fuel injector. A port fuel injector will have a fuel pressure of about 200 kPa, and a direct fuel injector will have a fuel pressure on the order of 5 to 10 MPa, since it is injecting directly into the high pressure cylinder gases. Digitally controlled fuel injectors were first patented in 1970, and were used on production vehicles in the United States beginning in 1982.

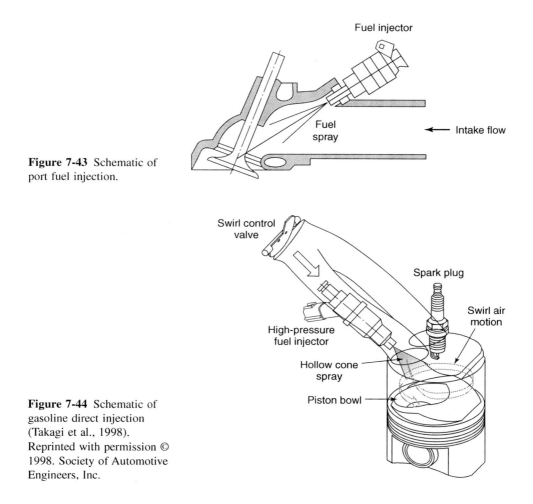

Figure 7-43 Schematic of port fuel injection.

Figure 7-44 Schematic of gasoline direct injection (Takagi et al., 1998). Reprinted with permission © 1998. Society of Automotive Engineers, Inc.

With a pintle fuel injector, the motion of a pintle nozzle opens the valve. At an appropriate time in the engine cycle, the engine control computer issues a square wave pulse to open and close the pintle nozzle. The pintle is rapidly lifted off its seat by a solenoid and the quantity of fuel injected increases more or less linearly with the duration of the open period, since the opening and closing times are much less than the open duration time. The variation of the open period over the pulse interval is shown for part and full load in Figure 5-25. Assuming quasi-steady flow through the fuel nozzle, one obtains for the mass of fuel injected in one period

$$m_f = \int_o^{\Delta t} \dot{m}_f \, dt = (2\rho_f \Delta P)^{1/2} \int_o^{\Delta t} A_f \, dt \tag{7.62}$$

where ΔP is the difference between the fuel delivery pressure and the air pressure upstream of the throttle, and A_f is the time dependent nozzle effective area. In terms of an average effective flow area of the injector nozzle

$$\overline{A}_f = \frac{1}{\Delta t} \int_o^{\Delta t} A_f \, dt \tag{7.63}$$

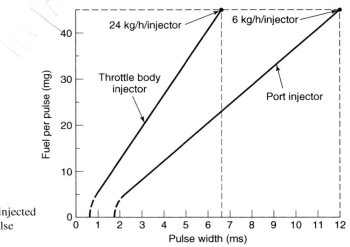

Figure 7-45 Mass of fuel injected as a function of injector pulse duration (Bowler, 1980).

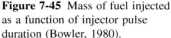

one has

$$m_f = (2\rho_f\,\Delta P)^{1/2}\,\overline{A}_f\,\Delta t \qquad (7.64)$$

and Δt is the open duration of the injection, typically 1 to 10 ms.

The prediction that the mass of fuel injected is linearly proportional to the open duration is borne out by experiment as evidenced by the results shown in Figure 7-45.

Diesel Engines

With diesel engines, fuel is sprayed directly into the cylinders at the time combustion is intended to occur. Diesel fuel injection systems operate at high pressures, on the order of 100 MPa, for two reasons:

1. the fuel pressure must be greater than the compression pressure in order to inject the fuel into the cylinder at the time combustion is to commence;

2. the fuel velocity relative to the air needs to be large so the atomized droplets will be small enough for rapid evaporation and ignition.

An example of a diesel common-rail system is shown in Figure 7-46. It is called common rail because one pump is used to deliver the fuel to all the injectors. The high pressure is generated mechanically using a camshaft. When the cam allows the injector plunger to rise, fuel enters the orifice at (A). Most of the fuel (80% or so) passes through the drain at (C) and returns to the tank. This bypass flow serves to cool the injector. Some of the fuel enters the cavity ahead of the plunger through the metering orifice (B). The greater the pump pressure is, the greater is the mass of fuel that enters the cavity. When the fuel is to be injected, the cam pushes down on the plunger, closing the metering orifice and compressing the fuel, causing it to discharge into the engine cylinder. Note that the only high pressure in the system is in the injector at this time; metering and distribution takes place at the low pressure in the common rail.

There are also fuel injection systems that employ a high pressure common rail. Figure 7-47 illustrates the differences with respect to a low pressure system. In this case the plunger is called a needle valve and it is lifted rather than forced down. This opens a flow path through the nozzle, and the fuel, which is already at a high pressure, discharges

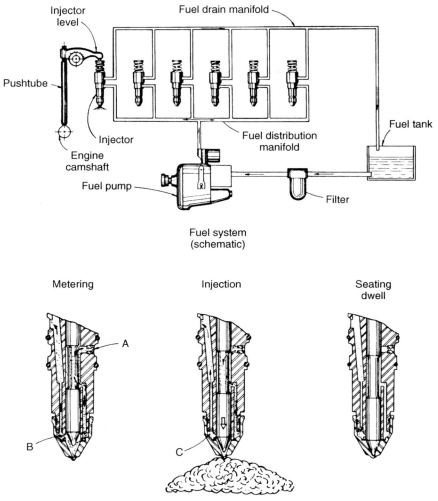

Figure 7-46 A common rail injection system used on diesel engines (Reiners et al., 1960). Reprinted with permission © 1960. Society of Automotive Engineers, Inc.

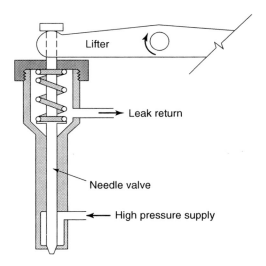

Figure 7-47 A high pressure common rail injector.

into the engine cylinder. Assuming quasi-steady flow of an incompressible fluid, one finds that the mass of fuel injected is

$$m_f = (2\rho_f \, \Delta P)^{1/2} \, \overline{A}_f \frac{\Delta\theta}{2\pi} \frac{1}{N} \tag{7.65}$$

Note that this expression is identical to Equation 7.64 except that Δt is expressed in terms of the crank angle change during the injection duration. It is clear, in this form, that in order to hold $\Delta\theta$ constant as engine speed varies, one must increase or decrease the fuel pressure to hold m_f constant. In fact, since typically the fuel injection pressure is large compared to the cylinder pressure, one must vary the fuel injection pressure P_f with the square of engine speed.

$$P_f \approx \Delta P \propto N^2, \quad \text{if} \quad m_f, \, \Delta\theta \text{ are constant} \tag{7.66}$$

Herein lies one basic problem with trying to build a diesel engine that will operate over a large speed range: if $N_{max}/N_{min} = 5$, then $P_{f,\,max}/P_{f,\,min} = 25$; furthermore, if at low speeds $P_{f,\,min} = 5$ MPa is needed to ensure good atomization and penetration into the combustion chamber, then $P_{f,\,max} = 125$ MPa. Pressures significantly higher than this can be dealt with but at great expense.

There are also diesel injection systems that use a positive displacement pump so that the mass injected is the independent variable and the fuel pressure adjusts itself accordingly. These so called jerk-pump systems utilize the principle depicted in Figure 7-48. A low pressure transfer pump fills the cavity ahead of a pumping plunger. A cam is configured to displace the plunger at the time injection is to occur. The plunger moves up, shuts off the inlet port, and, because the fuel is nearly incompressible, it rapidly increases the fuel pressure. The rise in fuel pressure creates a pressure imbalance on the needle in the injector nozzle, causing it to open and allowing fuel to discharge into the engine cylinder through the nozzle. Once the fuel pressure falls to some predetermined value, a spring forces the needle down shutting off the injector.

The mass of fuel injected is controlled by varying the displacement of the pumping plunger. One way in which this is done is shown in Figure 7-49. A helix is cut into the pumping plunger that reopens the inlet port at some intermediate position in the plunger's

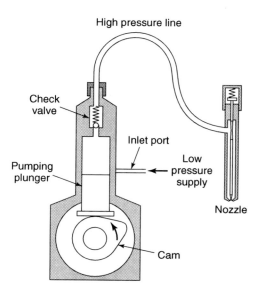

Figure 7-48 Essential features of a jerk-pump fuel injection system.

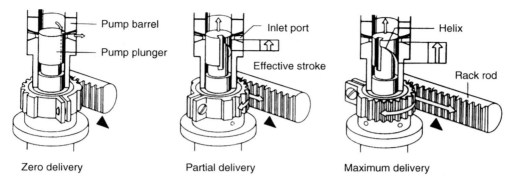

Zero delivery   Partial delivery   Maximum delivery

Figure 7-49 Use of a helix cut into the pumping plunger of a jerk-pump to vary the effective stroke. The rack rotates the helix with respect to the inlet ports to meter the fuel injected (Courtesy Robert Bosch Corporation).

stroke. A rack and pinion arrangement varies the effective stroke by rotating the plunger and therefore the position at which the port will reopen and thus dropping the fuel pressure.

An example of an electronically controlled unit fuel injector is shown in Figure 7-50. Metering is initiated by actuation of a solenoid operated valve. Closure of the solenoid valve initiates pressurization and fuel injection. The duration of valve closure determines the quantity of fuel injection. Figure 7-51 is a plot of needle lift as a function of crank angle for an electronic fuel injector. The injector pressure is about 50 MPa when the needle opens, and it increases to a maximum of about 85 MPa just before the needle closes.

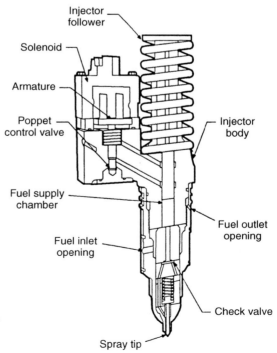

Figure 7-50 Diesel electronic unit injector (Merrion, 1994). Reprinted with permission © 1994. Society of Automotive Engineers, Inc.

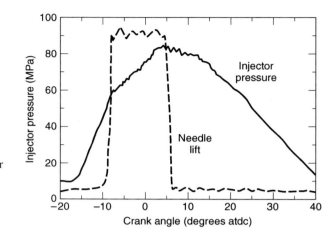

Figure 7-51 Diesel fuel injector pressure and needle lift (Espey and Dec, 1993). Reprinted with permission © 1993. Society of Automotive Engineers, Inc.

7.9 CARBURETION

Carburetors are used on spark ignition engines to control the fuel flow delivered to an engine so that it is proportional to the air flow. As shown in Fig. 7-52 and 7-53 carburetors are used for both liquid and gaseous fuels. With liquid fuels, they also serve to mix the fuel with the air by atomizing the liquid into droplets so that it will evaporate quickly. Liquid fuel carburetors were used on automobiles until the mid-1980s. Currently, they are primarily used on small (< 25 kW) engines.

The basic principle behind a liquid fuel carburetor is shown in Figure 7-52. Liquid fuel carburetors atomize the fuel by processes relying on the air speed being greater than the fuel speed at the fuel nozzle. The inlet air flows through a venturi nozzle. The pressure difference between the carburetor inlet and the nozzle throat is used to meter the fuel to achieve a desired air-fuel ratio. Therefore, the fuel is metered using the air flow as the independent variable. The mass flow between (1) and (4) is determined by the engine speed and throttle position. The pressure at (2) and the fuel-air ratio, m_f/m_a, are

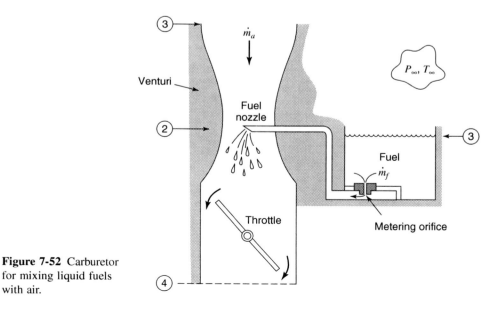

Figure 7-52 Carburetor for mixing liquid fuels with air.

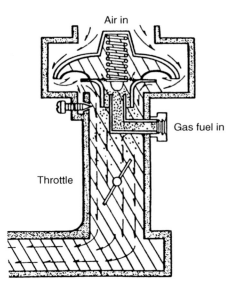

Figure 7-53 Carburetor for mixing gaseous fuels with air (Courtesy Impco, Inc.).

dependent variables that adjust themselves to match the mass flow $m_4 = m_f + m_a$ that the engine is demanding.

Assuming steady ideal gas flow through the carburetor, the air flow is

$$\dot{m}_a = \rho_\infty c_\infty A_a \sqrt{\frac{2}{\gamma - 1}\left[\left(\frac{P_2}{P_\infty}\right)^{2/\gamma} - \left(\frac{P_2}{P_\infty}\right)^{(\gamma+1)/\gamma}\right]} \qquad (7.68)$$

In Equation 7.68, c_∞ is the speed of sound in the atmospheric air and A_a is the effective flow area at the venturi throat. That flow area is less than that of a circle equal to the venturi throat diameter because of blockage by the fuel nozzle and boundary layers along the venturi walls.

The fuel flow is computed assuming the fuel is incompressible, in which case

$$\dot{m}_f = C_d A_o \sqrt{2\rho_f(P_\infty - P_2)} \qquad (7.69)$$

where C_d and the area A_o are the discharge coefficient and the flow area of the metering orifice, respectively. According to Obert (1973), the discharge coefficient of metering orifices used in carburetors is typically 0.75; it accounts for boundary layers in the orifice and for the small pressure drop from the orifice to the nozzle.

The maximum flow rate of air through a carburetor occurs when the flow chokes at the venturi nozzle. In this case one has, with application of Equations 7.8 and 7.9

$$\left(\frac{P_2}{P_\infty}\right)_c = \left(\frac{2}{\gamma + 1}\right)^{\gamma/(\gamma-1)} \qquad (7.70)$$

$$\dot{m}_{a,\,max} = \rho_\infty c_\infty A_a \left(\frac{2}{\gamma + 1}\right)^{(\gamma+1)/2(\gamma-1)} \qquad (7.71)$$

Equation 7.71 is useful in sizing a carburetor venturi; the effective area A_a is a function of the maximum air flow rate.

Let us call the ratio of the air flow to the critical or choked air flow, the carburetor demand D_c. It should be clear that $0 <= D_c <= 1$. Assuming $\gamma = 1.4$, which when substituted into Equations 7.68 and 7.71 and solved with Equation 7.69 for the fuel-air

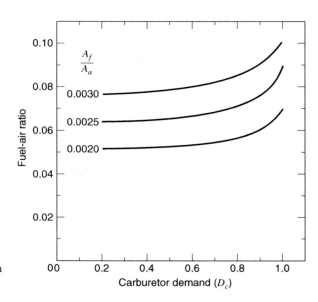

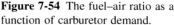

Figure 7-54 The fuel–air ratio as a function of carburetor demand.

ratio F yields

$$F = \frac{\dot{m}_f}{\dot{m}_a} = \frac{1.73}{D_c} \frac{\rho_f}{\rho_\infty} \frac{A_f}{A_a} \left[\frac{2(P_\infty - P_2)}{\rho_f c_\infty^2} \right] \tag{7.72}$$

By definition, the carburetor demand is then

$$D_c = 3.86 \left[\left(\frac{P_2}{P_\infty} \right)^{1.43} - \left(\frac{P_2}{P_\infty} \right)^{1.71} \right]^{1/2} \tag{7.73}$$

A graph of the fuel-air ratio as a function of carburetor demand is shown in Figure 7-54 assuming typical values for gasoline properties and different values of the effective area ratio A_f/A_a. The curves are based on Equation 7.72 for $P_\infty = 0.987$ bar, $\rho_f = 749$ kg/m^3, $c_\infty = 346$ m/s, and $\rho_\infty = 1.17$ kg/m^3. Note that for demands between 20 and 80%, the fuel-air ratio is a weak function of demand and its value is dependent primarily upon the geometric properties of the carburetor through the ratio A_f/A_a. At demands less than about 20%, the fuel-air ratio would in reality be much less than predicted by Equation 7.72 because of surface tension effects at the nozzle exit. The simple carburetor just described can be expected then to operate only over the range $0.20 < D_c < 0.80$.

7.10   REFERENCES

ADRIAN, R. (1991), "Particle-Imaging Techniques for Experimental Fluid Mechanics," *Ann. Rev. Fluid Mech.,* Vol. 23, p. 261–304.

AMSDEN, A. (1997), "KIVA-3V: A Block Structured KIVA Program for Engines with Vertical or Canted Valves," Los Alamos Report LA-13313-MS.

ANDERSSON, J., A. BENGTSSON, and S. ERIKSSON (1984), "The Turbocharged and Inter-Cooled 2.3 Liter Engine for the Volvo 760," SAE paper 840253.

ANNAND, W. J. D. and G. E. ROE (1974), *Gas Flow in the Internal Combustion Engine,* G. T. Foulis, Somerset, England.

BELAIRE, R. C., R. G. DAVIS, J. C. KENT, and R. J. TABACZYNSKI (1983), "Combustion Chamber Effects on Burn Rates in a High Swirl Spark Ignition Engine," SAE paper 830335.

BIANCHI, G., K. RICHARDS, and R. REITZ (1999), "Effects of Initial Conditions in Multidimensional Combustion Simulations of HSDI Diesel Engines," SAE paper 1999–01–1180.

BLAIR, G. P. and M. C. ASHE (1976), "The Unsteady Gas Exchange Characteristics of a Two-Cycle Engine," SAE paper 760644.

BLAIR, G. P. (1996), *Design and Simulation of Two Stroke Engines,* SAE International, Warrendale, Pennsylvania.

BLAIR, G. and F. DROUIN (1996), "Relationship Between Discharge Coefficients and Accuracy of Engine Simulation," SAE paper 962527.

BORETTI, A., M. BORGHI, and G. CANTORE (1994), "Numerical Study of Volumetric Efficiencies in a High Speed, Four Valve, Four Cylinder Spark Ignition Engine," SAE paper 942533.

BOWLER, L. (1980), "Throttle Body Fuel Injection (TBI)—An Integrated Engine Control System," SAE paper 800164.

ESPEY, C. and J. DEC (1993), "Diesel Engine Combustion Studies in a Newly Designed Optical-Access Engine Using High Speed Visualization and 2-D Laser Imaging," SAE paper 930971.

FLUENT USERS' MANUAL (1999), Fluent Incorporated, Hanover, New Hampshire.

GALPERIN, B. and S. ORSZAG (1993), *Large Eddy Simulation of Complex Engineering and Geophysical Flows,* Cambridge University Press, New York.

GASCHLER, E., W. EIB, and W. RHODE (1983), "Comparison of the 3-Cylinder DI-Diesel with Turbocharger or Comprex-Supercharger," SAE paper 830143.

HAN, Z. and R. REITZ (1995), "Turbulence Modeling of Internal combustion Engines using RNG k-ε Models," *Comb. Sci. and Tech;* Vol. 106, p. 207–295.

HEYWOOD, J. and E. SHER (1999), *The Two-Stroke Cycle Engine,* SAE International, Warrendale, Pennsylvania.

JASAK, H., J. LUO, B. KALUDERCIC, and A. GOSMAN (1999), "Rapid CFD Simulation of Internal Combustion Engines," SAE paper 1999-01-1185.

KAJIYAMA, K., K. NISHIDA, A. MURAKAMI, M. ARAI, and H. HIROYASU (1984), "An Analysis of Swirling Flow in Cylinder for Predicting D. I. Diesel Engine Performance," SAE paper 840518.

KAWASHIMA, J., H. OGAWA, and Y. TSURU (1998) "Research on a Variable Swirl Intake Port for 4-Valve High Speed DI Diesel Engines," SAE paper 982680.

LAUSTELA, E., U. GRIBI, and K. MOOSER (1995), "Turbocharging the Future Gas and Diesel Engines of the Medium Range," ASME ICE Conf; Vol 25-1, p. 15–21.

LIOU, T. M., M. HALL, D. A. SANTAVICCA, and F. N. BRACCO (1984), "Laser Doppler Velocimetry Measurements in Valved and Ported Engines," SAE paper 840375.

LIVENGOOD, J. C. and J. B. STANITZ (1943), "The Effect of Inlet Valve Design, Size and Lift on the Air Capacity and Output of a Four-Stroke Engine," NACA TN 915.

LIVENGOOD, J. C., C. F. TAYLOR, and P. C. WU (1958), "Measurements of Gas Temperatures by the Velocity of Sound Method," *SAE Trans.* Vol. 66, p. 683.

MERRION, D. (1994), "Diesel Engine Design for the 1990s," SAE SP-1011.

NOMA, K., Y. IWAMOTO, N. MURAKAMI, K. IIDA, and O. NAKAYAMA (1998), "Optimized Gasoline Direct Injection Engine for the European Market," SAE paper 980150.

OBERT, E. F. (1973), *Internal Combustion Engines and Air Pollution,* Harper & Row, New York, pp. 388–389.

OLSEN, D., P. PUZINAUSKAS, and O. DAUTREBANDE (1998), "Development and Evaluation of Tracer Gas Methods for Measuring Trapping Efficiency in 4-Stroke Engines," SAE paper 981382.

REINERS, N., R. SCHMIDT, and J. PERR (1960), "Cummins New PT Fuel Pump," SAE paper 258B.

REYNOLDS, A. (1974), *Turbulent Flows in Engineering,* John Wiley & Sons, New York.

SCHWEITZER, P. H. (1949), *Scavenging of Two-Stroke Diesel Engines,* Macmillan, New York.

SHER, E. (1990), "Scavenging the Two-Stroke Engine," *Prog. Energy Combust. Sci.,* Vol. 16, pp. 95–124.

SILVESTRI, J., T. MOREL, AND M. COSTELLO (1994), "Study of Intake System Wave Dynamics and Acoustics by Simulation and Experiment," SAE paper 940206.

SMITH, J. R. (1982), "Turbulent Flame Structure in a Homogeneous-Charge Engine," SAE paper 820043.

SORENSON, S. C. (1984), "Simulation of a Positive Displacement Supercharger," SAE paper 840244.

STAR-CD USERS' MANUAL (1999), Computational Dynamics, Inc., London, England.

TABACZYNSKI, R. J. (1976), "Turbulence and Turbulent Combustion in Spark Ignition Engines," *Prog. Energy Combust. Sci.,* 2, p. 143-165.

TABACZYNSKI, R. J. (1982), "Effects of Inlet and Exhaust System Design on Engine Performance," SAE paper 821577.

TAKAGI, Y., T. ITOH, and K. NAITOH, "Simultaneous Attainment of Low Fuel Consumption, High Output Power and Low Exhaust Emissions in Direct Injection SI Engines," SAE paper 980149.

TAYLOR, C. F. (1985), *The Internal Combustion Engine in Theory and Practice.* MIT Press, Cambridge, Massachusetts.

TAYLOR, C. F., J., LIVENGOOD, and D. TSAI (1955), "Dynamics of the Inlet System of a Four-Stroke Single-Cylinder Engine," *ASME Trans.,* Vol. 77, p. 1133.

UZKAN, T., C. BORGNAKKE, and T. MOREL (1983), "Characterization of Flow Produced by a High-Swirl Inlet Port," SAE paper 830266.

VECTIS USERS' MANUAL (1999), Ricardo Software, Inc., Burr Ridge, Illinois.

WATSON, N. and M. S. JANOTA (1982), *Turbocharging the Internal Combustion Engine,* John Wiley & Sons, New York.

WAVE USERS' MANUAL (1999), Ricardo Software, Inc., Burr Ridge, Illinois.

WEBER, H. (1995), *Shock Wave Engine Design,* John Wiley & Sons, New York.

WILCOX, D. (1994), *Turbulence Modeling for CFD,* DCW Industries, La Canada, CA.

WILSON, D. (1984), *The Design of High Efficiency Turbomachinery and Gas Turbines,* MIT Press, Cambridge, Massachusetts,

WINTERBONE, D. and R. PEARSON (1999), *Design Techniques for Engine Manifolds,* Society of Automotive Engineers, Warrendale Pennsylvanin.

7.11   HOMEWORK

7.1   Derive an expression for the volumetric efficiency of a supercharged engine, using an analysis similar to the derivation of Equation 7.27.

7.2   If an engine has a bore of 0.1 m, stroke of 0.08 m, inlet flow effective area of $4.0 \times 10^{-4}\,m^2$, and inlet temperature of 320 K, what is the maximum speed it is intended to be operated while maintaining good volumetric efficiency?

7.3   Explain how unburned fuel can appear in the exhaust during the intake and exhaust strokes.

7.4   A constant-volume cylinder (see Figure 7-a) contains air at $P_o = 50$ atm, $T_o = 298$ K. At time $t = 0$, a valve opens and closes 20 ms later. When the valve is open, its effective flow area is given by

$$\frac{A_f}{A_{f,\,max}} = \left[\sin\left(\frac{\pi t}{\tau}\right) \right]^{1/4}$$

where $A_{f,\,max} = 1.0\ cm^2$ and $\tau = 20 \times 10^{-3}$ s. Assuming the heat transfer coefficient between the gas and the vessel walls is $h = 100\ W/m^2\,K$, find the pressure and temperature as a function of time during the valve-open period. How much mass escapes the cylinder? The air may be assumed to have constant specific heats with $\gamma = 1.4$ and $M = 29.0$.

7.5   It was explained in Section 7.2 that because of the pressure drop across a valve, it is advantageous to close the intake valve after bottom dead center. Use the same logic to explain why exhaust valves are closed after top dead center and what the effect of engine speed is on the residual fraction.

7.6   Suppose an engine were constructed with variable valve timing, thus ensuring optimum timing at all speeds. Explain how the volumetric efficiency would depend on speed for wide-open throttle operation with short pipes and $Z < 0.6$.

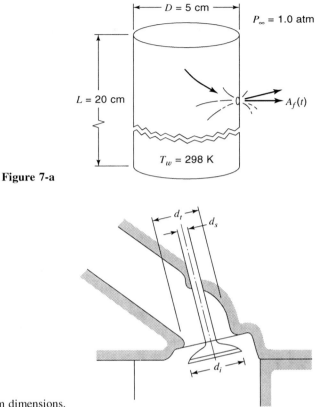

$D = 5$ cm

$P_\infty = 1.0$ atm

$L = 20$ cm

$A_f(t)$

$T_w = 298$ K

Figure 7-a

Figure 7-b Valve throat and stem dimensions.

7.7 Figure 7-b shows an inlet valve opened to $l/d_i = 0.25$. If the stem is chosen to be $d_s = 0.15 d_i$ and the throat of the port is $d_t = 0.85 d_i$, what would be the flow coefficient based purely on the geometrical blockage?

7.8 (**a**) Calculate the ratios of the inlet valve area to piston area for the three configurations, a, b, c in Figure 7-11 recommended by Taylor (1985) as being the maximum feasible for a flat cylinder head.

   (**b**) If the inlet Mach index in each case is held to $Z_i = 0.6$ and $c_i = 400$ m/s, $A_i = 0.35\, n_i$ $(\pi/4)\, d_i^2$ (n_i = number of intake valves), then what is the maximum piston speed in each case?

7.9 The inlet air flow in a single-cylinder four-stroke engine can be modeled as a Helmholtz resonator with an effective volume of

$$V_f = \frac{V_d}{2} \frac{r + 1}{r - 1}$$

The resonant tuning rpm (N_t) of the inlet pipe of length L_i and diameter D_i is predicted to be

$$N_t = \frac{15}{\pi} c_o \left(\frac{\frac{1}{4}\pi D_i^2}{L_i V_f} \right)^{1/2}$$

Where L_i is an effective length from the inlet valve to the atmosphere and D_i is an effective diameter that with L_i matches the inlet system volume.

Compare the predicted resonant tuning rpm of the Helmholtz resonator with the simple acoustic tuning rpm of Equation 7.38 and also the experimental results for maximum volumetric efficiency plotted in Figure 7-22. Assume D_i is equal to the inlet pipe diameter. Make a table of the tuning rpm versus tuning inlet pipe length for the five cases shown in Figure 7-22.

Assume $b = 83$ mm, $s = 106$ mm, $D_i = 0.05$ m, $r = 9$, $T_\infty = 300\ K$.

7.10 A number of cam shafts are available for an engine. These include

	TIMING (DEG)				
CAM	**IO** btc	**IC** abc	**EO** bbc	**EC** atc	**LIFT** mm
Factory	30	60	60	30	9.5
A	26	66	66	26	11.4
B	22	62	62	22	10.3

Discuss the effects these different cams might have, including duration and overlap effects.

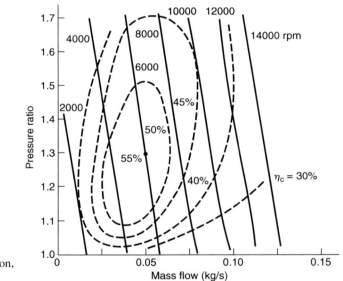

Figure 7-c Simulated supercharger performance (Sorenson, 1984).

7.11 A simulated Roots supercharger map is given in Figure 7-c. Match (i.e., find the resultant pressure ratio) this supercharger to a 2.0 liter, 4-stroke engine with the following volumetric efficiencies.

N (rpm)	$e_v(\%)$
1000	68
3000	75
6000	70

Find the power required to drive the supercharger at each condition as well as the outlet temperature. Choose a compressor speed equal to twice the engine speed. The procedure is as follows:

(a) Assume a pressure ratio.

(b) Read the compressor efficiency and mass flow rate.

(c) Solve for the temperature and density after compression (adiabatic values but not isentropic values).

(d) Calculate the engine mass flow rate with the density found in part (c) and the known volumetric efficiency.

(e) Iterate until the engine mass flow rate and compressor flow rate are equal.

(f) Calculate the compressor power.

7.12 A naturally aspirated four-cylinder, four-stroke gasoline engine has the following specifications.

$$V_d \quad 2316 \text{ cm}^3$$
$$b \quad 96 \text{ mm}$$
$$s \quad 80 \text{ mm}$$
$$r \quad 9.5$$
$$\dot{W}_b \quad 83 \text{ kW at } N = 90 \text{ rps}$$

A turbocharged version of the engine utilizes the compressor mapped in Figure 7-d. Estimate the brake power of the turbocharged engine at $N = 90$ rps if the compressor ratio is $P_2/P_1 = 1.5$. What is the compressor efficiency and speed? What is the heat transfer to the inter-cooler?

Make the following assumptions.

• For the naturally aspirated (NA) engine

Inlet manifold conditions: $T_i = 310$ K, $P_i = 1.0$ bar, $\phi = 1.0$.

Volumetric efficiency: $e_v = 0.84$.

Mechanical efficiency: $\eta_M = \text{bmep}/(\text{imep})_{\text{net}} = 0.90$.

• For the turbocharged (TC) engine

Aftercooled gas temperature: $T_i = 340$ K.

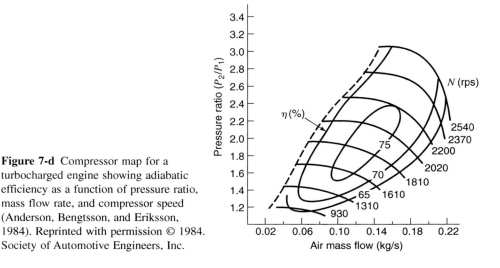

Figure 7-d Compressor map for a turbocharged engine showing adiabatic efficiency as a function of pressure ratio, mass flow rate, and compressor speed (Anderson, Bengtsson, and Eriksson, 1984). Reprinted with permission © 1984. Society of Automotive Engineers, Inc.

Volumetric efficiency: $e_v = 0.91$

Mechanical efficiency: $\eta_M = 0.88$.

- Net indicated power is proportional to airflow rate

$$\frac{(imep)_{net,TC}}{(imep)_{net,NA}} = \frac{\left(\dfrac{e_v P_i}{T_i}\right)_{TC}}{\left(\dfrac{e_v P_i}{T_i}\right)_{NA}}$$

In practice, the compression ratio was lowered to 8.7 to avoid knock and the engine produced 117 kW at 88 rps.

7.13   The carburetion analysis presented assumed that a steady flow existed. This is approximately true in carburetors serving four or more cylinders as at any point in the cycle there is an intake process occurring, Obert (1973). If there are less than four cylinders served by a carburetor, then the fraction of time an intake flow exists is given approximately by

$$t_f \approx n_c/4n$$

where n_c is the number of cylinders served by n carburetors. The average quasi-steady air flow rate through a carburetor when an intake flow is occurring is then

$$\dot{m}_a \approx \frac{1}{t_f}\frac{N}{2}e_v\rho_i V_d \tag{7.74}$$

where $t_f \leq 1$. Carburetor venturis are sized assuming the maximum quasi-steady flow is twice the average.

(a) Assuming $\rho_i = \rho_\infty$, $F = 0$, solve Equation 7.68 for the effective flow area of the air based on a demand of $D_c = 1.0$.

(b) Assume that there are n identical single-barrel carburetors so that

$$A_a = nC_d\frac{\pi}{4}d_v^2$$

and that the discharge coefficient is $C_d = 0.75$. Derive an expression for the venturi throat diameter d_v for an engine with n_c cylinders.

(c) Campbell (1971) offers the following practical hint in sizing carburetor venturis.

$$d_v\,(mm) = 20\sqrt{\frac{V_{d,1}}{1000} \times \frac{N}{1000}}$$

where

d_v = venturi diameter

$V_{d,1}$ = displacement volume of one cylinder (cm^3)

N = maximum engine speed (rpm)

He points out that this is based on a mean gas speed through the venturi of 110 m/s and that the venturi size is independent of the number of cylinders and the number of carburetors.

Compare Campbell's equation with your result assuming $P_\infty = 1$ bar, $T_\infty = 298$ K and show that venturi size is independent of the number of cylinders and independent of the number of carburetors if and only if Equation 7.74 is applicable.

Chapter **8**

Heat and Mass Transfer

8.1 INTRODUCTION

Satisfactory engine heat transfer is required for a number of important reasons, including material temperature limits, lubricant performance limits, emissions, and knock. Since the combustion process in an internal combustion engine is not continuous, as is the case for an external combustion engine, the component temperatures are much less than the peak combustion temperatures. However, the temperatures of certain critical areas need to be kept below material design limits. Aluminum alloys begin to melt at temperatures greater than 775 K, and the melting point of iron is about 1800 K. Differing temperatures around the cylinder bore will cause bore distortion and subsequent increased blow-by, oil consumption, and piston wear. Cooling of the engine is also required to prevent knock in spark ignition engines.

Exhaust system heat transfer is also an important factor in emissions and exhaust turbine performance. Satisfactory catalytic converter performance occurs above a threshold or light-off temperature. The threshold temperature (oxidation efficiency greater than 50%) for the catalyzed oxidation of hydrocarbon and carbon monoxide emissions is about 500 K, so that at exhaust temperatures less than 500 K, catalytic converter performance is adversely affected. In addition, the continued oxidation of hydrocarbons and other pollutants in the exhaust system is a function of the exhaust system temperature.

Heat transfer to the air flow in the intake manifold lowers the volumetric efficiency, since the density of the intake air is decreased. Plastic intake manifolds with reduced thermal conductivity (as well as reduced weight) are now being used to reduce intake air heating.

The heat transfer rate in an engine is dependent on the coolant temperature and the engine size, among other variables. There are complex interactions between various operational parameters. For example, as the temperature of the engine coolant decreases, the heat transfer to the coolant will increase, and the combustion temperature will decrease. This will cause a decrease in the combustion efficiency and an increase in the volumetric efficiency. It will also cause an increase in the thermal stresses in the cylinder sleeve, and increase the size of the radiator needed, since the coolant–ambient temperature difference will decrease. The formation of nitrogen oxides will decrease and the oxidation of hydrocarbons will decrease. The exhaust temperature will also decrease, causing a decrease in the performance of the catalytic converter and a turbocharger. For more detailed information about engine heat transfer, a comprehensive review is given in Borman and Nishiwaki (1987).

8.2 ENGINE COOLING SYSTEMS

There are two types of engine cooling systems used for heat transfer from the engine block and head; liquid cooling and air cooling. With a liquid coolant, the heat is removed through the use of internal cooling channels within the engine block, as shown schematically in

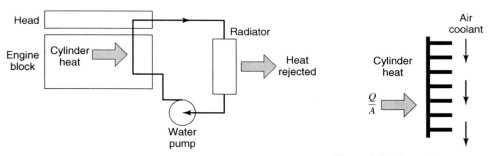

Figure 8-1 Liquid cooling system.   **Figure 8-2** Air cooling system.

Figure 8-1. With air as a coolant, the heat is removed through the use of fins attached to the cylinder wall, as shown schematically in Figure 8-2. Both types of cooling systems have various advantages and disadvantages. Liquid systems are much quieter than air systems, since the cooling channels absorb the sounds from the combustion processes. However, liquid systems are subject to freezing, corrosion, and leakage problems that do not exist in air systems.

As indicated in Figure 1-17 and Figure 8-1, the water cooling system is usually a single loop where a water pump sends coolant to the engine block, and then to the head. The coolant will then flow to a radiator or heat exchanger and back to the pump. The boiling temperature of the liquid coolant can be raised by increasing the pressure or by adding an additive with a high boiling point, such as ethylene glycol. During engine warm-up, a thermostatically controlled valve will recycle the coolant flow through the engine block, bypassing the heat exchanger. As the engine heats up, the valve will open up, and allow the coolant to flow to the radiator. The time required for engine warm-up to a steady state operating temperature depends on the engine size, speed, and load, and is typically of the order of 10 minutes for an automotive engine. Dual circuit cooling with separate circuits to the head and block has also been used.

The design of the liquid cooling passages in the engine block and head is done empirically. The primary design consideration is to provide for sufficient coolant flow at the high heat flux regions, such as the exhaust valves. Since the area between exhaust valves is difficult to cool, some automotive engine designs use only one exhaust valve to reduce the heating of the inlet air-fuel mixture, and thus increase the volumetric efficiency. A review of precision cooling considerations is given in Robinson et al. (1999).

The heat fluxes and surface temperatures near the exhaust manifold and port are high enough so that nucleate boiling can occur in the coolant at those locations. The boiling heat transfer coefficients are much larger than single phase forced convection, so that the surface temperatures will be correspondingly lower. For heat fluxes of the order of 1.5 MW/m^2, the resulting surface temperature of the cooling jacket will be about 20 to 30°C above the saturation temperature, which is typically 130°C (400 K). The nucleate boiling process is very complex, as bubbles formed on the cooling channel surface are swept downstream and then condense in cooler fluid.

Engines with relatively low power output, less than 20 kW, primarily use air cooling. Because the thermal conductivity of air is much less than that of water, air systems use fins to lower the air side surface temperature. For higher power output, an external cooling fan is used to increase the air side heat transfer coefficient. Aircraft engines are, for the most part, air cooled; supplying the required air flow is not a problem since the engine need not be enclosed and is usually located right behind a propeller. Engines that are operated for very short periods of time, such as engines used in 1/4-mile dragsters, do not use a cooling system, and use the thermal capacitance of the engine block to keep the gas side surface temperatures within limits.

Since about one third of the fuel energy is lost as heat transfer to the coolant, it would seem reasonable to try to reduce this heat loss, and thereby increase the efficiency of the engine. One way to reduce the heat flow to the coolant is to increase the thermal resistance of the engine block through the use of lower thermal conductivity materials, such as ceramics, or adding thermal insulation to the engine. Ceramic wall materials that can operate at higher temperatures and have a lower thermal conductivity than cast iron are silicon nitride and zirconia.

Experimental results from such engines show that a reduction in the coolant heat loss does not result in a corresponding increase in the efficiency of the engine (Sun et al., 1996). There are a number of reasons for this. First, since the internal combustion engine is a thermodynamic cycle, the heat to work conversion efficiency is limited by the second law. For the expansion stroke, the engine converts about 30% of the input heat to work, so at most, the same fraction of the coolant heat transfer could be converted to work. Second, by insulating the engine, the average cylinder temperature increases, and the exhaust temperature and enthalpy increase. The thermal energy that had been conducted to the coolant is now added to the exhaust stream. Third, since the penetration depth of the combustion heat flux is only about a millimeter, the coolant heat transfer is a relatively steady state process throughout the thermodynamic cycle, but the positive work is produced only during the expansion stroke. The coolant heat transfer that occurs during the other strokes is not available to be converted to work. Finally, the higher wall temperatures will heat the incoming gas during the intake stroke, lowering its volumetric efficiency. For spark ignition engines, the higher wall temperatures during the compression stroke can give rise to knock problems. Therefore, the majority of the engines that have employed increased cylinder thermal resistance are compression ignition engines.

8.3   ENGINE ENERGY BALANCE

An engine energy balance is obtained through experiments performed on instrumented engines. Figure 8-3 depicts an engine instrumented to determine the quantities of heat rejected to oil, water, and to the ambient air. Flow meters are installed in the water, and oil circuits and thermocouples measure the inlet and outlet temperatures. An energy balance applied to the water coolant and the oil flowing through the engine yields

$$\dot{Q}_{water} = (\dot{m} c_p)_{water} (T_3 - T_4) \qquad (8.1)$$
$$\dot{Q}_{oil} = (\dot{m} c_p)_{oil} (T_1 - T_2) \qquad (8.2)$$

Determining the heat loss to the ambient air is more involved. The first law applied to the engine system is

$$\dot{Q}_{amb} = (\dot{m} h)_{air} + (\dot{m} h)_{fuel} - (\dot{m} h)_{ex} \\ - \dot{Q}_{water} - \dot{Q}_{oil} - \dot{W}_{shaft} \qquad (8.3)$$

The mass flow rate of the exhaust is known in terms of the measured mass flow rates of air and fuel since

$$\dot{m}_{ex} = \dot{m}_{air} + \dot{m}_{fuel} \qquad (8.4)$$

The enthalpies of the exhaust, the air, and the fuel are based on the measured temperatures T_5, T_6, and T_7, respectively. The exhaust composition can be calculated theoretically from the known fuel-air equivalence ratio or it may be measured. In either case it is

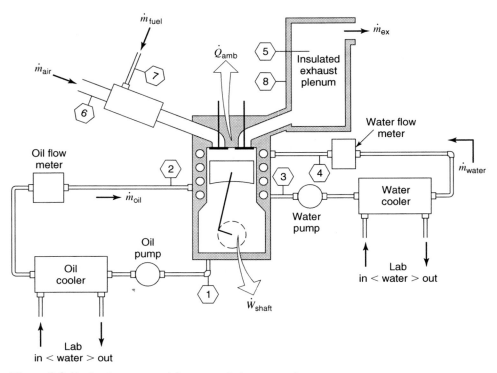

Figure 8-3 Engine instrumented for energy balance experiments.

important that the temperature T_5 corresponds to the mass averaged temperature of the exhaust; for this reason there is an insulated plenum that serves to mix the hot exhaust gas emitted early in the cycle with the cooler exhaust gas emitted later in the cycle. A further complication is that the thermocouple measurement must be corrected for radiation heat transfer to obtain the true gas temperature. An energy balance on the thermocouple tip yields

$$T_{ex} = T_5 + \frac{\varepsilon \sigma}{\boldsymbol{h}} (T_5^4 - T_8^4) \tag{8.5}$$

where ε is the emissivity of the thermocouple tip, σ is the Stefan-Boltzmann constant, $\boldsymbol{h}$ is the heat transfer coefficient at the tip (printed in bold to distinguish it from the symbol h for enthalpy), T_5 is the tip temperature, and T_8 is the exhaust plenum inner wall temperature.

In doing these energy balances, it is common practice to evaluate the maximum heat that can be recovered from the exhaust gas. This is computed from an energy balance on the exhaust where it is cooled to ambient temperature

$$\dot{Q}_{ex} = \dot{m}_{ex} \left[h_{ex}(T_{ex}) - (h_{ex}(T_{amb})) \right] \tag{8.6}$$

In evaluating the exhaust enthalpy at ambient temperature, equilibrium water quality should be used, as discussed in Chapter 4.

If Equation 8.6 is substituted into Equation 8.3, one obtains

$$\dot{Q}_{amb} = \dot{Q}_{in} - \dot{Q}_{ex} - \dot{Q}_{water} - \dot{Q}_{oil} - \dot{W}_{shaft} \tag{8.7}$$

Table 8-1   Energy Balance on a Medium Speed, Four-Stroke, Direct-Injection, Shallow-Bowl, Turbocharged Diesel Engine[a]

N (rpm)	bmep (bar)	$\dot{Q}_{\text{exhaust}}$	$\dot{Q}_{\text{water}}$	$\dot{Q}_{\text{oil}}$	$\dot{Q}_{\text{ambient}}$	$\dot{W}_{\text{shaft}}$	$\dot{W}_{\text{friction}}$	$\dot{Q}_{\text{loss}}$
500	9.90	0.459	0.118	0.037	0.069	0.317	0.100	0.124
500	3.52	0.437	0.108	0.065	0.092	0.298	0.178	0.087
400	3.50	0.432	0.151	0.074	0.026	0.315	0.092	0.159

Source: Whitehouse (1970–71).

[a]Engine: b = 304.8 mm, s = 381 mm, r = 12.85.

Ambient: T = 293 K, P = 1.01 bar.

$\dot{Q}_{\text{loss}} = \dot{Q}_{\text{water}} + \dot{Q}_{\text{oil}} + \dot{Q}_{\text{ambient}} - \dot{W}_{\text{friction}}$

All energy rates are normalized by the fuel rate, $\dot{Q}_{\text{in}}$.

where, by definition

$$\dot{Q}_{\text{in}} = (\dot{m}h)_{\text{air}} + (\dot{m}h)_{\text{fuel}} - \dot{m}_{\text{ex}}\, h_{\text{ex}}\, (T_{\text{amb}}) \tag{8.8}$$

Finally, if the fuel and air are at the ambient temperature, the engine runs lean or stoichiometric, and the ambient temperature and pressure are coincident with the reference temperature and pressure, then the inlet enthalpy is the product of the fuel flow rate and the fuel's stoichiometric heat of combustion.

Some results obtained by Whitehouse (1970–1971) for a medium speed diesel engine are given in Table 8-1. The diesel engine was a single cylinder test engine with a bore of 304.8 mm and a stroke of 381 mm. The table also gives heat equivalence of the friction so that one can ascertain how much of the heat lost to the ambient air, to the oil, and to the water, is from the working fluid and how much is from friction. The overall heat loss is the sum of the heat transfer to the water, oil, and ambient air minus the friction work. The terms in the table are normalized by the input energy of the fuel. Inspection of the shaft work term reveals that the engine has a brake thermal efficiency of about 30%. About 45% of the energy is rejected in the exhaust, 10 to 15% is dissipated by friction, and 10 to 15% is dissipated by heat loss.

We now compare the diesel engine results with energy balance results for spark ignition engines. Figure 8-4 shows the results of an energy balance on a small, spark-ignition automobile engine. This engine has an internal oil pump and the heat rejected to the oil is carried away partly by the coolant and partly by the heat lost to ambient air. As the load increases and the intake manifold pressure increases from P_i = 0.4 bar to P_i = 0.8 bar, the energy converted to shaft work increases from about 20 to 30%, the coolant load decreases from about 40 to 30%, the exhaust energy varies from about 30 to 35%, and the heat lost to ambient air decreases from about 10 to 5%. The energy dissipated by friction decreases from about 14 to 7% for the same loads. The total heat loss from the gas to the coolant and ambient air during the cycle is about 28 to 36%.

The diesel engine loses only about one half as much heat to the coolant and ambient air as the gasoline engine, yet their shaft efficiencies are about equal since the diesel engine has more exhaust energy than the gasoline engine. As discussed above, experiments with insulated engines show that a reduction in the coolant heat loss has a small impact on the shaft efficiency and that the heat no longer lost to the coolant mostly appears in the exhaust energy. If a turbocharger is used, the available portion of the exhaust energy can be converted to useful work.

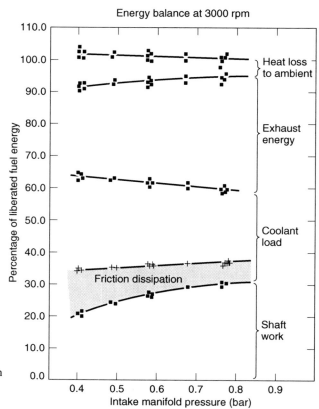

Figure 8-4 Energy balance on an automotive engine (Courtesy D. Brigham of Ford Motor Co.).

Figure 8-5 shows one cylinder used on a Pratt and Whitney Mark R-2800 air-cooled aircraft engine. Fins were used to enhance the heat transfer from the engine. Note that 42% of the heat loss is dissipated by fins in the vicinity of the exhaust valve. Results of an energy balance obtained on a single-cylinder version of this engine are given in Table 8-2 (Ryder, 1950).

Under cruise conditions the engine runs at 1800 rpm, with a bmep = 880 kPa and $\phi = 0.90$. During takeoff, the engine speed is 2700 rpm, with a bmep = 1385 kPa and $\phi = 1.65$; the engine is fueled extremely rich, and only about one half of the fuel's energy is released. This is done to utilize the liquid fuel's latent heat for cooling and to avoid knock limiting the power. Because the fuel consumed during takeoff is small compared to that used in the entire trip, the fact that fuel is wasted is of secondary concern. Notice

Table 8-2   Energy Balance on an Air-Cooled, Spark Ignition Aircraft Engine[a]

N (rpm)	bmep (bar)	Normalized by $\dot{Q}_{in}$					
		$\dot{Q}_{exhaust}$	$\dot{Q}_{oil}$	$\dot{Q}_{ambient}$	$\dot{W}_{shaft}$	ϕ	$\dot{Q}_{in}/\dot{m}_f q_c$
1800	8.75	0.44	0.09	0.18	0.29	0.90	0.98
2700	13.72	0.44	0.08	0.12	0.35	1.65	0.46

Source: Ryder (1950).

[a]Engine: $b = 146.1$ mm, $s = 152.4$ mm.

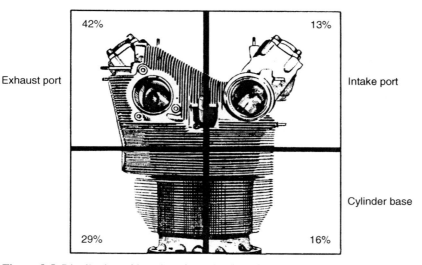

| 42% | 13% |

Exhaust port .... Intake port

Cylinder base

| 29% | 16% |

Figure 8-5 Distribution of heat loss from the fins of an air-cooled aircraft engine (Ryder, 1950). Reprinted with permission © 1950. Society of Automotive Engineers, Inc.

that during takeoff 35% of the heat released is used, but that the brake thermal efficiency of the engine is only $0.45 \times 35\% = 16.1\%$.

8.4 CYLINDER HEAT TRANSFER MEASUREMENTS

There are a wide range of temperatures and heat fluxes in an internal combustion engine. The values of local transient heat fluxes can vary by an order of magnitude depending on the spatial location in the combustion chamber and the crank angle. The source of the heat flux is not only the hot combustion gases, but also the engine friction that occurs between the piston rings and the cylinder wall. When an engine is running at a steady state, the heat transfer throughout most of the engine structure is steady. As will be shown, unsteady periodic effects are limited to a penetration layer about 1- to 5-mm thick at the gas-wall interface.

The maximum heat flux through the engine components occurs at fully open throttle and at maximum speed. Peak heat fluxes are on the order of 1 to 10 MW/m^2. The heat flux increases with increasing engine load and speed. The heat flux is largest in the center of the cylinder head, the exhaust valve seat and the center of the piston. About 50% of the heat flow to the engine coolant is through the engine head and valve seats, 30% through the cylinder sleeve or walls, and the remaining 20% through the exhaust port area.

The piston and valves, since they are moving, are difficult to cool, and operate at the highest temperatures. Temperature measurements indicate that the greatest temperatures occur at the top or crown of the piston, since it is in direct contact with the combustion gases. The crown temperatures can be as high as 550 K. The temperatures of the piston and valves depend on their thermal conductivity. As the thermal conductivity increases, the conduction resistance decreases, resulting in lower surface temperatures. For the same speed and loading, aluminum pistons are about 40 K cooler than cast iron pistons. The main cooling paths for the piston are conduction through the piston rings to the cylinder wall and conduction through the piston body to the air-oil mist on the underside of the piston, as shown in Figure 8-6. About half of the heat rejected to the cylinder wall from the piston is from cylinder friction.

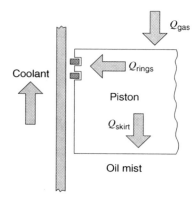

Figure 8-6 Piston cooling paths.

The main cooling path for the exhaust valves is through the valve seat, since the exhaust valve is closed for about three strokes of the four-stroke cycle. The cooling mechanism is thermal contact conduction. The conductance depends on the maximum cylinder pressure, which compresses the valve onto the valve seat. Values of the thermal contact conductance between 5,000 and 35,000 W/mK have been measured (Wisniewski, 1998). Hollow valve stems partially filled with sodium have been used to increase the effective axial thermal conductivity of the exhaust valve. The sodium melts at 370 K, so at temperatures greater than 370 K, there is internal natural conduction in the axial direction inside the valve stem.

Resolution of the instantaneous heat transfer at the cylinder surface can be achieved by inserting a surface thermocouple into the engine structure, configured in a design originated by Bendersky (1953). The essential features of a surface thermocouple used for this purpose are shown in Figure 8-7. Within the plug are two thermocouple junctions, one at the surface and one at a depth Δx from the surface. The basic idea is that according to Fourier's law for small Δx,

$$q'' = -k\frac{\Delta T}{\Delta x} \tag{8.9}$$

The criterion for small Δx is that it be small compared to the penetration layer δ. Unfortunately it is just not practical to build a plug with $\Delta x \ll \delta$. Therefore, instead, one solves the heat conduction equation between the two thermocouples, assuming that it is one dimensional.

The measured surface temperature variation in time is fitted by a Fourier series

$$T(0, t) = \overline{T}(0) + \sum_{i=1}^{N}\left[A_i \cos{(i\omega t)} + B_i \sin{(i\omega t)}\right] \tag{8.10}$$

to determine the coefficients A_i and B_i. The heat flux is then

$$q'' = -k\left\{\frac{\Delta \overline{T}}{\Delta x} + \sum_{i=1}^{N}\sqrt{\frac{i\omega}{2\alpha}}\left[(B_i - A_i)\sin{(i\omega t)} + (B_i + A_i)\cos{(i\omega t)}\right]\right\} \tag{8.11}$$

The term ω is one half the engine frequency for a four-stroke engine and equal to the engine frequency for a two-stroke engine. The curve fit, Equation 8.10 is applied to one engine cycle.

In order to calculate the total heat loss from the combustion chamber, one must evaluate the heat transfer at every point in the combustion chamber and integrate, as indicated

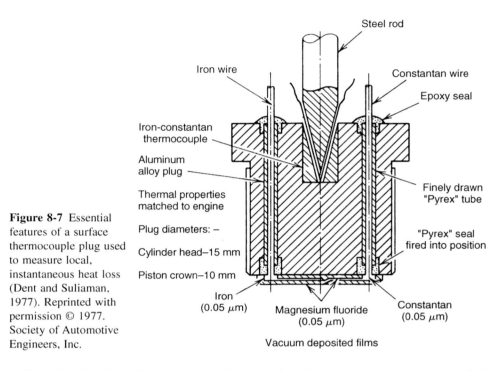

Figure 8-7 Essential features of a surface thermocouple plug used to measure local, instantaneous heat loss (Dent and Suliaman, 1977). Reprinted with permission © 1977. Society of Automotive Engineers, Inc.

by Equation 8.12. For this reason a number of surface thermocouples should be installed into the engine.

$$\dot{Q}(t) = \int_{A(t)} q''(t)dA \tag{8.12}$$

Figure 8-8 shows the location of five surface thermocouples in a four-stroke, propane fueled, spark ignition engine used by Alkidas and Myers (1982). There are four in the

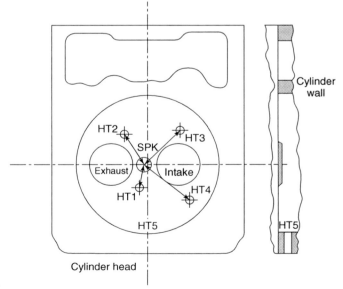

Figure 8-8 Five locations (HT1, HT2, ...) of surface thermocouples in a spark ignition engine studied by Alkidas and Myers (1982).

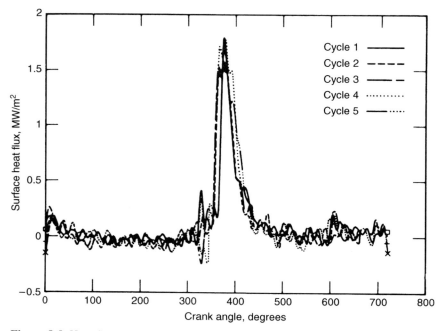

Figure 8-9 Heat flux histories of five consecutive cycles at location HT1 defined in Fig. 8-8. N = 1500 rpm, ϕ = 0.87, MBT (Alkidas and Myers, 1982).

head (HT1, HT2, HT3, and HT4) and one (HT5) on the major thrust side of the cylinder. The combustion chamber is disk shaped and has a centrally located spark plug. Results obtained at the location HT1 are given in Figure 8-9 for five consecutive cycles. The cycle to cycle variations noted in Figure 8-9 are caused by cycle to cycle variations in arrival times of the turbulent flame at the position HT1. Notice that most of the heat transfer occurs early in the expansion stroke at 360 to 420 degrees crank angle. Negative flux can occur late in the expansion stroke and during the intake stroke. Results obtained for other thermocouple locations are shown in Figure 8-10. That the flame arrives at position HT5 later than at the other positions is clearly indicated.

8.5 HEAT TRANSFER MODELING

The heat transfer processes in an internal combustion engine can be modeled with a variety of methods. These methods range from simple thermal networks to multidimensional differential equation modeling.

Thermal network models, using resistors and capacitors, are very useful for rapid and efficient estimation of the conduction, radiation, and convection heat transfer processes in engines. Using a thermal network, the significant resistances to heat flow, and the effects of changing material thermal conductivity, thickness, and coolant properties can be easily determined. A simple four node series network, which includes convection and conduction resistances shown in Figure 8-11 is an illustration of the steady state heat transfer from the engine cylinder gas to the coolant. This series path is composed of convection through the cylinder gas boundary layer, conduction across the cylinder head wall, and convection through the coolant liquid boundary layer. The cylinder gas boundary layer insulates the cylinder wall from the high temperature cylinder gases. Thermal networks are primarily used for convection and conduction heat transfer, as the radiation heat transfer

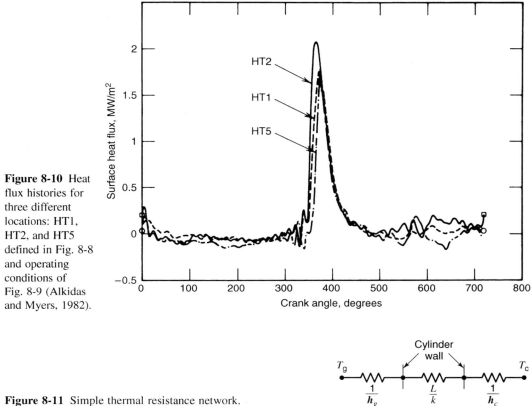

Figure 8-10 Heat flux histories for three different locations: HT1, HT2, and HT5 defined in Fig. 8-8 and operating conditions of Fig. 8-9 (Alkidas and Myers, 1982).

Figure 8-11 Simple thermal resistance network.

equation needs to be linearized to conform to the resistance model. Using Fourier's equation, the conduction resistance is

$$R_{cond} = \frac{\Delta T}{Q/A} = \frac{L}{k} \qquad (8.13)$$

and using Newton's equation, the convection resistance is

$$R_{conv} = \frac{\Delta T}{Q/A} = \frac{1}{h} \qquad (8.14)$$

Examples of resistor-capacitor thermal networks applied to engine warm-up and steady-state operation are given in Shayler et al. (1993) and Bohac et al. (1996).

We now consider the unsteady nature of the heat flux from the combustion gas to the cylinder wall. The cylinder wall has a periodic heat flux on the gas side and a constant surface temperature on the coolant side. The problem posed requires solution of the heat conduction equation.

$$\frac{\partial T}{\partial t} = \alpha \frac{\partial^2 T}{\partial x^2} \qquad (8.15)$$

subject to the following boundary conditions

$$-k \frac{\partial T}{\partial x} = q_0'' + q_1'' \sin(\omega t) \quad \text{at } x = 0 \qquad (8.16)$$

$$T = T_L \quad \text{at } x = L \qquad (8.17)$$

as well as an initial condition

$$T = T_i(x) \quad \text{at } t = 0 \tag{8.18}$$

An exact solution can be written in closed form but it is quite cumbersome and as a result, no more illustrative than a computer solution. Fortunately, an approximate solution can be derived for the practical case where

$$\omega t \gg 1 \quad \text{and} \quad \frac{\omega L^2}{2\alpha} \gg 1 \tag{8.19}$$

In this case the temperature field is given by

$$T = T_L + \frac{q_0''}{k}(L - x) + \frac{q_1''}{\left(\dfrac{\alpha}{\omega}\right)^{1/2}} \exp\left[\left(-\frac{\omega}{2\alpha}\right)^{1/2} x\right] \sin\left(\omega t - \left(\frac{\omega}{2\alpha}\right)^{1/2} x - \frac{\pi}{4}\right) \tag{8.20}$$

Inspection of this solution shows that:

1. The surface temperature at $x = 0$ oscillates with the same frequency as the imposed heat flux but with a phase difference of $\pi/4$;

2. The amplitude of the oscillations decays exponentially with the distance x from the surface; the amplitude is reduced to 10% of that at the surface at a distance given by

$$\delta = -\ln(0.10)\left(\frac{2\alpha}{\omega}\right)^{1/2} = 2.3\left(\frac{2\alpha}{\omega}\right)^{1/2} \tag{8.21}$$

For a two-stroke engine operating at 2000 rpm ($\omega = 209 \text{ s}^{-1}$) and made of cast iron ($\alpha = 21 \times 10^{-6} \text{ m}^2/\text{s}$) this length is rather small, $\delta = 1.0$ mm. For aluminum, $\delta = 2.2$ mm and for partially stabilized zirconia (a ceramic used in prototype insulated engines) $\delta = 0.7$ mm.

The penetration distance δ is a measure of how far into the material fluctuations about the mean heat flux penetrate. For distances x greater than δ, the temperature distribution is more or less steady and driven only by the time average heat flux. Since the length δ is rather small compared to the dimensions (wall thickness, bore, etc.) over which conduction heat transfer occurs, two simplifications can be made:

1. Conduction heat transfer in the various parts can be assumed steady and driven by the average flux;

2. Heat transfer from the gas can be coupled to the conduction analysis accounting for capacitance only in a penetration layer of thickness δ in series with a resistance computed or measured for a steady state.

A five mode thermal network for a cylinder wall is given in Figure 8-12. The modeling of the penetration layer can be complicated by the presence of an oil film or deposits. Fortunately an accurate model is not required as the fluctuations about the mean $\overline{T}_\delta$ tend to be small compared to the gas-penetration depth temperature difference $T_g - \overline{T}_\delta$. For an engine operated at a steady state, the penetration layer is thin because the engine frequency, which dictates the frequency components of the heat flux imposed on the gassolid interfaces, is rather high. On the other hand, in the case of an engine being accelerated or decelerated, the penetration layer is thicker because lower frequency components are added to the heat flux that are characteristic of the rates of change of engine

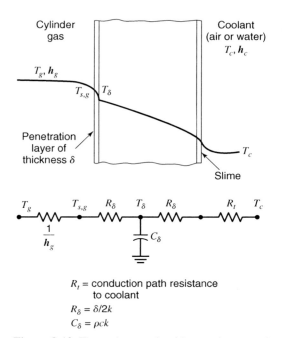

R_t = conduction path resistance
    to coolant
$R_\delta = \delta/2k$
$C_\delta = \rho ck$

Figure 8-12 Thermal network with capacitance node
for penetration layer.

speed. For example, one could define a characteristic time τ by

$$\tau^{-1} = \frac{1}{\omega}\frac{d\omega}{dt} \tag{8.22}$$

In the case where $\tau = 5s$, the penetration layer would be

$$\delta = 2.3\,(2\alpha\tau)^{1/2} \tag{8.23}$$

which is $\delta = 33$ mm for cast iron and no longer small compared to the typical dimensions over which the heat is transferred.

Determination of the temperature profile of an engine component such as the piston requires solution of the three-dimensional heat conduction equation. As an illustration, a case study of heat transfer in a piston will be presented (Li, 1982). Figure 8-13 shows how a piston can be divided into a number of elements for analysis. Only one quadrant of the piston in the x-y plane needs to be considered as the piston is, in this case, symmetrical. As mentioned earlier, the piston can be treated as steady and driven by an average heat flux since the penetration layers are small. The mean cylinder gas temperature is computed using a cycle simulation to predict instantaneous gas temperatures and then integrated according to

$$\overline{T}_g = \frac{1}{4\pi\overline{h}_g}\int_0^{4\pi} h_g T_g\,d\theta \tag{8.24}$$

where h_g is the instantaneous heat transfer coefficient (the determination of which is the subject of the next section). Likewise an average heat transfer coefficient

$$\overline{h}_g = \frac{1}{4\pi}\int_0^{4\pi} h_g\,d\theta \tag{8.25}$$

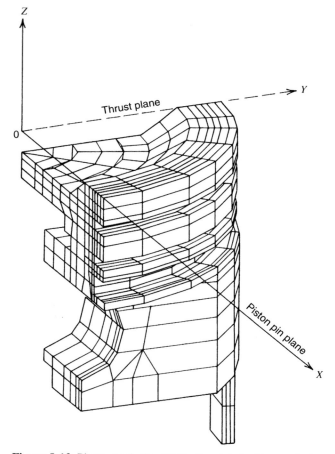

Figure 8-13 Piston mesh (Li, 1982). Reprinted with permission
© 1982. Society of Automotive Engineers, Inc.

is used in estimating the heat transfer coefficients on the crown of the piston in contact
with the cylinder gas. Results obtained for a 2.5 liter, four-cylinder engine at WOT are
given in Table 8-3. Notice that both the mean gas temperature and the mean heat transfer coefficient increase with engine speed. The mean gas temperature increases because
there is less time for the gases to lose heat as engine speed increases; whereas the mean
heat transfer coefficient increases because of increased gas motion at higher speeds.

Results obtained for a dished and slotted piston run in a 2.5 liter four-cylinder engine
at 4600 rpm and wide open throttle are given in Figure 8-14. The values associated with

Table 8-3   Variation of Mean Gas Temperature and Heat Transfer Coefficient at the Top of the
Piston with Engine Speed[a]

Engine Speed (rpm)	$\overline{T}_g(K)$	$\overline{h}_g(W/m^2K)$
2400	1263	1820
3600	1310	2430
4600	1335	2800

Source: Li (1982).

[a]A dished piston running in a 2.5 L engine WOT.

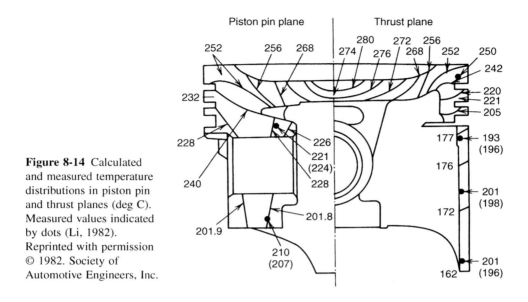

Figure 8-14 Calculated and measured temperature distributions in piston pin and thrust planes (deg C). Measured values indicated by dots (Li, 1982). Reprinted with permission © 1982. Society of Automotive Engineers, Inc.

the dots are measured temperatures, and the isotherms shown are the results of computations. The agreement obtained is reasonable except in the skirt area where the computed temperatures are too low. The calculated results show that three areas are particularly important in dissipating the piston heat input: (1) the ring groove surfaces, (2) the underside of the dome, and (3) the upper portion of the pin bearing surface. From the ring grooves, heat flows into the rings, through the bore, and is eventually absorbed by the coolant. From the underside of the dome and the surface of the pin bearing, the heat is convected into an air-oil mist and is eventually absorbed by the oil in the sump.

The results of similar calculations are summarized in Figure 8-15 to show the effect of engine speed at wide open throttle on piston temperature. Temperatures in the piston are determined by the average heat flow into the piston and the effectiveness with which the heat can be dissipated to the oil and the coolant. As speed increases, the heat flow increases, whereas the overall heat transfer coefficients to the coolant and oil change little; thus piston temperature increases.

The finite heat release model introduced in Chapter 2 can be modified to include the differential heat transfer dQ_w to the cylinder walls, if the instantaneous average cylinder heat transfer coefficient $\boldsymbol{h}_g(\theta)$ and engine speed N are known. The finite heat release equation, Equation 2.30, with the addition of wall heat transfer is:

$$\frac{dP}{d\theta} = \frac{\gamma - 1}{V}\ Q_{\text{in}}\frac{dx_b}{d\theta} - \frac{dQ_w}{d\theta} - \gamma\frac{P\,dV}{V\,d\theta} \tag{8.26}$$

The heat transfer rate at any crank angle θ to the exposed cylinder wall at an engine speed N is determined with a Newtonian convection equation:

$$\frac{dQ_w}{d\theta} = \boldsymbol{h}_g(\theta)A_w(\theta)\,(T_g(\theta) - T_w)/N \tag{8.27}$$

The cylinder wall temperature T_w in the above equation is the area-weighted mean of the temperatures of the exposed cylinder wall, the head, and the piston crown. The heat transfer coefficient $\boldsymbol{h}_g(\theta)$ is the instantaneous area averaged heat transfer coefficient. As given in Equation 8.28, the exposed cylinder area $A_w(\theta)$ is the sum of the cylinder bore area,

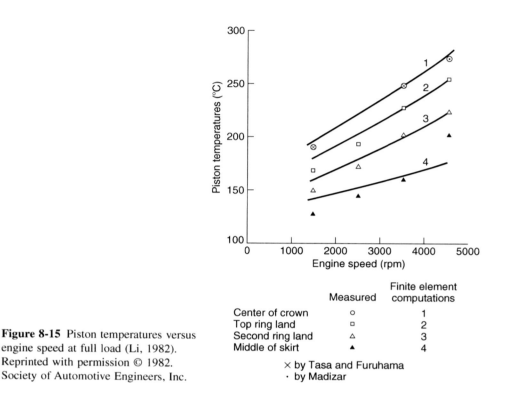

Figure 8-15 Piston temperatures versus engine speed at full load (Li, 1982). Reprinted with permission © 1982. Society of Automotive Engineers, Inc.

	Measured	Finite element computations
Center of crown	o	1
Top ring land	□	2
Second ring land	△	3
Middle of skirt	▲	4

× by Tasa and Furuhama
· by Madizar

the cylinder head area, and the piston crown area, assuming a flat cylinder head.

$$A_w(\theta) = A_{\text{wall}} + A_{\text{head}} + A_{\text{piston}} = \pi b y + \frac{\pi}{2} b^2 \qquad (8.28)$$

where

   y = exposed cylinder wall height = $a + l - \left[(l^2 - a^2 \sin^2 \theta)^{1/2} + a \cos \theta \right]$

   a = crank radius (s/2)

   l = connecting rod length

A more rigorous heat transfer analysis is zonal modeling, in which the cylinder volume is divided into individual control volumes, each with its own thermodynamic properties and heat transfer coefficient. For example, a two zone model separating the cylinder gases into unburned and burned gas fractions, with the moving flame separating the two zones is given in Krieger and Borman (1996). A four zone model consisting of the central core zone, a squish zone, a head recess zone, and a piston recess zone was used by Tillock and Martin (1996). With zonal modeling, the characteristic length and velocity are zone dependent. The characteristic velocity is usually taken as an effective velocity with components from the mean and turbulent flow field.

As the number of zones increases to length scales that are much less than the cylinder bore, the modeling is termed multidimensional. With multidimensional models, the mass, momentum, and energy conservation equations take the form of partial differential equations, which are solved numerically. Detailed turbulence and reaction rate models are also required. For example, the use of turbulent heat transfer models in the multidimensional KIVA code is given in Reitz (1991).

8.6   HEAT TRANSFER CORRELATIONS

Engine heat transfer data is correlated with the fluid conditions using two nondimensional parameters, the Nusselt and Reynolds numbers. The heat transfer coefficient of interest can be extracted from the Nusselt number. The Nusselt number, Nu, is the ratio of the convection to the conduction heat transfer over the same temperature difference, and expressed as:

$$Nu = \frac{hL}{k} \qquad (8.29)$$

where h is a heat transfer coefficient, L is a length scale such as the cylinder bore, and k is the working fluid thermal conductivity. The heat transfer coefficient varies with position in the cylinder and is time-dependent. The Reynold number, Re, a ratio of the inertial to viscous fluid forces, is expressed as:

$$Re = \frac{\rho UL}{\mu} \qquad (8.30)$$

where ρ is the fluid density, U is a characteristic velocity, and μ is the dynamic viscosity. Values of the thermal conductivity and viscosity are given in Appendix A. Because of the approximate nature of the correlations, it has been suggested (Krieger and Borman, 1966) that the use of air data for the thermal conductivity and viscosity of the combustion gases is adequate.

The Reynolds and Nusselt numbers vary both in time, i.e., with crank angle and with location in the engine. Since many energy balance calculations only require an average Nusselt number, averaging processes over space and/or time are used. The gas properties are evaluated at the appropriate mean effective cylinder gas temperature, which can be obtained using the ideal gas equation for known cylinder pressure and volume.

The characteristic gas velocity in the cylinder depends on a number of parameters, such as the piston speed, the amount of swirl and tumble present, the amount of combustion, and the level of turbulence. Since the gas velocity in the cylinder scales with the piston speed, the mean piston speed is often chosen as a first order estimate of the characteristic gas velocity in the cylinder for the Reynolds number. The mean piston speed $\overline{U}_p$ is

$$\overline{U}_p = 2 N s \qquad (8.31)$$

In terms of the mass flow rate into the engine per unit piston area, the Reynolds number is

$$Re = \frac{(\dot{m}_a + \dot{m}_f) b}{A_p \mu_g} \qquad (8.32)$$

A correlation for the overall engine heat transfer coefficient, h_o, between the cylinder gas and the coolant, is that of Taylor (1985). The Nusselt number in the Taylor correlation implicitly includes the conduction and radiation heat transfer components, and the Reynolds number is based on the fuel-air mass flow rate. For two- or four-stroke engines, compression or spark ignited, the correlation is

$$Nu = 10.4 \, Re^{0.75} \qquad (8.33)$$

The overall heat flux using the piston area $A_p = 1/4 \, \pi \, b^2$ as the reference area is

$$\frac{Q}{A_p} = h_o (\overline{T}_g - T_c) \qquad (8.34)$$

EXAMPLE 8.1 *Overall Heat Transfer Coefficient*

Compute the overall engine heat transfer coefficient h_o and the overall heat flux for a single cylinder engine with a 0.1-m bore and stroke operating at 1000 rpm, average combustion gas temperature of 1000 K, coolant temperature of 350 K, and fuel-air flowrate of 2×10^{-2} kg/s. Assume $k = 0.06$ W/m K and $\mu = 20 \times 10^{-6}$ Ns/m².

SOLUTION The Reynolds number is

$$Re = \frac{\dot{m}\, b}{A_p \mu_g} = 1274$$

The overall average heat transfer coefficient is found using the Taylor correlation:

$$\frac{h_o\, 0.1}{0.06} = 10.4\,(1274)^{3/4} = 2218$$

$$U = 1330 \text{ W/m}^2 \text{ K}$$

The average heat transfer per unit area from the cylinder to the coolant is therefore

$$Q/A = h_o\,(\overline{T}_g - T_c) = (1330)(1000 - 350) = 0.86 \text{ MW/m}^2$$

The cylinder instantaneous average heat transfer coefficient, $h_g(\theta)$, between the cylinder gas and wall is an input to global, i.e., single zone cycle calculations, such as the finite heat release model represented by Equation 8.26. There are two correlations that are used, the Annand and the Woschni correlation.

The Annand (1963) correlation was developed from cylinder head thermocouple measurements of instantaneous heat flux. It uses a constant characteristic velocity, the mean piston speed, and a constant characteristic length, the cylinder diameter. The properties, such as the thermal conductivity and viscosity, to be used in the correlation are the zone averaged instantaneous values. The instantaneous gas density can be determined from the known constant charge mass and instantaneous cylinder volume. The Annand correlation is

$$Nu = a\, Re^{0.7} \tag{8.35}$$

where the constant, a, is 0.49 for a four-stroke engine, and 0.26 for a two-stroke engine.

Another correlation for the instantaneous cylinder average heat transfer coefficient is due to Woschni (1967). The Woschni correlation was developed using a heat balance analysis on a diesel engine and uses a constant characteristic length and a variable characteristic velocity to account for the gas motion induced by combustion. The Woschni correlation is

$$Nu = 0.035\, Re^{0.8} \tag{8.36}$$

The characteristic gas velocity in the Woschni correlation is proportional to the mean piston speed during intake, compression, and exhaust. During combustion and expansion, it is assumed that the gas velocities are increased by the pressure rise resulting from combustion, so the characteristic gas velocity has both piston speed and cylinder pressure rise terms.

$$U = 2.28\, \overline{U}_p + 0.00324\, T_o \frac{V_d}{V_o} \frac{\Delta P_c}{P_o} \tag{8.37}$$

where

$\overline{U}_p$ = mean piston speed (m/s)

T_o = temperature at intake valve closing (K)

V_o = cylinder volume at intake valve closing (m^3)

V_d = displacement volume (m^3)

ΔP_c = instantaneous pressure rise due to combustion (kPa)

P_o = pressure at intake valve closing (kPa)

The pressure rise due to combustion is the cylinder pressure in the firing engine minus the cylinder pressure in the motored engine at the same crank angle. The latter can be estimated by use of the isentropic relation $PV^\gamma = P_o V_o^\gamma = $ constant. Equation 8.37 is applicable when the valves are shut. When the valves are open, the gases are accelerated because of the flow into or out of the cylinder. In this case Woschni uses

$$U = 6.18 \, \overline{U}_p \qquad (8.38)$$

In dimensional form, assuming $k \sim T^{0.75}$, and $\mu \sim T^{0.62}$ (Heywood, 1988), the Woschni correlation is

$$h_g = 3.26 \, P^{0.8} \, U^{0.8} \, b^{-0.2} \, T^{-0.55} \qquad (8.39)$$

where the units of h_g, P, U, b, and T are in W/m^{2}K, kPa, m/s, m, and K, respectively. The constants in Woschni's correlation were determined by matching experimental results from a particular engine. When applied to any other engine, the constants are estimates at best and it is not uncommon to find engineers adjusting them to better suit their own engine. For example, an empirical formula for the instantaneous heat transfer coefficient for a spark ignition engine is given in Han et al. (1997) as:

$$h_g = 687 \, P^{0.75} \, U^{0.75} \, b^{-0.25} \, T^{-0.465} \qquad (8.40)$$

where

$$U = 0.494 \, \overline{U}_p + 0.73 \times 10^{-6}(PdV + VdP) \qquad (8.41)$$

EXAMPLE 8.2   *Effect of Cylinder Heat Transfer*

Compare the mean effective pressure, the net work and indicated thermal efficiency for an engine modeled with and without cylinder heat transfer. Use the Woschni heat transfer correlation. The engine has a bore of 0.1 m, stroke of 0.1 m, and connecting rod length of 0.15 m, with a compression ratio of 10. The engine speed is 3000 rpm. The burn initiation is -25 degrees atdc, the burn duration is 70 degrees crank angle and the Weibe parameters are $a = 5$ and $n = 3$. The fuel is octane with an equivalence ratio of 1.0, and the inlet temperature and pressure are 300 K and 1 bar. The average cylinder wall temperature is 400 K.

SOLUTION   The web software accompanying the text contains an applet, *Finite Heat Release with Heat Transfer,* that solves the finite heat release equation, Equation 8.26, for the cylinder pressure, temperature, and work, for a given set of engine and fuel conditions during the compression and expansion strokes. The valve of the lower heat of combustion q_c is obtained using Table 4.1 for the fuels octane, diesel, methane, propane, methanol, and ethanol. Equations 2.64 and 2.65 are used in the applet to account for the effect of the equivalence ratio ϕ on the average specific heat ratio γ and the heat

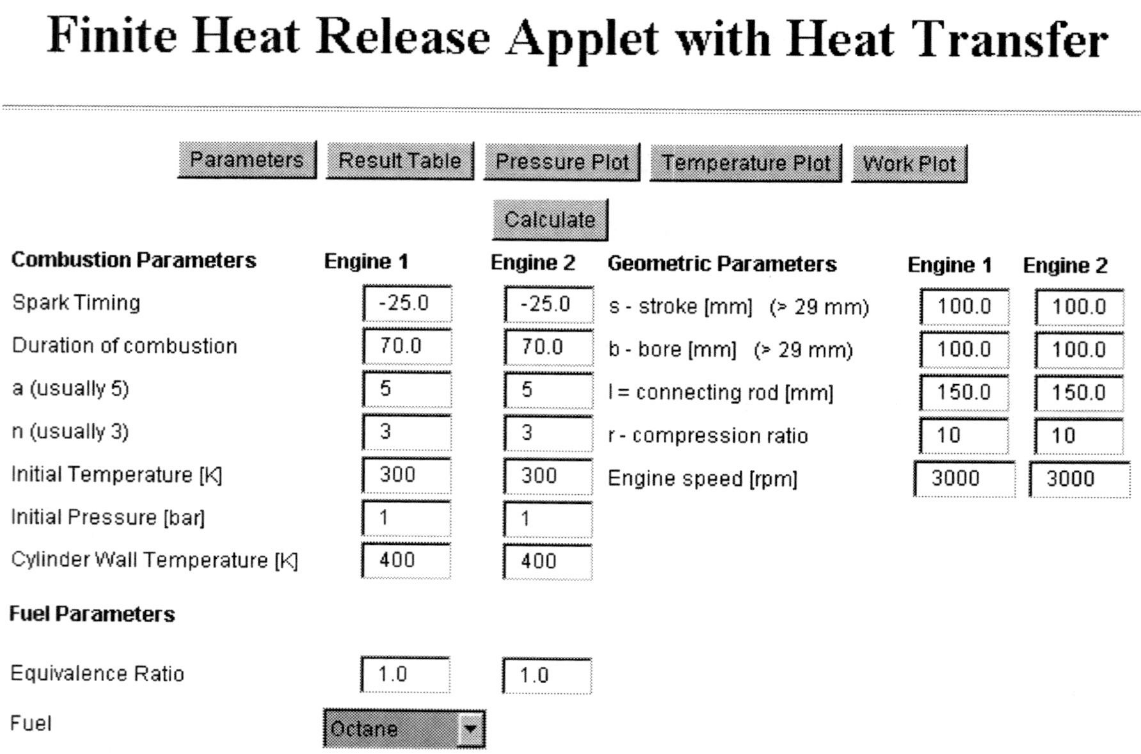

Finite Heat Release Applet with Heat Transfer
Parameters
Result Table
Pressure Plot
Temperature Plot
Work Plot
Calculate
Combustion Parameters
Engine 1
Engine 2
Spark Timing -25.0 -25.0
Duration of combustion 70.0 70.0
a (usually 5) 5 5
n (usually 3) 3 3
Initial Temperature [K] 300 300
Initial Pressure [bar] 1 1
Cylinder Wall Temperature [K] 400 400
Geometric Parameters
Engine 1
Engine 2
s - stroke [mm] (> 29 mm) 100.0 100.0
b - bore [mm] (> 29 mm) 100.0 100.0
l = connecting rod [mm] 150.0 150.0
r - compression ratio 10 10
Engine speed [rpm] 3000 3000
Fuel Parameters
Equivalence Ratio 1.0 1.0
Fuel
Octane

Figure 8-16 Input for finite heat release with heat transfer applet.

input Q_{in}. The applet uses the Woschni correlation for the cylinder wall heat transfer. The applet also computes the engine performance without wall heat transfer for comparison purposes.

The above engine parameters are entered into the *Finite Heat Release* applet as shown on Figure 8-16. The temperature profiles are shown in Figure 8-17. The heat transfer to the cylinder walls lowers the cylinder gas temperature by about 200 K

2500
1250
0
T (K)
w/o heat transfer
w/heat transfer
−180
−90
0
90
180
Crank angle (degrees)

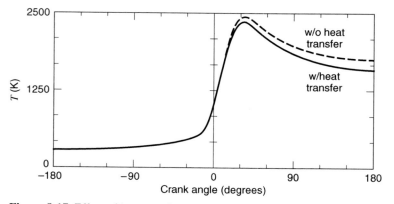

Figure 8-17 Effect of heat transfer on cylinder gas temperatures for Example 8-2.

Table 8-4   Effect of heat transfer (HT) on predicted engine performance

	w/HT	w/o HT
Indicated Work [J]	1087	1150
Indicated Power [kW]	26.9	28.4
Mean Effective Pressure [bar]	13.8	14.6
Thermal Efficiency	0.37	0.39
Q_{in} [J]	2933	2933

throughout the combustion and expansion stroke. As shown in Table 8-4, the mean effective pressure decreases from about 14.6 to about 13.8 bar. The corresponding decrease in the indicated net work is from 1150 to 1087 J and thermal efficiency is from 39% to 37%.

EXAMPLE 8.3   *Comparison of Annand and Woschni Correlations*

Compare the Annand and Woschni correlations for an engine operating at 1000 rpm and 2000 rpm. Assume that the engine has a bore and stroke of 0.1 m, and connecting rod length of 0.15 m. The fuel is octane, with an equivalence ratio of 1.0. The engine is unthrottled, with inlet pressure of 1 bar, and temperature of 300 K. The spark ignition is at −25 degrees atdc, and the burn duration is 70 degrees crank angle. The average cylinder wall temperature is 400 K.

SOLUTION   The Annand and Woschni heat transfer coefficients are computed and compared in the *Heat Transfer Coefficient Applet* (Figure 8-18). For two given set of engine conditions, the applet computes the Annand and Woschni heat transfer coefficients, heat flux, and heat transfer rate as a function of crank angle for the compression and expansion strokes. In addition, the applet uses the finite heat release analysis with heat transfer to determine overall parameters such as work, power, imep, and indicated thermal efficiency.

The heat transfer coefficients are plotted in Figure 8-19 for the two engine speeds. The maximum heat transfer coefficient increases from about 1500 to 1750 W/m²K as the engine speed is increased from 1000 to 2000 rpm. The instantaneous Woschni and Annand heat flux is plotted in Figure 8-20 for an engine speed of 1000 rpm, and in Figure 8-21 for an engine speed of 2000 rpm. The maximum heat flux for the Woschni correlation is about 2 MW/m² for the engine speeds considered. The work, power, imep, and thermal efficiency are compared in Table 8-5.

Convective heat loss is an important consideration in exhaust pipe and port design, especially for engines with exhaust turbines or catalytic converters. The ports are relatively short and curved, with unsteady flow, so the flow will not be fully developed. The heat flux is relatively high due to high exhaust gas velocities and temperatures.

Correlations have been developed for the average and the instantaneous heat transfer in the exhaust system. Malchow et al. (1979) obtained the following correlation for the average heat loss in a straight circular exhaust pipe, where D is the exhaust pipe diameter, and L is the pipe length;

$$Nu = 0.0483 \, Re^{0.783} \quad (D/L = 0.3) \tag{8.42}$$

Heat Transfer Coefficient Applet

Comparision of Woschni and Annand

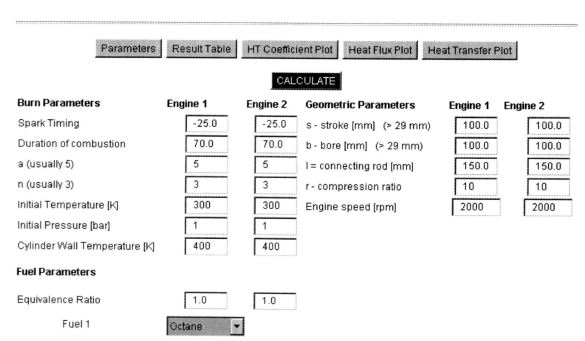

Figure 8-18 Input for heat transfer coefficient applet.

Equation 8.42 is compared in Figure 8-22 to computations based on the steady-state correlation for developing turbulent flow in a smooth pipe (Incropera and DeWitt, 2001):

$$Nu = 0.023\, Re^{0.8}\, Pr^{0.33}(1 + D/L)^{0.7} \qquad (8.43)$$

In Equation 8.43, Pr is the Prandtl number and L is the pipe length. Notice that the heat transfer in the exhaust pipe is about 50% greater than would exist in a steady flow in the same pipe.

Hires and Pochmara (1976) correlated experimental results for the instantaneous heat loss from 10 different exhaust port designs. The suggested correlation equation is:

$$Nu = 0.158\, Re^{0.8} \qquad (8.44)$$

Table 8-5 Comparison of Annand and Woschni heat transfer correlations

	1000 rpm (Woschni)	1000 rpm (Annand)	2000 rpm (Woschni)	2000 rpm (Annand)
Indicated Work [J]	1012	949	1058	975
Indicated Power [kW]	8.43	7.91	17.64	16.26
Mean Effective Pressure [bar]	12.75	11.96	13.33	12.30
Thermal Efficiency	0.34	0.32	0.36	0.33
Q_{in} [J]	2933	2933	2933	2933

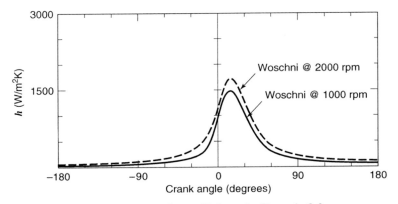

Figure 8-19 Woschni heat transfer coefficients for Example 8-3.

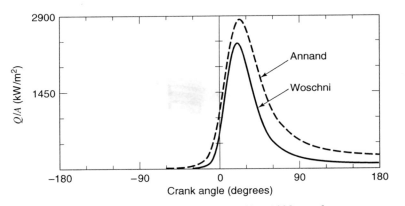

Figure 8-20 Annand and Woschni heat flux at $N = 1000$ rpm for Example 8-3.

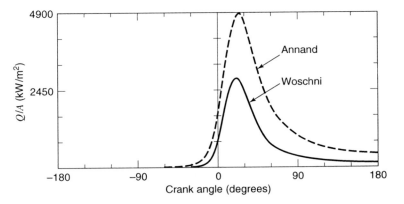

Figure 8-21 Annand and Woschni heat flux at $N = 2000$ rpm for Example 8-3.

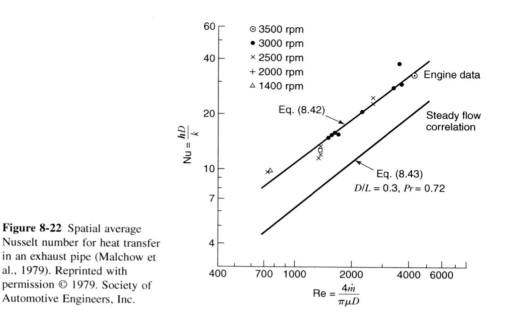

Figure 8-22 Spatial average Nusselt number for heat transfer in an exhaust pipe (Malchow et al., 1979). Reprinted with permission © 1979. Society of Automotive Engineers, Inc.

where their Reynolds number is defined as

$$Re = \frac{\dot{m}\,d}{\mu A} \tag{8.45}$$

and $\dot{m}$ is the instantaneous flow rate, d is the throat diameter, A is the exit cross section, and μ is the exhaust gas viscosity. Some of the ports studied are shown in Figure 8-23. In an exhaust port, it appears that the heat transfer coefficient is about eight times what it would be in a steady flow in the same port. This is probably caused by increased turbulence generated by flow separation at the valve.

In that neither Equation 8.42 nor 8.44 includes information about the valve, such as lift or valve seat angle, they are applicable to other engines only in so far as they are geometrically similar. As discussed in Chapter 7, at low valve lifts the port flow is in the form of a jet, and for larger lifts, the port flow is in the form of developing pipe flow. The valve geometry and lift have been incorporated into heat transfer correlations by Caton and Heywood (1981).

8.7 RADIATION HEAT TRANSFER

During the combustion process in an engine, high temperature gases and particulate matter radiate to the cylinder walls. In a spark ignition engine the fraction of the gaseous and particulate matter radiation is very small in comparison to the convection heat transfer to the cylinder wall. The flame front propagates quickly across the combustion chamber through a fairly homogeneous fuel-air mixture. Most of the gaseous radiation is in narrow bands from the H_2, CO_2, and H_2O molecules.

In a compression ignition engine, fuel burns in a turbulent diffusion flame formed around the fuel spray in the region where the equivalent ratio is close to 1. Soot particles are formed as an intermediate step in the compression ignition process. The radiation from the soot particles during their existence is significant, about 20 to 40% of the total heat transfer to the cylinder wall (Dent and Sulaiman, 1977). In contrast to the gaseous radiation, the soot particles radiate over the entire spectrum. The radiant heat

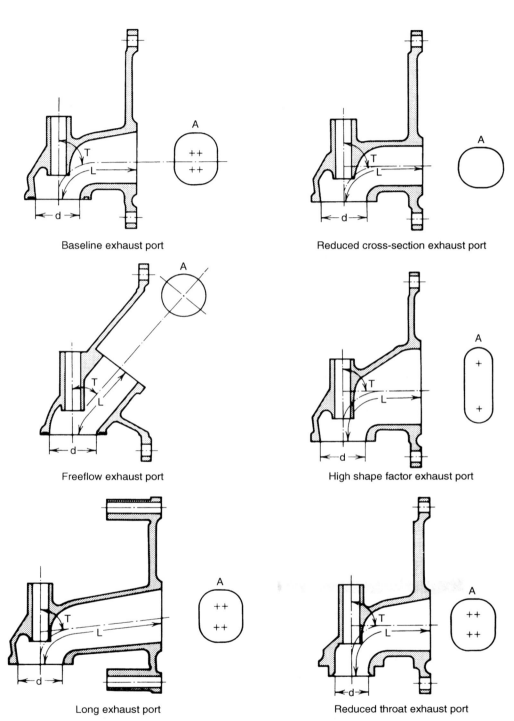

Baseline exhaust port

Reduced cross-section exhaust port

Freeflow exhaust port

High shape factor exhaust port

Long exhaust port

Reduced throat exhaust port

Figure 8-23 Various exhaust ports (Hires and Pochmara, 1976). Reprinted with permission © 1976. Society of Automotive Engineers, Inc.

transfer from the flame will reduce its temperature, which affects the local rate of NO formation. The Woschni correlation, since it is based on heat flux, does include the radiation heat transfer to the cylinder wall. The Annand correlation does not include the radiation heat flux.

Radiation heat transfer in participating media, such as the case in a combusting gas-particulate mixture, is modeled with the radiation transfer equation (RTE). The equation includes the radiant energy absorbed, emitted, and scattered along a given solid angle direction:

$$\frac{dI}{ds} = -(K + \sigma)I + KI_b + \frac{\sigma}{4\pi}\int P(\Omega,\Omega')\,I(\Omega')\,d\Omega' \tag{8.46}$$

where I is the radiant intensity in the direction of a solid angle Ω, s is the distance in that direction, I_b is the black body intensity, K is an extinction coefficient, σ is a scattering coefficient, P is the phase function or probability for scattering from solid angle Ω' into solid angle Ω. There are a variety of numerical methods for solution of the radiative transfer equation including flux methods, Monte Carlo techniques, the discrete ordinates method, and the discrete transfer method. The discrete ordinates method discretizes the radiative transfer equation for a set of finite solid angle directions. The resulting discrete ordinates equations are solved along the solid angle directions using a control volume technique.

The radiation transfer equation has been incorporated into multi-dimensional CFD codes, such as KIVA. Blunsdon et al. (1992) applied the discrete transfer method to the CFD code KIVA for the simulation of diesel combustion, and also modeled the radiation from combustion products in a spark ignition engine (Blunsdon et al., 1993). Using the discrete ordinates method, Abraham and Magi (1997) computed the radiant heat loss in a diesel engine. Inclusion of the radiant heat loss reduced the peak temperature by about 10% relative to a nonradiative computation, lowering the predicted frozen NO concentrations.

The exhaust system operates at a temperature high enough so that radiation heat transfer from the exhaust system to the environment is significant. At full load with a stationary engine on a test stand, it is possible to make the exhaust system glow red, which indicates the radiation emission is in the visible wavelength range. In many engines a radiation shield is used to reduce the radiation heat transfer from the exhaust manifold to the engine block and head through the exhaust manifold gasket.

8.8   MASS LOSS OR BLOWBY

There are three primary reasons for an interest in blowby. It influences (1) the gas pressure acting on the rings which influences the friction and wear characteristics; (2) the indicated performance; and (3) the hydrocarbon emissions.

Typically a ring pack consists of two compression rings and an oil ring. The pressure drop across the oil ring is generally negligible. Such a ring pack is represented in Figure 8-24. A one-dimensional representation of the ring pack is also shown; it consists of three plenums in series through passages whose sizes are dependent upon the ring gaps, the piston to cylinder wall clearance, and any ring tilt present. The volumes are all time dependent: V_0 changes because of piston motion; V_1 changes because of ring motion; and V_2 changes because of piston motion (including those of the other cylinders in a multicylinder engine).

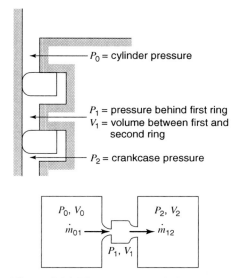

Figure 8-24 Ring pack and 1-D flow model of blowby.

Figure 8-25 shows the results of measurements made for the ring gas pressures in a two-stroke diesel engine (Ruddy, 1979). Notice that at about 70 degrees after top dead center, the pressure between the rings is greater than the cylinder pressure which, if the flows are quasi-steady, means the flow has reversed itself. The figure also shows pressures computed assuming the flows are one-dimensional, quasi-steady, and isentropic as they pass from one plenum to another. The mass flow from one plenum to another is governed by Equation 7.14. The requisite stagnation properties and the specific heat ratio are

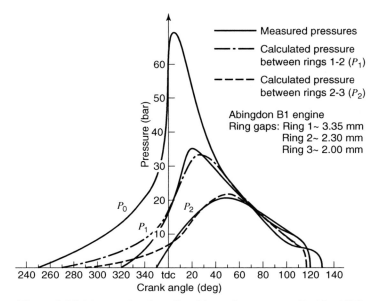

Figure 8-25 Measured and predicted inter-ring pressure (Ruddy, 1979).

based on current values in the upstream plenum. The throat area is proportional to the ring gap and the bore to cylinder clearance. The constant of proportionality, which depends at least on the Reynolds number of the flow and probably on whether or not the ring is tilted, is not known with certainty. It is of order unity and its magnitude is fixed matching between some measured and predicted data, such as the average blow-by rate and the interring pressure distribution.

It is typical in such computations to assume that the gas between rings is at a temperature equal to the average of the piston temperature and the cylinder liner temperature. In so doing, there is no need to solve the energy equation for each plenum. The equation of continuity for mass conservation is applied to each plenum where the mass flows in and out are determined as just described. By simultaneously integrating the resultant ordinary differential equations, one obtains the mass within each plenum. These are coupled to equations of motion for each ring, thereby obtaining the plenum volumes. With the plenum volumes and the mass contained therein, one then uses an equation of state to compute the pressure in each plenum.

It was mentioned that blowby influences the hydrocarbon emissions. During compression and the early stages of combustion, unburned fuel and air is being compressed into the plenums between the rings. As mentioned, the gases rapidly equilibrate thermally to the environment and are thus at the average of the piston and liner temperature. In fact, the heat transfer is so effective that the flame propagating in the cylinder is extinguished when it tries to propagate into the spaces between the rings. The unburned fuel and air pushed into the ring pack remains unburned. Soon after the blowby flow reverses itself, unburned fuel and air emerges from the ring pack back into the cylinder. Since this occurs late in the expansion stroke, the burned gases in the cylinder are relatively cold, and thus a large part of this reemerging fuel and air will not be oxidized as it mixed with the in-cylinder combustion products. Thus, unburned fuel or hydrocarbons will be expelled from the engine during the exhaust process. Namazian and Heywood (1982) estimate that anywhere from 2 to 7% of the fuel is wasted in this way. It is interesting to note that this is another advantage of diesel engines since in diesel engines mostly air will be compressed into the ring pack.

Figure 8-26 shows a sampling valve mounted in the piston of a 1.3 liter, four-cylinder, gasoline engine by Furuhama and Tateishi (1972). This valve can be opened for 1.25 ms at the same angle in consecutive cycles. Gases are withdrawn for analysis from the space between the top land and the cylinder for a variety of angles during the cycle. Results obtained at wide open throttle and 2000 rpm are given in Figure 8-27.

During the compression stroke, concentrations of oxygen and hydrocarbons (as n-hexane equivalent) are high because fuel and air is entering the ring pack. Likewise carbon dioxide and carbon monoxide are low in concentrations attributable to the residual gas content. About 15 degrees before top dead center, there is a sudden drop in oxygen and hydrocarbon concentrations as burned gases are beginning to enter the ring pack. About 30 degrees after top dead center, it appears that the unburned gases that entered earlier reemerge though diluted somewhat by the burned products that entered the ring pack.

By writing conservation equations for each species in the burned and unburned gases, Furuhama and Tateishi (1972) were able to formulate and integrate a set of ordinary differential equations for the gas composition in the top land as a function of time. The computed results at wide open throttle, shown in Figure 8-27 agree well with the experimental values. In the case of the hydrocarbons, the agreement is not as good and only the trend appears reliable. The discrepancy is attributable in part to the computation and in part to the instrumentation, both of which assume there is only one type of hydrocarbon

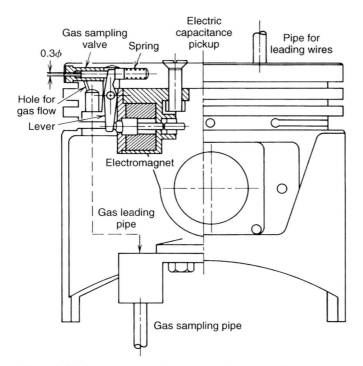

Figure 8-26 Gas sampling valve installed in a piston for
sampling gases at the top land (Furuhama and Tateishi, 1972).

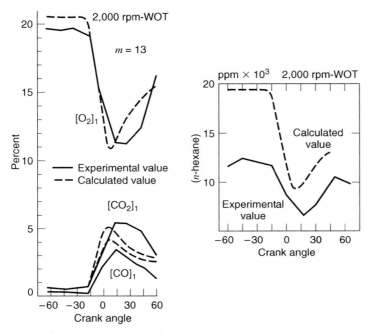

Figure 8-27 Measured and calculated gas composition at top land of a
gasoline engine at WOT (Furuhama and Tateishi, 1972).

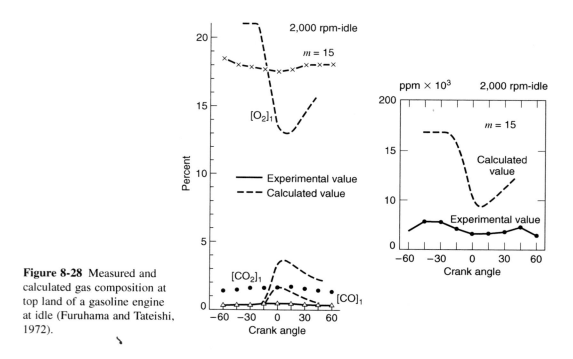

Figure 8-28 Measured and calculated gas composition at top land of a gasoline engine at idle (Furuhama and Tateishi, 1972).

(n-hexane) present when in fact there are 10 to 20 different species of significance. At the light loads, the computations and experiments show no correspondence whatsoever as shown in Figure 8-28.

8.9 REFERENCES

ABRAHAM, J. and V. MAGI (1997), "Application of the Discrete Ordinates Method to Compute Radiant Heat Loss in a Diesel Engine," *Num. Heat Transfer,* Part A (31), p. 597–610.

ALKIDAS, A. C. and J. P. MYERS (1982), "Transient Heat-Flux Measurements in the Combustion Chamber of a Spark Ignition Engine," *ASME J. Heat Transfer,* Vol. 104, p. 62–67.

ANNAND, W. J. D. (1963), "Heat Transfer in the Cylinders of a Reciprocating Internal Combustion Engine," *Proc. Instn. Mech. Engrs.,* 177, p. 973.

BENDERSKY, D. (1953), "A Special Thermocouple for Measuring Transient Temperature," *Mech. Eng.,* 75, p. 117.

BLUNSDON, C., W. MALALASEKERA, and J. DENT (1992), "Application of the Discrete Transfer Model of Thermal Radiation in a CFD Simulation of Diesel Engine Combustion and Heat Transfer," SAE paper 922305.

BLUNSDON, C., J. DENT, and W. MALALASEKERA (1993), "Modeling Infrared Radiation from the Combustion Products in a Spark Ignition Engine," SAE paper 932699.

BOHAC, S., D. BAKER, and D. ASSANIS (1996), "A Global Model for Steady State and Transient S. I. Engine Heat Transfer Studies," SAE paper 960073.

BORMAN, G. and K. NISHIWAKI (1987), "Internal-Combustion Engine Heat Transfer," *Prog. Energy Combustion Sci.,* 13, p. 1–46.

CATON, J. A. and J. B. HEYWOOD (1981), "An Experimental and Analytical Study of Heat Transfer in an Engine Exhaust Port," *Int. J. Heat Mass Trans.,* 24 (4), p. 581–595.

DENT, J. C. and S. J. SULAIMAN (1977), "Convective and Radiative Heat Transfer in a High Swirl Direct Injection Diesel Engine," SAE paper 770407.

FURUHAMA, S. and Y. TATEISHI (1972), "Gases in Piston Top-Land Space of Gasoline Engine," *Trans. Soc. Automotive Eng. of Japan,* 4, p. 30–39.

HAN, S., Y. CHUNG, Y. KWON, and S. LEE (1997), "Empirical Formula for Instantaneous Heat Transfer Coefficient in Spark Ignition Engine," SAE paper 972995.

HEYWOOD, J. (1988), *Internal Combustion Engine Fundamentals,* McGraw-Hill, New York.

HIRES, S. D. and G. L. POCHMARA (1976), "An Analytical Study of Exhaust Gas Heat Loss in a Piston Engine Exhaust Port," SAE paper 760767.

INCROPERA, F. and D. DEWITT, (2001), *Fundamentals of Heat and Mass Transfer,* John Wiley and Sons, New York.

KRIEGER, R. and G. BORMAN (1966), "The Computation of Apparent Heat Release for Internal Combustion Engines," *ASME Proc. of Diesel Gas Power,* paper 66-WA/DPG-4.

LI, C. (1982), "Piston Thermal Deformation and Friction Considerations," SAE paper 820086.

MALCHOW, G., S. SORENSON, and R. BUCKIUS (1979), "Heat Transfer in the Straight Section of an Exhaust Port of a Spark Ignition Engine," SAE paper 790309.

NAMAZIAN, M. and J. B. HEYWOOD (1982), "Flow in the Piston-Cylinder-Ring Crevices of a Spark-Ignition Engine: Effect on Hydrocarbon Emissions, Efficiency and Power," SAE paper 820088.

REITZ, R. (1991), "Assessment of Wall Heat Transfer Models for Premixed Charge Engine Combustion Computations," SAE paper 910267.

ROBINSON, K., N. CAMPBELL, J. HAWLEY, and D. TILLEY (1999), "A Review of Precision Engine Cooling," SAE paper 1999–01–0578.

RUDDY, B. (1979), "Calculated Inter-Ring Gas Pressures and Their Effect Upon Ring Pack Lubrication," DAROS Information, 6, 2–6, Sweden.

RYDER, E. A. (1950), "Recent Developments in the R-4360 Engine," *SAE Quart. Trans.,* 4(4), p. 559.

SHAYLER, P., S. CHRISTIAN, and T. MA (1993), "A Model for the Investigation of Temperature, Heat Flow, and Friction Characteristics During Engine Warm-up," SAE paper 931153.

SUN, X., W. WANG, D. LYONS, and X. GAO (1993), "Experimental Analysis and Performance Improvement of a Single Cylinder Direct Injection Turbocharged Low Heat Rejection Engine," SAE paper 930989.

TAYLOR, C. F. (1985), *The Internal Combustion Engine in Theory and Practice,* MIT Press, Cambridge, Massachusetts.

TILLOCK, B. and J. MARTIN (1996), "Measurement and Modeling of Thermal Flows in an Air-Cooled Engine," SAE paper 961731.

WHITEHOUSE, N. D. (1970–1971), "Heat Transfer in a Quiescent Chamber Diesel Engine," *Proc. Instn. Mech. Engrs.,* 185, p. 963–975.

WISNIEWSKI, T. (1998), "Experimental Study of Heat Transfer on Exhaust Valves of 4C90 Diesel Engine," SAE paper 981040.

WOSCHNI, G. (1967), "A Universally Applicable Equation for the Instantaneous Heat Transfer Coefficient in the Internal Combustion Engine," SAE paper 670931.

8.10   HOMEWORK

8.1   Practical applications of Equation 8.7 are limited because the heat loss to ambient air is determined by the small differences between much larger numbers. Suppose each term on the right-hand side can be determined to within ±5%. What tolerances could then be attached to $\dot{Q}_{amb}$? For nominal values use the results given in Table 8-1. The most probable error is computed from the square root of the sum of the squares of the errors of the RHS terms.

8.2   Estimate the heat transfer rate (kW) at which an automotive radiator would operate if an engine produces 75 kW brake power. What is the temperature change (°C) of the coolant as it passes through the radiator if its flow rate is 20 gpm? What is the temperature change of the coolant as it passes through the engine?   *Fig 8.4   estimate Q_cut*

8.3   If an engine cylinder head is changed from iron ($k = 60$ W/m K) to aluminum ($k = 170$ W/m K), estimate the change in the heat flux from the head. What is the change in the gas side cylinder head temperature? $T_c = 350 K$ $h_c = 1000 \frac{W}{m^2 \cdot k}$

8.4 At an engine speed of 1000 rpm, what is the approximate penetration depth of the temperature fluctuations in **(a)** cast iron block and **(b)** aluminum engine block?

8.5 Determine the effect of engine speed on the overall engine heat transfer coefficient of Example 8.1. Plot h_o versus N for $1000 < N < 6000$ rpm. Assume $e_v = 0.9$.

8.6 With reference to Equation 8.33, the heat transfer coefficient h_o increases with engine speed to the 0.75 power, whereas the time available to lose heat decreases with engine speed. What is the net effect of increasing engine speed on engine thermal efficiency?

8.7 Show that as far as engine speed is concerned, the indicated thermal efficiency should have the following dependence

$$(1 - \eta/\eta_{\text{otto}}) \sim N^{-0.25}$$

since the ratio η/η_{otto} is dominated by heat loss. (Hint: see Figure 4-15.)

8.8 If $k \sim T^{0.75}$, and $\mu \sim T^{0.62}$, how does the heat transfer coefficient of Equation 8.36 vary with pressure and temperature?

8.9 How would you expect the leading coefficient of Equation 8.38 to change if Woschni had included the effect of the intake valve size?

8.10 Compare the mean effective pressure, the net work and indicated thermal efficiency for an engine modeled with and without cylinder heat transfer. Use the Woschni heat transfer correlation. The engine has a bore of 0.085 m, stroke of 0.08 m, and connecting rod length of 0.12 m, with a compression ratio of 11. The engine speed is 5000 rpm. The burn initiation is at -10 degrees atdc, the burn duration is 60 degrees, and the Weibe parameters are $a = 5$ and $n = 3$. The fuel is propane with an equivalence ratio of 0.95, and the inlet temperature and pressure are 300 K and 1 bar. The average cylinder wall temperature is 400 K.

8.11 Compare the Annand and Woschni correlations for an engine operating at **(a)** 2000 rpm and **(b)** 4000 rpm. Assume that the engine has a bore and stroke of 0.085 m, and connecting rod length of 0.12 m. The fuel is methane, with an equivalence ratio of 0.95. The engine is unthrottled, with inlet pressure of 1 bar, and temperature of 320 K. The spark ignition is at -15 degrees atdc, and the burn duration is 70 degrees. The average cylinder wall temperature is 400 K.

8.12 What is the average heat transfer coefficient for an exhaust system that has an average mass flow rate of 0.08 kg/s at a temperature of 700 K and a pipe diameter of 0.045 m?

Chapter 9

Combustion and Emissions

9.1 INTRODUCTION

In this chapter we examine combustion and emissions processes in spark and compression ignition engines. The combustion processes that occur in each of these types of engines are very different. A spark ignition engine has a turbulent mixture of fuel and air, which once ignited by a spark, sustains a reaction process that propagates a flame in the form of a thin wrinkled sheet across the cylinder. The heat release starts relatively slowly, and increases to its largest value near the end of the process. On the other hand, a compression ignition engine has separate fuel and air streams that combust as they are mixed together. The chemical reaction, which produces a diffusion flame, takes place at the interface between the fuel and the air. The heat release begins at a relatively high value, and then decreases as the available oxygen is depleted.

The combustion product emissions from internal combustion engines of nitrogen oxide (NO_x), carbon monoxide (CO), hydrocarbons (HC), particulates (PM), and aldehydes are a significant source of air pollution. The internal combustion engine is the source of roughly half of the NO_x, CO, and HC pollutants in our air. These pollutants have various adverse health and environmental effects. For example, NO_x reacts with water vapor to form nitric acid, and reacts with solar radiation to form ground level ozone, both of which cause respiratory system problems. Carbon monoxide has an affinity for hemoglobin about 200 times that of oxygen, so it can interfere with oxygen distribution throughout a person's circulatory system. Finally, hydrocarbons can cause cellular mutations and also contribute to the formation of ground level ozone.

Combustion chemistry in internal combustion engines is very complex, and depends on the type of fuel used in the combustion process. Models for the reaction pathways for the oxidation of a hydrocarbon fuel such as paraffin, C_nH_{2n+2}, a major component of gasoline, can include at least 200 different reactions. The hydrocarbon reactions are generally grouped into three distinct steps. For example, the first step in the breakdown of a paraffin is via reactions with O and H atoms to form olefins (C_nH_{2n}) and hydrogen. The second step is the oxidation of the olefins to form CO and hydrogen. The third and last step is the oxidation of CO to form CO_2. Most of the energy release occurs during the last step, a step independent of the molecular weight. Consequently, hydrocarbon paraffins of different molecular weight have very similar heats of combustion. Detailed information about hydrocarbon chemical kinetics is given in Westbrook and Dryer (1984); and books by Borman and Ragland (1998), and Turns (1996) cover combustion chemistry and kinetics in internal combustion engines from an engineering perspective.

9.2 COMBUSTION IN SPARK IGNITION ENGINES

Flow Visualization

In the late 1930s, Rassweiler and Withrow (1938) performed a classic flow visualization experiment. They modified an L-head cylinder of a spark ignition engine so that a quartz window could be installed allowing an unobstructed view of the entire combustion space.

Using high-speed motion photography, they were able to record the combustion process. Figure 9-1 shows a typical sequence obtained illustrating flame propagation in a homogeneous charge, spark ignition engine. For the combustion process shown in Figure 9-1, ignition occurs at $\theta_s = -25°$. Notice that no flame is visible until $\theta = -16°$ which is 9° later. That 9° period is called the ignition delay, $\Delta\theta_{id}$. Once formed, the flame spreads like a spherical wave into the unburned gas with a ragged surface because of turbulence. The end of combustion at $\theta = +21°$ is determined from simultaneous measurement of cylinder pressure. Notice that the burned gases remain luminous even after combustion is complete.

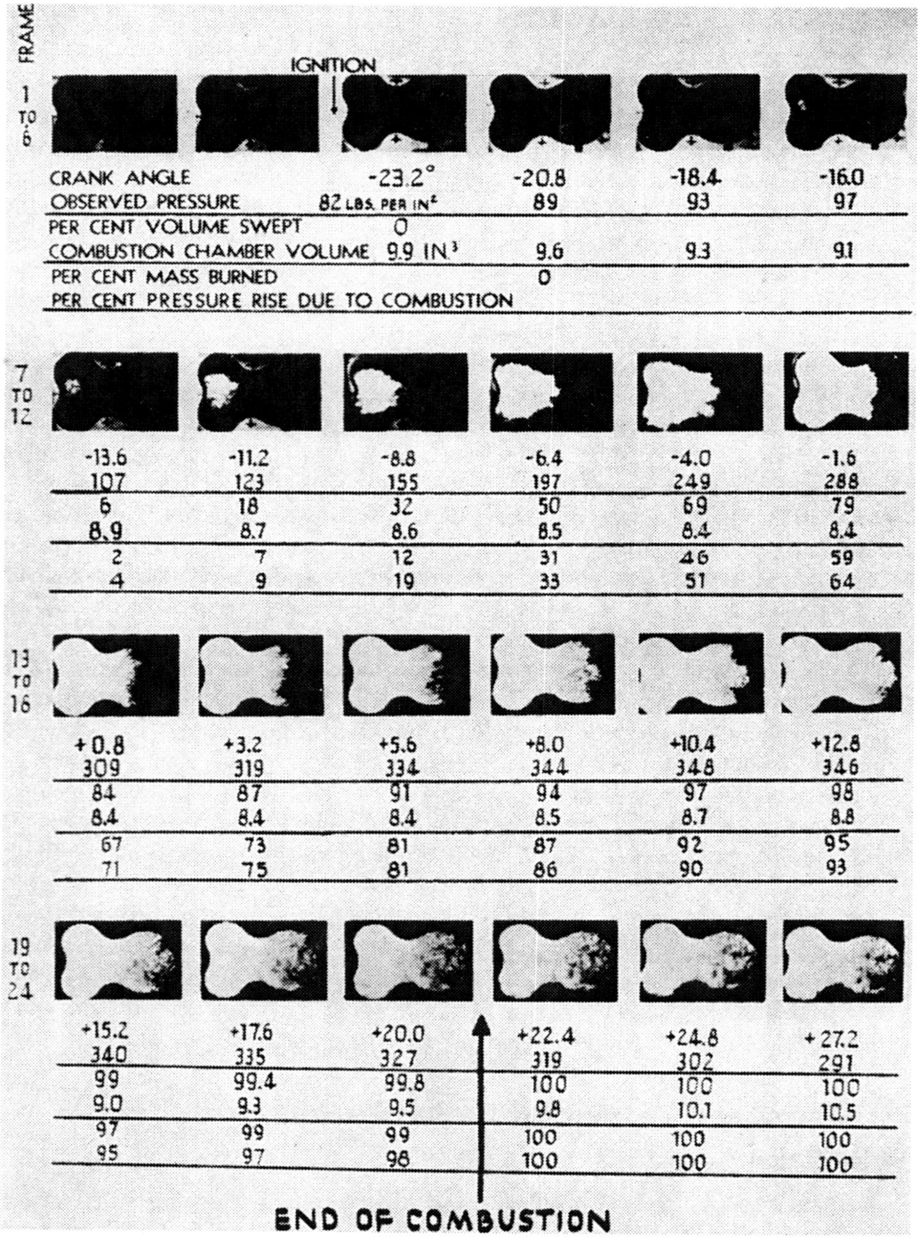

Figure 9-1 Visualization of spark ignition combustion (Rassweiler and Withrow, 1938). Reprinted with permission © 1980. Society of Automotive Engineers, Inc.

Shadowgraph photography is a method of flow visualization that shows contrasts due to differences in density of the flow. It does not record light emitted by the flame; rather it records light transmitted through and refracted by the gases. Figure 9-2 shows shadowgraph sequences for lean and slightly rich combustion. Again, a ragged edge wave

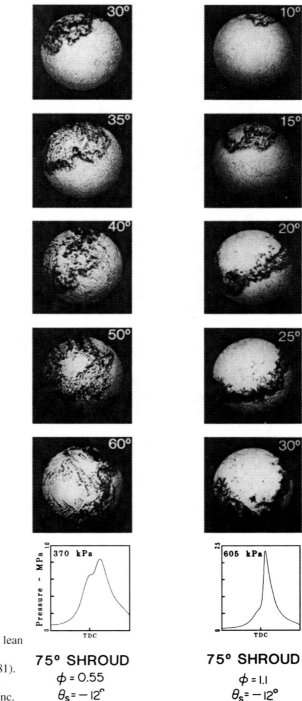

Figure 9-2 Laser shadow graph of lean ($\phi = 0.55$) and rich ($\phi = 1.1$) combustion (Witze and Vilchis, 1981). Reprinted with permission © 1981. Society of Automotive Engineers, Inc.

75° SHROUD
$\phi = 0.55$
$\theta_s = -12°$

75° SHROUD
$\phi = 1.1$
$\theta_s = -12°$

is seen propagating into the unburned mixture. Ignition delay (in degrees of crank angle), is on the order of 10° for the rich case and 20° for the lean case. At 20° and 25° after ignition in the rich case, the width of the flame front is clearly discernible, one of the advantages shadow photography offers over ordinary photography. The width is more difficult to discern in the lean case because it is two or three times thicker. Thus, a completely burned region does not appear until approximately 40° after ignition. At this time, the whitest region is burned gas, the greyish region in front of the flame is unburned gas, and the highly convoluted dark and white region is a mixture of burned, burning, and unburned gas.

Combustion Modeling

There are a number of models of the homogeneous charge combustion process in spark ignition engines. The models range in rigor from simple two zone models which divide the combustion chamber into burned and unburned zones separated by a turbulent flame front to differential equation models that are solved numerically. The combustion parameters incorporated into these models include the laminar flame speed and thickness, the turbulent flame speed, and the turbulence intensity.

The laminar flame speed s_l is a well-defined characteristic of a fuel-air mixture. It represents the speed at which a one-dimensional laminar flame propagates into the unburned gas under adiabatic conditions. As shown in Figure 9-3, the concentration of reactants decreases across the flame front, and the temperature of the mixture increases across the flame front. There is a preheat zone in front of the flame in which the temperature of the reactants is raised to the ignition temperature by conduction heat transfer from the flame front into the unburned region, and a reaction zone which contains the flame front where the combustion takes place. As the temperature rises in the reaction zone, the chemical reactions, which depend exponentially on temperature, increase until the reactants are consumed and their concentration decreases to zero, forming the downstream side of the flame front.

The laminar flame speed depends on the pressure, temperature, and composition of the unburned gas. Some of these dependencies are illustrated in Figures 9-4 and 9-5 in which data from Metghalchi and Keck (1982) is compared to predictions of Westbrook and Dryer (1980). The laminar flame speed, or burning velocity, shows a maximum for slightly rich mixtures, is a strong function of unburned gas temperature, and is a weak

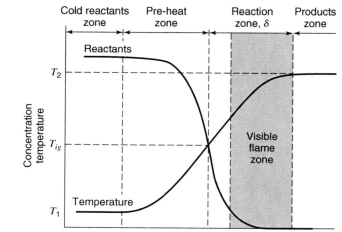

Figure 9-3 Temperature and concentration profiles of laminar flames (Borman and Ragland, 1999).

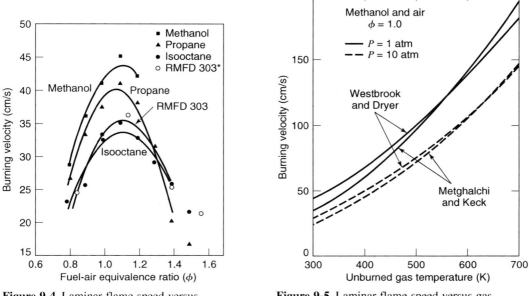

Figure 9-4 Laminar flame speed versus equivalence ratio (Metghalchi and Keck, 1982).

Figure 9-5 Laminar flame speed versus gas temperature (Metghalchi and Keck, 1982).

function of pressure. It also decreases linearly with the residual fraction and the exhaust gas recirculation (see Homework Problem 9-1). The dependence of the laminar flame speed on temperature is due to the exponential relation between reaction kinetics and temperature.

There are three regimes for turbulent flames. The regimes are wrinkled laminar flame, flamelets in eddies, and distributed reaction. The characteristics of the regimes are outlined in Table 9-1. Internal combustion engines operate in the wrinkled laminar flame and flamelets in eddies regimes, depending on the engine speed (Abraham et al., 1985).

In the wrinkled laminar flame regime, the flame thickness δ_l is thinner than the smallest turbulent eddy thickness η, and the turbulent intensity u_t is of the same order as the laminar flame speed s_l. The effect of turbulence in the cylinder therefore is to wrinkle and distort the laminar flame front. In the flow field, the turbulent vortices spread ignition sites via a ragged edge wave emerging from the spark plug. For the turbulent flow conditions of Example 7-3, the scale of the wrinkles is about 1 mm, and the flame is less than 0.01-mm thick. The turbulent flame speed u_t can be 3 to 30 times the laminar flame speed, depending on the turbulent intensity. The relationship between the turbulent flame speed and the laminar flame speed is modeled as

$$s_t/s_l = a(u_t/s_l)^b \tag{9.1}$$

The position of the flame front moves irregularly, making the time average flame profile appear relatively thick, forming a "turbulent flame brush."

Table 9-1   Turbulent Flame Regimes

Wrinkled laminar flame	$\delta_l < \eta$	$u_t \sim s_l$
Flamelets in eddies	$\eta < \delta_l < l$	$u_t \gg s_l$
Distributed reactions	$\delta_l > l$	

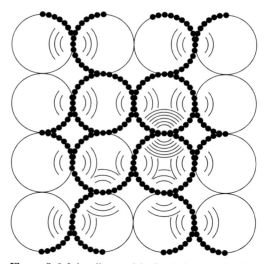

Figure 9-6 Ink roller model of turbulent combustion.

A convenient way to conceptualize the flame propagation in the wrinkled laminar regime is in terms of ink rollers. The ink roller model is shown in Figure 9-6. Imagine a bunch of cylindrical rolls as depicted to represent eddies of a similar diameter in the turbulent flow field. Now consider ignition as being analogous to continuously depositing a stream of ink at the periphery of one roll. The rollers are rotating, and as a result, the ink spreads. A ragged edge wave emerges from the initial deposition site. The speed of the propagation is proportional to the velocity at the edge of the vortices. The front will take on a thickness determined by the speed of the rollers, their size, and the rate at which ink seeps into the rolls. In the flow field the flame thickness will depend on the vorticity, the eddy sizes, and the laminar flame-spreading rate. Note that the eddies in the flow field are more likely to resemble a mesh of spaghetti than perfectly aligned ink rollers.

Since the turbulence intensity is proportional to the engine speed, at higher engine speeds the turbulent flame region can transition from the wrinkle sheet to the flamelets in eddies regime. In the flamelets in eddies regime, the flame thickness is greater than the small eddy thickness η, but less than the integral thickness l, and the turbulent intensity is much greater than the laminar flame speed. The increased wrinkling can result in the creation of pockets of unburned gas mixture. In this regime, the burning rate is controlled by the turbulent mixing rate, i.e., the integral length scale, not the chemical reaction rate.

The combustion also depends on the combustion chamber geometry. To illustrate the effect of combustion chamber geometry, consider two limiting cases of combustion: (1) in a sphere centrally ignited and (2) in a tube ignited at one end. Assume that the sphere and the tube have the same volume. In each case the flame will propagate as a ragged spherical front of radius r_f from the spark plug. In the sphere the area of the front grows as r_f^2. Thus, the entrainment rate gets faster and faster as the flame grows. On the other hand, in the tube, the flame front will initially grow as r_f^2, but it will soon hit the walls and be constrained to be more or less constant from then on. Thus, combustion in a sphere can be expected to burn faster; that is, it will take less time to burn the charge. The maximum cylinder pressure occurs at about the time that the flame reaches the cylinder wall. This is also the point of largest flame surface area, with the maximum flow of unburned gases into the flame.

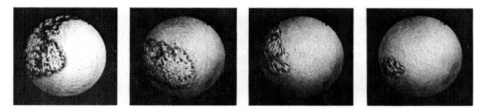

Figure 9-7 Examples of cyclic variation in the combustion process. The flames shown are for identical conditions of shroud angle $\Gamma = 45$ deg and $\phi = 0.55$ at 30 deg after ignition at $\theta_s = 12$ deg btdc (Witze and Vilchis, 1981). Reprinted with permission © 1981. Society of Automotive Engineers, Inc.

There are cycle-to-cycle variations in the flame propagation caused by the random features of the flow field. Figure 9-7 shows flames from four different cycles in an engine under identical operating conditions. It is not difficult to imagine that the leftmost picture coincided with the spark igniting a region of high shear near the perimeter of a vortex, so that turbulent spreading began immediately. In the rightmost picture the spark ignited the center of a vortex so that laminar spreading had to precede turbulent spreading.

9.3 ABNORMAL COMBUSTION (KNOCK) IN SPARK IGNITION ENGINES

Knock is the term used to describe a pinging noise emitted from homogeneous charge, spark ignition engines. It is caused by autoignition of the unburned or end gas ahead of the flame, creating pressure waves that travel through the combustion gases. Compression of the end gas by expansion of the burned part of the charge raises its temperature to the autoignition point. Whether or not an engine will knock depends on the engine design, the engine operating parameters (such as operation at wide open throttle), and the fuel type. The occurrence of knock puts a constraint on engine design since it limits the compression ratio and thus the engine efficiency.

High speed Schlieren photographs (see Figure 9-8) reveal that under knocking conditions, the flame spread occurs much faster than normal. This is because the unburned gas involved is at an elevated temperature, so the laminar flame speed is substantially increased. More importantly, however, several autoignition sites appear almost simultaneously. The attendant rapid fluctuations in pressure can be a serious problem as they can disrupt the cylinder thermal boundary layers causing higher piston surface temperatures, resulting in surface erosion and failure. The cylinder pressure profiles for normal and knocking combustion are shown in Figure 9-9. Note the high frequency ($5-10$ kHz) pressure fluctuations resulting from the autoignition.

Using a single-cylinder research engine, the unburned end gas in a high swirl, homogeneous charge engine has been isolated in the center of the combustion chamber by simultaneous ignition at four equally spaced spark plugs mounted in the cylinder wall. Figure 9-8 shows the dramatic change in the Schlieren pattern just before and just after ignition. It took about 2 ms for the flames to spread from the spark plugs to the positions shown in the photograph just before knock; whereas it took only 0.1 ms to propagate through the end gas once autoignition occurred. Neither shock nor detonation waves were observed.

In these experiments, temperature measurements have been made of the end gas using a laser-based technique. For temperatures less than 1100 K, coherent antistokes Raman spectroscopy (CARS) is used, and at the higher temperatures spontaneous Raman scattering is used. The principles underlying these techniques are discussed at length in Greenhaugh

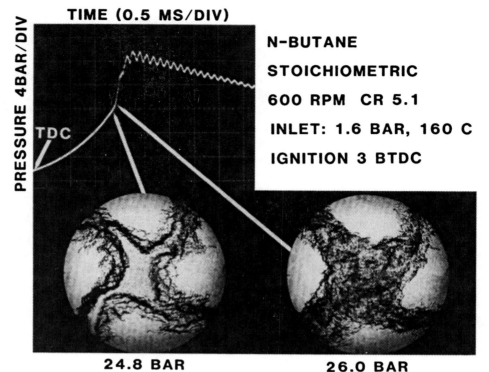

Figure 9-8 Schlieren photographs of knock process (Smith et al., 1984).

(1988) and Eckbreth (1988). The results, shown in Figure 9-10, show that the temperature, like the pressure, undergoes an abrupt change in the rate of change at the knock point. They also show that the temperature continues to rise even after the 0.1 ms required for the homogeneous ignition sites and the flame propagation to have consumed the end gas. Clearly oxidation is not complete in the after-knock photograph of Figure 9-8.

To provide a standard measure of a fuel's knock characteristics, a scale has been devised in which fuels are assigned an octane number. As discussed in Chapter 10, the octane number referred to on gasoline pumps as $(R + M)/2$ is the average of the research (R) and motor (M) method octane numbers. It is also called the antiknock index (AKI). A standardized engine operated under specified operating conditions is used to measure the octane number of a fuel. The engine employed is a Cooperative Fuel Research (CFR) engine that features a variable compression ratio. To measure knock, an American Society for Testing Materials (ASTM) knock meter that responds to rate of pressure rise is used. Two sets of operating conditions are employed: the motor and the research method, as described in Table 9-2.

Table 9-2 Octane Number Measurement Conditions

	Motor	Research
Inlet temperature (K)	422	325
Jacket temperature (K)	373	373
Speed (rpm)	900	600
Humidity (mass fraction)	0.0036 to 0.0072	0.0036 to 0.0072

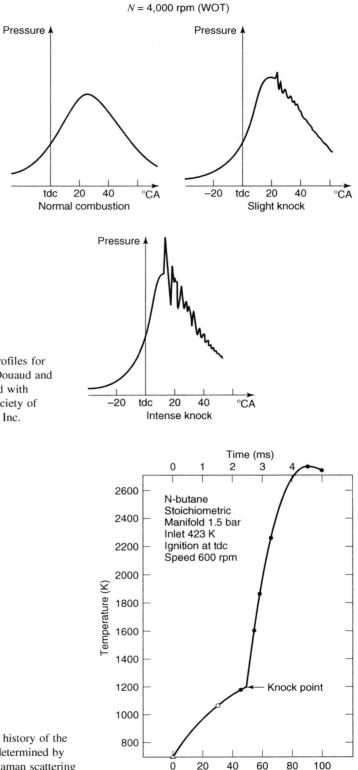

Figure 9-9 Pressure profiles for knocking conditions (Douaud and Eyzat, 1977). Reprinted with permission © 1977. Society of Automotive Engineers, Inc.

Figure 9-10 Temperature history of the end gas in Figure 9-8 as determined by CARS and spontaneous Raman scattering (Smith et al., 1984).

The procedure to measure the octane number of a test fuel is as follows:

- Run the CFR engine on the test fuel at either the motor or research operating conditions.
- Slowly increase the compression ratio until the standard amount of knock occurs.
- At that compression ratio, run the engine on blends of the reference fuels isooctane and *n*-heptane.
- The octane number is the percentage of isooctane in the blend that produces the standardized knock at that compression ratio.

One measure of an engine's octane requirement is its knock limited indicated mean effective pressure (klimep). The greater the klimep, the smaller the octane requirement. Knock limited imep is measured by increasing the inlet pressure P_i (which will increase the cylinder charge density and temperature) until knock occurs; the imep at that condition is the klimep. Figure 9-11 shows experimental results that show klimep decreases with increasing coolant temperature. Similar results are obtained with increasing inlet air temperature. Both results are to be expected since chemical reaction rates are accelerated strongly by increases in temperature.

Figure 9-11 also illustrates two problems with the octane number scale:

1. At low coolant temperatures *di*-isobutylene performs better than isooctane (implying the octane number is greater than 100).

2. The relative ranking of isooctane and *di*-isobutylene depends on coolant temperatures; whereas if the octane scale were decoupled from engine design, making the assigned number a fuel property, the fuel with the greater octane number would always yield the largest klimep.

The former problem is dealt with by extrapolation. A performance number defined as the ratio of klimep for the fuel in question to klimep of isooctane is used for this purpose. The latter problem is dealt with by using two standard operating conditions (research and motor) and reporting an average number. These shortcomings should be kept in mind; they are easy to forget because of the great utility of the octane number scale.

Typical results obtained for varying fuel-air ratio are shown in Figure 9-12. Notice that near-stoichiometric mixtures have the lowest klimep (therefore the highest octane requirement). Notice too that maximum klimep is attained with very rich ($\phi \sim 1.6$) mixtures. Therefore, to obtain maximum power from an engine, one should run very rich, near $\phi = 1.6$, with a compression ratio and inlet pressure such that imep is equal to klimep.

Knock occurs if there is enough time for sufficient autoignition precursors to form. Thus, at high engine speeds one might not expect knock to be a problem since there is less time available for the precursors to form. On the other hand, as engine speed increases, there is less heat loss from the gases so that gas temperatures will be higher. This accelerates the precursor formation rate so that less time is required to form a concentration high enough for autoignition to occur. As a result of these and other competing effects, some engines show a klimep increasing with speed, and others show a decrease.

One way to model engine knock is to suppose that there exists a critical mass fraction of precursors that if attained anywhere within the charge will cause autoignition. Let x_p denote the mass fraction of precursors and x_c denote the critical value. Knock will then occur prior to the end of normal combustion if the integrated rate of formation equals the critical mass fraction:

$$\int_0^t \frac{dx_p}{dt} \, dt = x_c \tag{9.2}$$

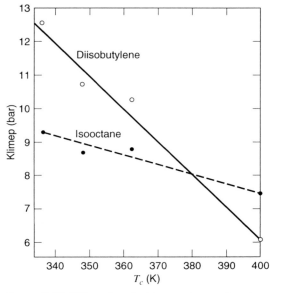

Figure 9-11 Effect of coolant temperature on knock limited imep: T_c is the temperature of the entering coolant (Hesselberg and Lovell, 1951). Reprinted with permission © 1951. Society of Automotive Engineers, Inc.

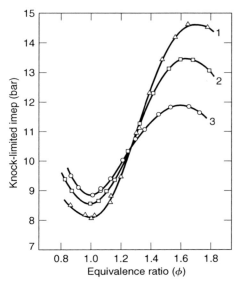

Figure 9-12 Effect of fuel-air ratio on knock limited imep for three aircraft fuels (Cook et al., 1944).

The lower limit of integration is chosen to coincide with the closing of the intake valve, thus ignoring precursor formation in the intake manifold. The non-dimensional rate of formation of precursors is represented by an equation of the following form

$$\frac{1}{x_c}\frac{dx_p}{dt} = A\, P^n \exp\left(\frac{-B}{T}\right) \tag{9.3}$$

where the empirical constants A, B, and n are determined from a set of experimental results. Like constants in algebraic burning laws, these constants will vary from engine to engine and fuel to fuel.

It is instructive to consider the dependence of the precursor formation rate on temperature. For isooctane at $\phi = 1.0$, $P = 10$ bar, the results shown in Table 9-3 are obtained using the constants A, B, and n of Douaud and Eyzat (1978). Notice that the reaction rate is an extremely strong function of the temperature. Indeed, at temperatures

Table 9-3   Precursor Formation Rate Versus Temperature

$\dfrac{1}{x_c}\dfrac{dx_p}{dt}$ (s^{-1})	T (K)
2.8×10^{-5}	300
5.9×10^{-1}	500
4.2×10^{1}	700
4.5×10^{2}	900
2.0×10^{3}	1100

characteristic of the intake manifold, the rate of formation of precursors is negligible. Since the combustion occurs over times of order 10^{-2}s, not until the rates approach $100s^{-1}$ or $T \sim 900$ K will knock occur with isooctane.

To incorporate Equation 9.2 into an arbitrary heat release model of combustion one defines the extent of reaction ζ as the ratio of the precursor mass fraction to the critical mass fraction.

$$\xi = x_p / x_c \tag{9.4}$$

and

$$\frac{d\xi}{dt} = \frac{1}{x_c} \frac{dx_p}{dt} \tag{9.5}$$

Equation 9.5 is integrated simultaneously with the energy and continuity equations discussed in Chapter 4. If at any time prior to the end of combustion ξ reaches 1, knock is said to occur and the remaining unburned gas burns instantaneously. An example of the incorporation of a knock model into an engine performance simulation code is given in Ho et al. (1996).

9.4   COMBUSTION IN COMPRESSION IGNITION ENGINES

Flow Visualization and Diagnostics

Flow visualization and diagnostic techniques are significant tools used to better understand diesel combustion processes. The application of optical diagnostics to diesel combustion is constrained by the need to maintain realistic combustion chamber geometry, while maintaining satisfactory optical access. An example of a test engine modified for optical access by researchers at Sandia National Laboratory is shown in Figure 9-13 (Dec and Espey, 1995). The engine uses an extended piston with a piston crown window. This single cylinder research engine is based on a typical commercial, heavy duty diesel engine, and has a stroke of 140 mm and a bore of 152 mm. Windows at the top of the cylinder provide for laser access along the axis of the fuel spray, which in this engine is 14° from the horizontal.

The piston crown in Figure 9-13 illustrates the trade-off between optical access and combustion bowl geometry. When a flat piston crown window and a flat cylinder head window are installed, the combustion bowl has a flat "pancake" bottom allowing the laser sheet to be viewed from above and below throughout the bowl and the squish region. Other optical engines have been built using more realistic bowl geometries at the expense of increased optical distortion in the images.

Figure 9-15 shows a high-speed cinematography sequence of the luminous combustion in the direct injection engine of Figure 9-13. The high-speed cinematography data show the fuel jet penetration and spread of the luminous combustion zones. The first luminosity is seen about 5° after the start of injection. The rapid appearance of widespread combustion indicates that the ignition occurs at multiple points throughout the jet. The luminosity is yellow, suggesting fuel rich combustion. After a few more degrees, the burning fuel jets contact the edge of the combustion bowl, and then spread along the circumference into the space between the jets and downward into the bottom of the bowl. For the case shown in Figure 9-15, about half of the fuel is injected after the burning fuel jet reaches the edge of the bowl.

High-speed cinematography is a qualitative measurement technique, since the image is integrated along the line of sight of an optically thick medium; and it is also not species specific, so additional diagnostic techniques have been developed. Various laser-based

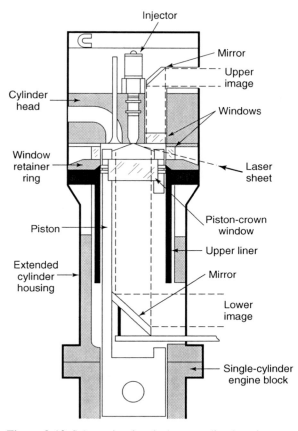

Figure 9-13 Schematic of optical-access diesel engine showing the laser sheet along the fuel jet axis (Dec and Espey, 1995). Reprinted with permission © 1995. Society of Automotive Engineers, Inc.

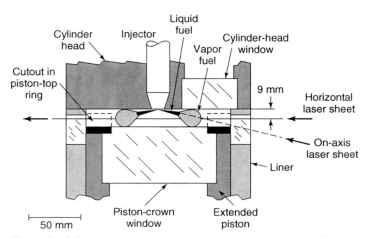

Figure 9-14 Side view of optical access combustion chamber (Dec and Espey, 1995). Reprinted with permission © 1995. Society of Automotive Engineers, Inc.

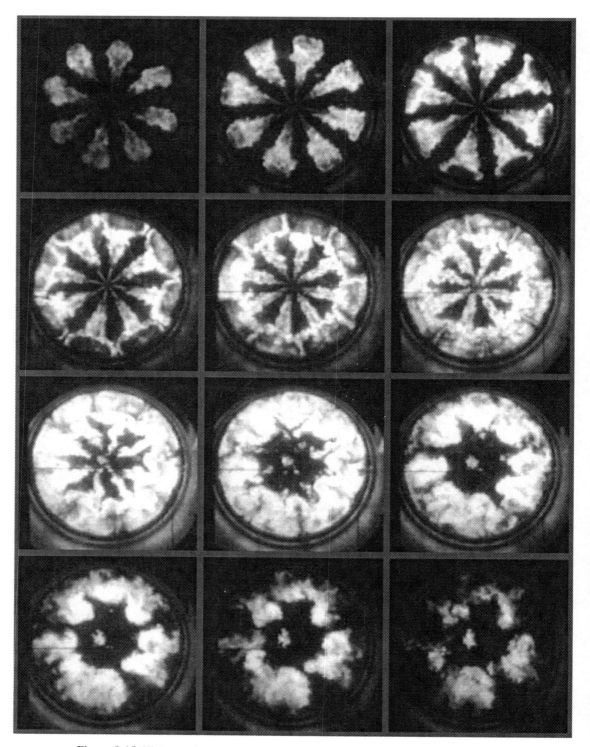

Figure 9-15 High-speed photographic sequence of the luminosity of a diesel flame (Espey and Dec, 1993). Reprinted with permission © 1993. Society of Automotive Engineers, Inc.

combustion diagnostic techniques developed by combustion researchers have been applied to diesel engines in order to obtain more detailed and species specific information about the combustion processes taking place in a diesel engine. The techniques include laser light scattering, and laser-induced incandescence and fluorescence.

With a scattering technique laser light is elastically scattered by fuel droplets and/or soot particles. The scattering distribution and intensity depends on the particle size. Mie scattering is defined as elastic scattering from particles whose diameter is of the same order of magnitude or smaller as the light wavelength. Elastic scattering of light from molecules or small particles with diameters much smaller than the wavelength of the laser light is termed Rayleigh scattering. Liquid spray patterns have been determined via measurements of Mie scattering. The vapor phase fuel-air mixture patterns and temperature fields have been determined through Rayleigh scattering measurements.

Laser light is also used to induce incandescence and fluorescence of given species. Both relative and absolute soot concentrations have been determined using laser-induced incandescence (LII). Planar laser-induced fluorescence (PLIF) has been used to determine polyaromatic hydrocarbon (PAH) concentrations, which are precursors to soot, OH distributions, and NO distributions. The OH radical distribution provides information about the location and intensity of a diffusion flame. The NO radicals indicate the location of NO_x production in the cylinder. For further information, a review of various optical diagnostic techniques is given in Eckbreth (1988).

Combustion Modeling

The diesel combustion process has been classified into three phases: ignition delay, premixed combustion, and mixing controlled combustion. In a diesel engine a low volatility fuel must be converted from a cold liquid into a finely atomized state, vaporized, and its temperature raised to a point to support autoignition. This time interval between the start of ignition and the start of combustion is termed the ignition delay. Once regions of vapor-air mixture form around the fluid jet as it is first injected into the cylinder is at or above the autoignition temperature, it will spontaneously ignite. The combustion of this initial vapor-air mixture is termed the premixed combustion phase. Finally, the fuel in the main body of the fuel jet will mix with the surrounding air and burn, which is called the mixing controlled combustion phase. Combustion occurs at a rate limited by the rate at which the remaining fuel to be burned can be mixed with air.

Early models of diesel combustion assumed that a burning diesel jet was composed of a dense fuel-rich core surrounded by a uniformly leaner fuel-air mixture, as shown in Figure 9-16. With reference to the models used for steady spray combustion in furnaces and gas turbines, the diesel fuel autoignition and premixed combustion phases were also assumed to occur in a diffusion flame in the near stoichiometric ($\phi \sim 1$) regions between the rich ϕ_R and the lean ϕ_L limits, at the outer edge of the jet. Soot was assumed to form in a narrow region on the fuel-rich side of the diffusion flame.

Recent laser sheet diagnostic experiments in diesel engines have indicated that the combustion process in diesel engines is different than that in furnaces and gas turbines. Dec (1997) has proposed an alternative conceptual model based on laser sheet experimental results. The Dec model features two stages of fuel oxidation for both of the premixed and mixing controlled combustion phases. The first stage is partial oxidation of the fuel in a rich premixed reaction, and the second stage is combustion of the fuel-rich, partially oxidized products of the first stage in a near stoichiometric diffusion flame.

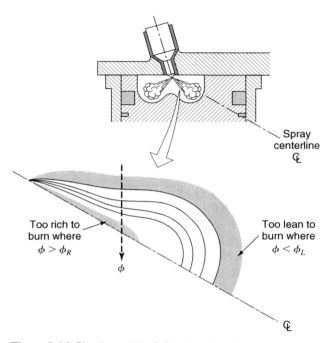

Figure 9-16 Simple model of diesel combustion.

This conceptual model is shown schematically in Figure 9-17. Figure 9-17 is a temporal sequence showing the progressive changes during the injection process. Significant events in the evolution of the jet state are drawn at successive degrees after the start of injection (ASI). Six parameters are shown in Figure 9-17; the liquid fuel, the vapor-air mixture, the PAHs, the diffusion flame, the chemiluminscence emission region, and the soot concentration.

At 1.0° in Figure 9-17, near the beginning of the ignition delay phase, as the liquid fuel is injected into the cylinder, it entrains hot cylinder air along the sides of the jet, leading to fuel evaporation. Note that throughout the injection process, the liquid length portion of the jet remains relatively constant. There is limited penetration of the fuel droplets into the combustion chamber. The penetration depth of the liquid jet has been found to be dependent on the volatility of the fuel injector hole size, fuel and cylinder air temperature, and relatively insensitive to the injection pressure (Siebers, 1998).

At about 4.0° ASI, a vapor head vortex is beginning to form in the leading portion of the jet downstream of the liquid jet. The bulk of the vaporized fuel is in the head of the jet. The fuel vapor-air mixture region in the head vortex is relatively uniform, has a well-defined boundary separating it from the surrounding air, and has an equivalence ratio between 2 and 4 throughout its cross section. At about 5.0° ASI, premixed combustion begins in the head vortex. As a consequence of the high equivalence ratio, the initial premixed combustion is fuel-rich with a temperature of about 1600 K, and produces PAHs and soot. The soot concentration is fairly uniform throughout the jet cross section.

At about 6.5° ASI, a turbulent diffusion flame forms at the edge of the jet around the products of the initial premixed stage. This turbulent diffusion flame begins the transition to the mixing controlled phase, and is near stoichiometric. The diffusion flame causes the formation of larger soot particles at the jet periphery. The soot concentration

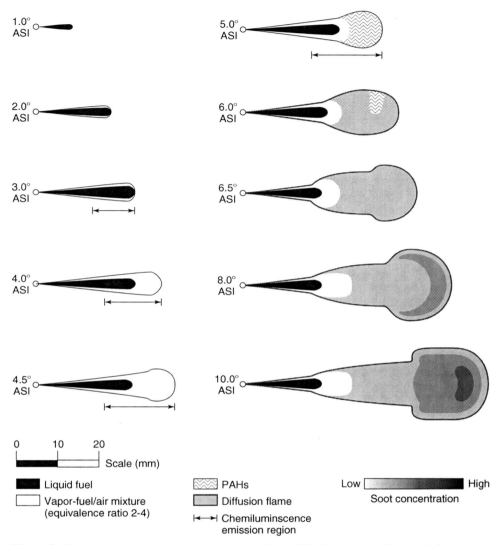

Figure 9-17 Detailed model of diesel combustion (Dec, 1997). Reprinted with permission © 1997. Society of Automotive Engineers, Inc.

continues to increase throughout the head vortex region at the head of the jet. Since the head vortex of the jet is composed of recirculating gases, the soot particles also recirculate and grow in size.

At about 8° ASI, the jet reaches a quasi-steady condition in which the general features of the jet do not change significantly as it expands across the combustion chamber. The combustion is in the mixing controlled phase. The fuel first passes through a very fuel rich ($\phi > 4$) premixed reaction stage and then burns out in the turbulent diffusion flame at the edge of the jet.

Most of the soot is burned with the fuel at the diffusion flame. The fraction of soot that is not oxidized becomes an exhaust emission. NO_x is formed in the high temperature regions at the diffusion flame where both oxygen and nitrogen are available, and in the postcombustion hot gas regions.

The quantity of fuel burned in each of the premixed and mixing controlled phases is not only influenced by the engine and injector design, but also by the fuel type and the load. For example, aromatic hydrocarbons have chemical bonds that are difficult to break and result in a long ignition delay. If this fuel is injected rapidly enough to mix completely with air before autoignition occurs, it will all burn rapidly when ignition occurs in the premixed phase, producing a large rate of change of pressure and a high peak pressure. On the other hand, the chemical bonds of some fuels are easily broken. Ignition delay is then short, and with a slow injection most of the fuel to be burned is injected after autoignition occurs. Little fuel burns in the premixed phase and most burns at a rate limited by the rate of mixing with the cylinder air. At idle, most of the fuel injected in small bore diesel engines is burned in the premixed phase. As the load increases, the injection duration increases, and the relative size of the mixing controlled phase increases relative to the premixed phase.

Since diesel combustion is heterogenous, numerical models need to be multidimensional to account for the spatial variations in temperature and species concentrations. The numerical code KIVA (Amsden et al., 1989), developed at Los Alamos National Laboratory, is a public domain three-dimensional CFD program that has been used by a number of research groups to model diesel combustion. Reitz and co-workers at the University of Wisconsin (for example, see Rutland et al., 1994, and Kong et al., 1995) have added a number of improvements to the original KIVA model that incorporate more realistic analysis of the fuel spray breakup, vaporization, spray-wall impingement, wall heat transfer, ignition, combustion, and pollutant formation.

Diesel Centane Number

Diesel fuels are compared using an ignition delay criterion and classified by cetane number. The cetane number is measured using a standard CFR prechamber engine that features a variable compression ratio. It is operated under the standard operating conditions shown in Table 9-4. The compression ratio is adjusted until the ignition delay is 13° with the test fuel. At that compression ratio, reference fuels are blended to again produce an ignition delay of 13°. Then

$$\text{Cetane No.} = \% \text{ hexadecane} + 0.15 \cdot (\% \text{ heptamethylnonane}) \qquad (9.6)$$

The name cetane is derived from the fact that many persons refer to hexadecane as *n*-cetane. Originally, the cetane scale assigned a value of zero to α-methylnaphthalene as a reference fuel. Later the poor reference fuel was changed to heptamethylnonane, probably because it is cheaper, and assigned a cetane number of 15, so that results obtained in the past were still valid. The octane number and the cetane number of a fuel are inversely correlated, as shown in Figure 9-18. Gasoline is a poor diesel fuel and vice versa.

A low cetane number fuel will mix more completely with the cylinder air before burning, so that the local equivalence ratio of the initial premixed burn will be less ($\phi \sim 3$) than the local equivalence ratio ($\phi \sim 4$) for a higher cetane number.

Table 9-4   Cetane Number Measurement Conditions

Inlet temperature	339 K
Jacket temperature	373 K
Speed	900 rpm
Injection timing	13° btdc

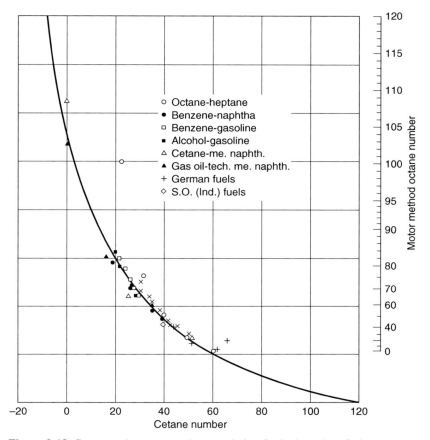

Figure 9-18 Cetane and octane number correlation for hydrocarbon fuels (Taylor, 1985). Copyright © 1985 MIT Press.

9.5   THERMODYNAMIC ANALYSIS

Spark Ignition Engines

Measurements of the cylinder pressure versus crank angle in spark ignition engines can be incorporated into a first law analysis to compute the mass fraction burned x_b, the ignition delay, and the combustion duration θ_d. If one assumes thermal equilibrium at each crank angle, a uniform mixture, and ideal gas behavior, the first law for a single zone (see Equation 8.26) is

$$\frac{dQ}{d\theta} = \frac{\gamma}{\gamma - 1} P \frac{dV}{d\theta} + \frac{1}{\gamma - 1} V \frac{dP}{d\theta} + \frac{dQ_w}{d\theta} \tag{9.7}$$

Equation 9.7 can be solved numerically to obtain the net heat release per unit crank angle $dQ/d\theta$. The mass fraction burned x_b is the normalized integral of the heat release

$$x_b(\theta) = \frac{\displaystyle\int_{\theta_o}^{\theta} \frac{dQ}{d\theta} \, d\theta}{\displaystyle\int_{\theta_s}^{\theta_e} \frac{dQ}{d\theta} \, d\theta} \tag{9.8}$$

For increased accuracy, the crevice flow and the variation of the specific heat ratio with composition, pressure, temperature, and equivalence ratio should be taken into account, as done in the heat release analysis of Cheung and Heywood (1993). The temperature dependence of the specific heat ratio is usually assumed to be linear:

$$\gamma = 1.40 - 7.18 \times 10^{-5} \, T \, (\text{K}) \tag{9.9}$$

The single zone analysis can be extended to two zones by assuming that the charge can be split into a burned and an unburned zone. The unburned zone include gas ahead of the flame and unburned gas within the flame. The burned zone includes gas behind the flame and burned gas within the flame. Thus, the highly convoluted flame structure observed via flow visualization is accounted for, and the analysis is limited in principle only by the assumption that the mass of gas actually reacting is small. In practice the analysis is limited further by imprecise estimates of the heat and mass loss as well as experimental error in the pressure measurement.

Typical results obtained by this method are shown in Figure 9-19. The engine used was a CFR engine operating at 1600 rpm. For conditions near stoichiometric, the ignition delay is approximately 10° and the combustion duration is approximately 40°. Consistent with observations made via flame photography, the ignition delay and combustion duration both increased as the mixture is leaned out from stoichiometric. For the particular engine and fuel in Figure 9-19, the lean flammability limit is near $\phi = 0.68$. Under these conditions, the ignition delay and combustion duration are quite long, about 20° and 70°after spark, respectively. That the mass-fraction-burned curves so deduced do not go to one and remain constant is caused by imprecision in the model and the experimental measurements. Results are usually deemed satisfactory if the curves level off at $x_b = 1.00 \pm 0.05$. To achieve this one has to account for blowby, heat loss, and incomplete combustion.

A model (Tabaczynski et al., 1980) has been developed to predict mass-fraction-burned curves from fundamental quantities such as the laminar flame speed of the fuel and the turbulence intensity of the flow. Key to the analysis are the "ink roller" assumptions that ignition sites are spread by turbulence and the laminar burnup of material between

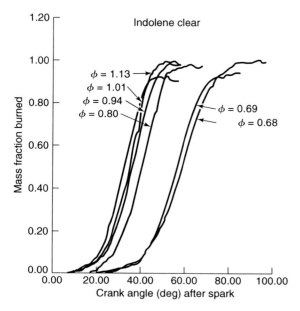

Figure 9-19 Mass fraction burned (x_b) versus crank angle degrees (θ) and equivalence ratio (ϕ) for indolene: CFR engine, 1600 rpm, $r = 6.9$, $T_i = 355$ K, imep $= 3.7$ bar (inlet pressure varies) (LoRusso, 1976).

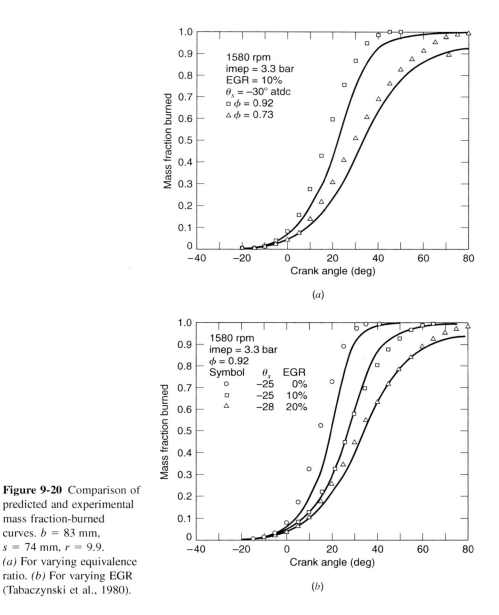

Figure 9-20 Comparison of predicted and experimental mass fraction-burned curves. $b = 83$ mm, $s = 74$ mm, $r = 9.9$. *(a)* For varying equivalence ratio. *(b)* For varying EGR (Tabaczynski et al., 1980).

shear layers occurs. Figure 9-20 shows comparisons of measured and predicted mass-fraction-burned curves for varying equivalence ratios and amounts of exhaust gas recirculation. The good agreement realized is also obtained for variations in spark advance, engine speed, and load (imep).

Two aspects of the mass-fraction-burned curve that are used to characterize the combustion are the ignition delay and the combustion duration. Figure 9-21 shows results obtained by Young (1980) for ignition delay, defined in this case as the angle between the time of spark firing and the time at which 1% of the mass fraction is burned. The ignition delay varies mainly with spark timing, residual fraction, and equivalence ratio. The ignition delay increases with spark advance because the laminar flame speed drops as a result of lower temperatures at the time of spark, but it is not the sole effect, for the

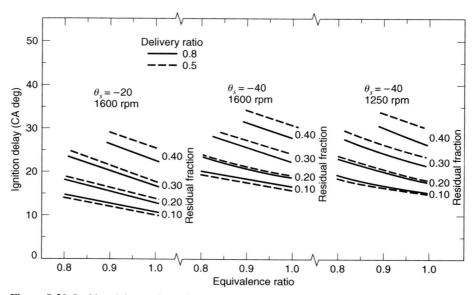

Figure 9-21 Ignition delay as determined by a regression analysis applied to three differently shaped combustion chambers on an automotive engine. Adapted from Young (1980).

turbulent field is also different. Likewise the ignition delay increases as the mixture is diluted either by leaning the charge or recirculating the exhaust. The change is proportionately less than the change in laminar flame speed. The combustion models indicate that this is because the combustion influences the turbulence as the flame grows.

The combustion duration results obtained by Young (1980) are shown in Figure 9-22. The combustion duration in this case is defined as the angle change from 1% to 90% burned fraction. Like ignition delay, the combustion duration depends on the equivalence ratio, the residual fraction, and the spark timing. However, it shows a stronger dependence on engine speed and delivery ratio. The combustion duration depends on the laminar flame speed, the turbulence intensity of the flow, and the combustion chamber geometry.

Minimizing the combustion duration in an engine requires a high turbulence intensity (which is often achieved at the expense of volumetric efficiency), a flame area that increases with distance from the spark plug, and a centrally located plug to minimize flame travel. As one expects, minimizing the combustion duration maximizes the work done, since the combustion approaches constant volume, and it also lowers the octane required. Figure 9-23 shows experimental results for three different combustion chamber shapes, that confirm these expectations.

For use in engine simulation programs, the mass-fraction-burned profile is fitted with a Wiebe function equation, Equation 2.22, restated here as Equation 9.10:

$$\frac{dQ}{d\theta} = na \frac{Q_{in}}{\theta_d} \left(\frac{\theta - \theta_s}{\theta_d} \right)^{n-1} \exp\left[-a\left(\frac{\theta - \theta_s}{\theta_d} \right)^n \right] \qquad (9.10)$$

Compression Ignition Engines

Diesel engine combustion analysis is performed using an energy balance analysis to determine either the effective heat release or the effective fuel injection rate. Typical calculations for a direct injection (DI) engine use Equation 9.7, which assumes homogeneous

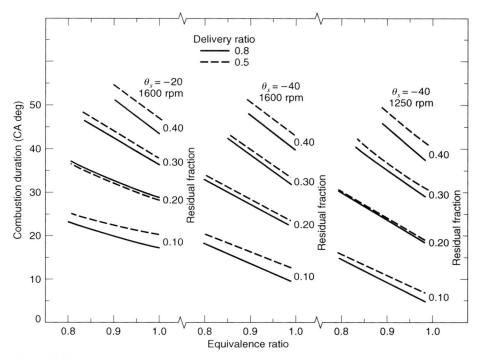

Figure 9-22 Combustion duration as determined by a regression analysis applied to a low squish, open-chamber automobile engine. Adapted from Young (1980).

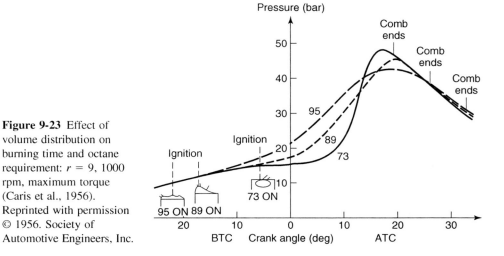

Figure 9-23 Effect of volume distribution on burning time and octane requirement: $r = 9$, 1000 rpm, maximum torque (Caris et al., 1956). Reprinted with permission © 1956. Society of Automotive Engineers, Inc.

conditions throughout the combustion chamber during the injection and combustion process, and ideal gas behavior. The heat release in indirect injection engines (IDI) is modeled with an energy balance applied to both the main chamber and the prechamber, so that pressure data is required for both chambers.

The effective net heat release (J/deg) for low and high load conditions for the direct injection diesel engine described in Section 9.4 is plotted in Figures 9-24 and 9-25 as a function of crank angle. Also plotted is the cylinder pressure and the injector needle lift.

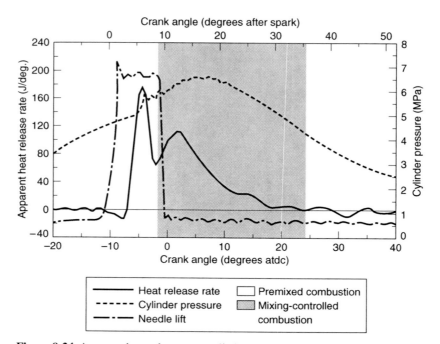

Figure 9-24 Apparent heat release rate, cylinder pressure, and injector needle lift for low fuel load ($\phi = 0.25$) (Dec, 1997). Reprinted with permission © 1997. Society of Automotive Engineers, Inc.

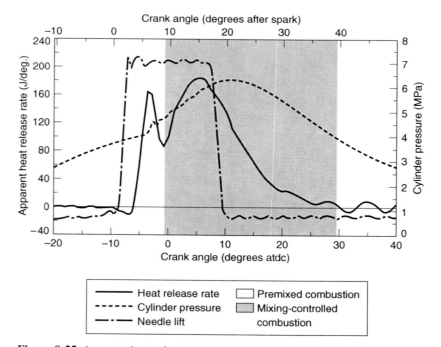

Figure 9-25 Apparent heat release rate, cylinder pressure, and injector needle lift for high fuel loading ($\phi = 0.43$) (Dec, 1997). Reprinted with permission © 1997. Society of Automotive Engineers, Inc.

The injection duration for the low load case is about 10 degrees, resulting in an overall equivalence ratio of 0.25. The injection duration for the high load case is about 20°, giving an overall equivalence ratio of 0.43.

The double peak shape of the heat release profile in Figures 9-24 and 9-25 is characteristic of diesel combustion. The first peak occurs during the premixed combustion phase and results from the rapid combustion of the portion of the injected fuel that has vaporized and mixed with the air during the ignition delay period. Note that the heat release curve in the premixed combustion phase is relatively independent of the load, since the initial mixing is independent of the injection duration. The second peak occurs during the mixing controlled combustion. The heat release during this phase depends on the injection duration. As the injection duration is increased, the amount of fuel injected increases, thus increasing the magnitude and duration of the mixing controlled heat release.

A dual Wiebe function (see Figure 9-26), which has two peaks, has been used to fit diesel combustion heat release data (Miyamoto et al., 1985). The dual equation, with seven parameters, is

$$\frac{dQ}{d\theta} = a\left(\frac{Q_p}{\theta_p}\right)(m_p)\left(\frac{\theta}{\theta_p}\right)^{m_p-1} \exp\left[-a\left(\frac{\theta}{\theta_p}\right)^{m_p}\right]$$
$$+ a\left(\frac{Q_d}{\theta_d}\right)(m_d)\left(\frac{\theta}{\theta_d}\right)^{m_d-1} \exp\left[-a\left(\frac{\theta}{\theta_d}\right)^{m_d}\right]$$

(9.11)

The subscripts p and d refer to the premixed and mixing controlled combustion portions, respectively. The parameter a is a nondimensional constant, θ_p and θ_d are the burning duration for each phase, Q_p and Q_d are the integrated energy release for each phase, and m_p and m_d are nondimensional shape factors for each phase. The adjustable parameters are selected using a least squares fit.

Miyamoto et al. (1985), for the specific direct (DI) and indirect injection (IDI) diesel engines tested in their experiments, reported that the m_p, m_d, and θ_p parameters were essentially independent of engine speed, load, and injection timing. The fitted values of these parameters were: $a = 6.9$, $m_p = 4$, $m_d = 1.5$ (DI) or 1.9 (IDI), and $\theta_p = +7°$.

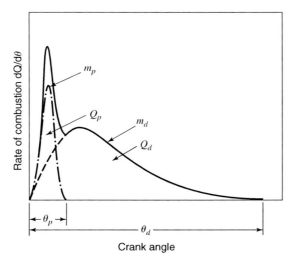

Crank angle

Figure 9-26 Dual Wiebe function for diesel heat release (Miyamoto et al., 1985). Reprinted with permission © 1985. Society of Automotive Engineers, Inc.

The effective diesel fuel injection rate is also obtained using the energy equation. The effective fuel injection rate is based on the assumptions that the chamber mixture is homogeneous and in thermodynamic equilibrium. Therefore the different liquid and vapor fuel fractions are not included at this level of modeling. The open system first law for the combustion chamber, with the injected fuel now explicitly included is

$$-\dot{Q}_l - P\dot{V} = \frac{d}{dt}(mu) - \dot{m}_f h_f \tag{9.12}$$

and the mass conservation equation is

$$\frac{dm}{dt} = \dot{m} - \dot{m}_f \tag{9.13}$$

In Equations 9.12 and 9.13, $\dot{m}_f$ is the fuel injection rate, h_f is the enthalpy of the injected fuel, and $\dot{Q}_l$ is the heat loss rate. With the above assumptions, the ideal gas equation in differential form is

$$P\dot{V} + V\dot{P} = RT\dot{m} + Rm\dot{T} \tag{9.14}$$

If dissociation is neglected, the internal energy is a function of temperature and equivalence ratio only.

$$u = u(T, \phi) \tag{9.15}$$

Differentiation of Equation 9.15 with respect to time gives

$$\dot{u} = \frac{\partial u}{\partial T}\dot{T} + \frac{\partial u}{\partial \phi}\dot{\phi} \tag{9.16}$$

If the mass of air in the cylinder is constant, with no residual fuel in the chamber at the beginning of injection, the overall equivalence ratio increases solely due to the fuel injection, and in differential form is

$$\dot{\phi} = \phi \frac{\dot{m}_f}{m_f} \tag{9.17}$$

Finally, combining Equations 9.12 through 9.17 leads to

$$\dot{m}_f = \frac{-\dot{Q}_l - \left(1 + \frac{c_v}{R}\right)P\dot{V} - \frac{c_v}{R}V\dot{P}}{u - h_f - c_v T + \frac{m}{m_f}\frac{\partial u}{\partial \phi}} \tag{9.18}$$

Equations 9.14, 9.17, and 9.18 are a set of ordinary differential equations that when numerically integrated using measured values for P, $\dot{P}$, V, and $\dot{V}$ yield T, $\dot{T}$, ϕ, $\dot{\phi}$, m_f, and $\dot{m}_f$ as functions of time. At each time step, an equilibrium combustion product numerical routine gives the required partial derivatives of the internal energy. The heat loss $\dot{Q}_1$ is computed at each time step from an appropriate model.

Results obtained by Kreiger and Borman (1966) for a heat release computation which includes the relatively small effects of dissociation are given in Figure 9-27. The cylinder pressure, cylinder pressure gradient, and effective fuel injection rate (mg/deg) are plotted as a function of crank angle. The effective fuel injection rate curve is double peaked,

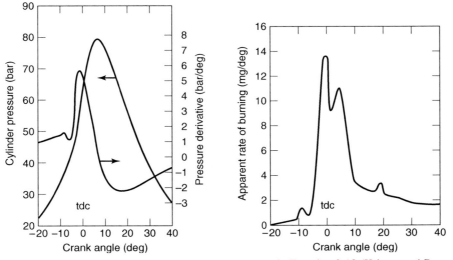

Figure 9-27 Engine data and the deduced burning rate via Equation 9-18 (Krieger and Borman, 1966). Reprinted with permission © 1966. Society of Automotive Engineers, Inc.

similar to the effective heat release rate. The area under the curve is approximately equal to the actual mass of fuel injected.

9.6   NITROGEN OXIDES

Nitrogen oxides (NO_x) are formed throughout the combustion chamber during the combustion process due to the reaction of atomic oxygen and nitrogen. The reactions forming NO_x are very temperature dependent, so the NO_x emissions from an engine scale proportionally to the engine load, and NO_x emissions are relatively low during engine start and warmup. In spark ignition engines, the dominant component of NO_x is nitric oxide, NO, as the concentration of nitric dioxide, NO_2, is of the order of only 1% to 2%.

Three reaction mechanisms that produce NO are the thermal or Zeldovich mechanism, the Fenimore or prompt mechanism, and the N_2O intermediate mechanism. For internal combustion engines, the most significant is the Zeldovich mechanism (Zeldovich, 1946) in which NO is formed in the high temperature burned gases left behind by the flame front. The prompt mechanism occurs within the flame front, and is relatively small if the volume of the high temperature burned gases is much larger than the instantaneous volume of the flame front, as is usually the case in internal combustion engines.

The following three chemical equations form the extended Zeldovich reaction (Miller and Bowman, 1989):

$$O + N_2 \rightleftharpoons NO + N \tag{9.19}$$

$$N + O_2 \rightleftharpoons NO + O \tag{9.20}$$

$$N + OH \rightleftharpoons NO + H \tag{9.21}$$

The first reaction, Equation 9.19, is a nitrogen dissociation reaction triggered by an oxygen atom. This reaction is endothermic with activation energy of +75.0 kcal, so that it is the controlling reaction. In the second reaction, Equation 9.20, a nitrogen atom reacts exothermically ($+31.8$ kcal) with an oxygen molecule to form nitric oxide and an oxygen atom. The third reaction, Equation 9.21, is an exothermic ($+49.4$ kcal) reaction between a nitrogen atom and a hydroxide radical which forms nitric oxide and a hydrogen atom.

The forward rate constants for these reactions are as follows (Borman and Ragland, 1998):

$$k_1 = 1.8 \times 10^{11} \exp(-38,370/T)$$
$$k_2 = 1.8 \times 10^7 \exp(-4,680/T) \tag{9.22}$$
$$k_3 = 7.1 \times 10^{10} \exp(-450/T)$$

where the units are $m^3/kmol$-s.

Using the chemical reactions given, one can write the following expression for the rate of change of nitric oxide concentration:

$$\frac{d}{dt}[NO] = +k_1[O][N_2] - k_{1r}[NO][N] + k_2[N][O_2]$$
$$- k_{2r}[NO][O] + k_3[N][OH] - k_{3r}[NO][H] \tag{9.23}$$

where the brackets denote concentrations in units of molecules/m^3 and the additional subscript r on the rate constants denotes the reverse rate constant.

To apply Equation 9.23 two approximations are introduced.

- The C-O-H system is in equilibrium and is not perturbed by N_2 dissociation.
- N atoms change concentration by a quasi-steady process.

The first approximation means simply that given the pressure, temperature, equivalence ratio, and residual fraction of a fluid element, one simply computes the equilibrium composition to determine the concentrations of N_2, O_2, O, OH, and H. The second approximation means that one can solve for the N atom concentration by setting the rate of change of N atoms to zero:

$$\frac{d}{dt}[N] = +k_1[N_2][O] - k_{1r}[N][NO] - k_2[N][O_2]$$
$$+ k_2[NO][O] - k_3[N][OH] + k_{3r}[NO][H] \tag{9.24}$$
$$= 0$$

With these two approximations it can be shown that

$$\frac{dx_{NO}}{dt} = \frac{60}{\rho}(1 - \alpha^2)\frac{R_1}{1 + \alpha R_1/(R_2 + R_3)} \tag{9.25}$$

where α is the ratio of the nitric oxide mass fraction to its equilibrium value

$$\alpha = \frac{x_{NO}}{x_{NO,e}} \tag{9.26}$$

and R_i ($i = 1, 2, 3$) is a forward rate of reaction at equilibrium, labeled with the subscript e.

$$R_1 = k_1[O]_e[N_2]_e$$
$$R_2 = k_2[N]_e[O_2]_e \tag{9.27}$$
$$R_3 = k_3[N]_e[OH]_e$$

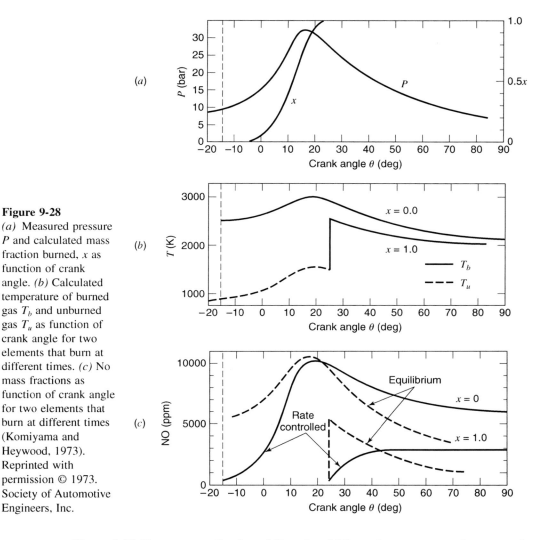

Figure 9-28
(a) Measured pressure *P* and calculated mass fraction burned, *x* as function of crank angle. *(b)* Calculated temperature of burned gas T_b and unburned gas T_u as function of crank angle for two elements that burn at different times. *(c)* No mass fractions as function of crank angle for two elements that burn at different times (Komiyama and Heywood, 1973). Reprinted with permission © 1973. Society of Automotive Engineers, Inc.

Figure 9-28 illustrates application of Equation 9.25 to a homogeneous-charge, spark ignition engine. The curves in *a* show the cylinder pressure and mass fraction burned at different crank angles. The curves in *b* show temperature-time histories (time is proportional to crank angle) of the first element to burn ($x = 0$) and the last element to burn ($x = 1$). The computation assumes that fluid elements retain their identity, which is to say no mixing among the fluid elements occurs.

Each element burns to its adiabatic flame temperature based on the unburned gas temperature at the time it burned. Once burned, an element's temperature tracks the pressure, as it is more or less isentropically compressed or expanded. Notice that the first element to burn is compressed considerably; each subsequent element to burn is compressed less, and the last element to burn undergoes no compression. As a result the first element to burn is hotter than all the rest, and the last element to burn is the coolest.

The curves in *c* illustrate how the nitric oxides vary with time in the different fluid elements. The dashed curves correspond to the equilibrium concentration based on the local temperature, pressure, equivalence ratio, and residual mass fraction. The solid curves are computed by integrating Equation 9.25.

At the time an element burns, its nitric oxide concentration is close to zero but finite because of the residual gas present. Since the chemistry is not fast enough to assume the process is quasi-static, it is rate controlled. Once the element is burned, the calculated equilibrium concentration is high, whereas the actual concentration is low. Notice that each element tries to equilibrate; if the equilibrium concentration is higher than the actual concentration, then nitric oxides are forming, whereas they decompose if the equilibrium concentration is less than the actual concentration.

The chemical reaction rates increase strongly with temperature. As a result there are large differences between the nitric oxide concentrations in the first and last elements. Furthermore, it can be seen that when the temperatures drop to about 2000 K, the decomposition rate becomes very slow and for practical purposes it may be said that the nitric oxides freeze at a concentration greater than the equilibrium values. The total amount of nitric oxide that appears in the exhaust is computed by summing the frozen mass fractions for all the fluid elements.

$$\bar{x}_{NO} = \int_0^1 x_{NO}\, dx \tag{9.28}$$

Comparisons between some measured exhaust concentrations with predictions made using the procedure described are given in Figure 9-29. The figure shows the agreement is quite good. It also shows a result typical of all engines, nitric oxides are maximized with mixtures slightly lean of stoichiometric. Recall that increased temperatures favor nitric oxide formation and that burned gas temperatures are maximized with mixtures that are slightly rich. On the other hand, there is little excess oxygen in rich mixtures to dissociate and attach to nitrogen atoms to form nitric oxide. The interplay between these two effects results in maximum nitric oxides occurring in slightly lean mixtures, where there is a slight excess of oxygen atoms to react with the nitrogen atoms.

At this point it is useful to discuss the mixing of the burned gases. Fluid elements are mixed with one another via turbulence. If the rate of mixing is faster than the rate at which burned gas is produced, then the burned gas can be assumed to be homogeneous and characterized by a single temperature. If the mixing is slow, then the burned gas must

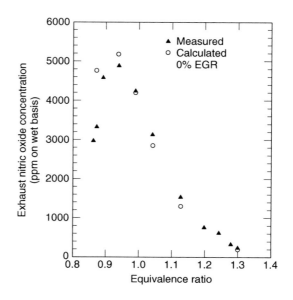

Figure 9-29 Measured and calculated exhaust NO concentration as function of equivalence ratio with no EGR (Komiyama and Heywood, 1973). Reprinted with permission © 1973. Society of Automotive Engineers, Inc.

be treated as an ensemble of fluid elements at different temperatures. Experimentally it is observed that there are different temperature fluid elements in the burned gases but that the differences are smaller than predicted, as in Figure 9-28. Thus it can be stated that mixing occurs, but it is not complete during the combustion.

It can be shown using the analyses in Chapter 4 that the energy of the burned gas is a nearly linear function of temperature (i.e., the specific heat varies little over the range of temperatures encountered in the burned gas), so that for computing cylinder pressure, the overall average gas temperature can be used. The same cannot be said of nitric oxides since the chemical rates are nonlinear functions of temperature. Since the computations done for Figure 9-28 did not admit to any mixing and ignored the temperature gradients due to wall boundary layers, the good agreement shown is somewhat fortuitous. The state of the art requires one to account for these effects to realize good agreement under all circumstances. With diesel engines, one has to further account for variations in equivalence ratio from fluid element to fluid element.

In conclusion, some experimental results are presented to illustrate how nitric oxides in the exhaust depend on various engine parameters. The trends, although typical, are by no means universal, especially for diesel engines. Figures 9-30 to 9-33 for homogeneous-charge, spark ignition engines lead to the following observations:

- The dependence on spark timing and inlet pressure is strong for lean mixtures and weak for rich mixtures.
- Nitric oxides are maximum for slightly lean mixtures.
- The dependence on engine speed cannot be stated simply; factors to consider are the variation in the combustion duration and heat loss with engine speed.
- Increased coolant temperature or the presence of deposits each reduce heat loss and increase the nitric oxides.

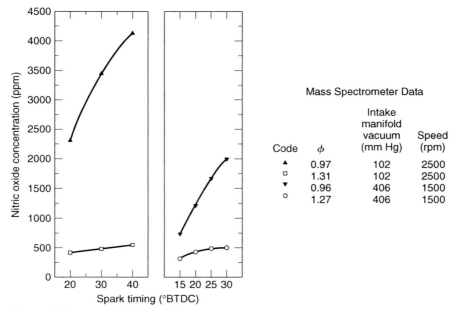

Figure 9-30 Advancing timing increases NO (Huls and Nickol, 1967). Reprinted with permission © 1967. Society of Automotive Engineers, Inc.

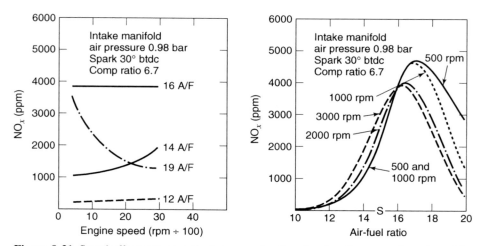

Figure 9-31 Speed effects are complicated as combustion duration (in terms of crank angle) increases slightly with speed and heat loss per unit mass decreases slightly with speed (Nebel and Jackson, 1958). Reprinted with permission © 1958. Society of Automotive Engineers, Inc.

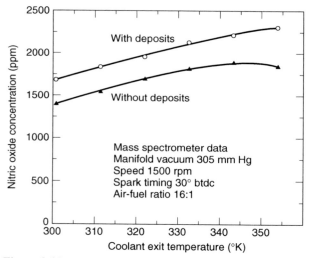

Figure 9-32 Increased coolant temperature or presence of deposits reduce heat losses, both of which increase NO (Huls and Nickol, 1967). Reprinted with permission © 1967. Society of Automotive Engineers, Inc.

- Dilution of the charge by residual gas, explicitly via exhaust gas recirculation, implicitly via throttling, or by moisture in the inlet air, reduces the nitric oxides.

The last four observations just cited also apply to diesel engines. Figure 9-34 illustrates the effects of injection timing and fuel-air equivalence ratio. The results show

- The nitric oxides are a strong function of injection timing.
- For direct injection engine (DI), nitric oxides increase with load (bmep).

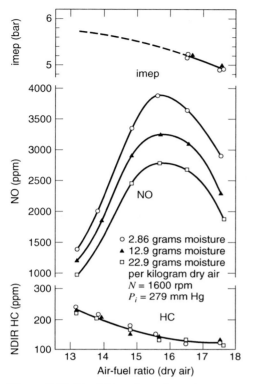

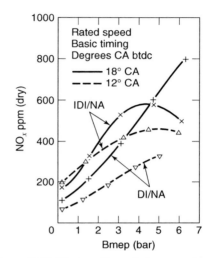

Figure 9-33 Humidity reduces burned gas temperatures and consequently NO emissions (Robinson, 1970). Reprinted with permission © 1970. Society of Automotive Engineers, Inc.

Figure 9-34 Nitric oxide emissions as a function of load for a naturally aspirated direct-injection and an indirect-injection engine (Pischinger and Cartellieri, 1972). Reprinted with permission © 1972. Society of Automotive Engineers, Inc.

- For indirect injection engines (IDI), nitric oxides are maximum at loads slightly less than full.

9.7   CARBON MONOXIDE

Carbon monoxide appears in the exhaust of rich-running engines since there is insufficient oxygen to convert all the carbon in the fuel to carbon dioxide. The most important engine parameter influencing carbon monoxide emissions is the fuel-air equivalence ratio. All other variables cause second order effects. Thus, results obtained when varying fuel-air ratio are more or less universal. Typical results are shown in Figure 9-35. Notice that at near stoichiometric conditions carbon monoxide emission is a highly nonlinear function of equivalence ratio. Under these circumstances, in multicylinder engines, it becomes important to ensure that the same fuel-air ratio is delivered to each cylinder. If half the cylinders run lean and the other half run rich, then the lean cylinder produces much less CO than the rich cylinders. The average CO emission of such an engine would correspond not to the average equivalence ratio but to an equivalence ratio richer than average, producing more CO than is necessary.

The carbon monoxide emissions from engines with rich mixtures could be predicted from the combustion product coefficients listed in Table 3-4 if the exhaust gases were in complete thermochemical equilibrium. It has already been mentioned that the C-O-H

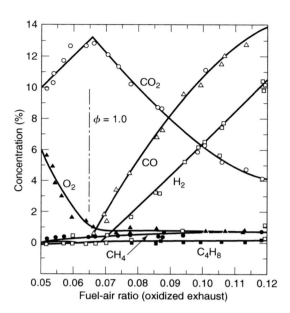

Figure 9-35 Exhaust gas composition versus oxidized or measured fuel–air ratio for supercharged engine with valve overlap; fuel C_8H_{18} (Gerrish and Meem, 1943).

system is more or less in equilibrium during combustion and expansion to the point where the nitric oxide chemistry freezes. Thus, whether it is a lean- or rich-running engine, one can compute the carbon monoxide concentration during these times using equilibrium assumptions.

Late in the expansion stroke, with temperatures down to around 1800 K, the chemistry in C-O-H systems starts to become rate limited and is generally frozen by the time blowdown finishes. The rate controlling reactions in the C-O-H systems are three-body recombination reactions such as

$$H + H + M \rightleftharpoons H_2 + M \tag{9.29}$$

$$H + OH + M \rightleftharpoons H_2O + M \tag{9.30}$$

$$H + O_2 + M \rightleftharpoons HO_2 + M \tag{9.31}$$

In addition, the CO destruction reaction is

$$CO + OH \rightleftharpoons CO_2 + H \tag{9.32}$$

Results obtained by using an unmixed model for the burned gas and accounting for these rate limiting reactions are illustrated in Figure 9-36. In these plots x is the fraction of the total charge burned when an element is burned and z is the mass fraction that has left the cylinder at the time an element leaves the cylinder. Because that time is unknown, results are given for several values of z for each element. Gas that leaves early ($z \ll 1$) cools more rapidly than gas that leaves last ($z \sim 1$). The results show that gases that burn early carry more CO into the exhaust than gases that burn later. They also show the fortuitous fact that the frozen concentrations are close to the equilibrium concentrations that exist in the cylinder at the time the exhaust valve opens. This suggests an approximation that is often used in practice, to assume that the C-O-H system is in equilibrium until the exhaust valve opens at which time it freezes instantaneously.

In lean-running engines there appears to be an additional source of CO caused by the flame-fuel interaction with the walls, the oil films, and the deposits. Under these circumstances

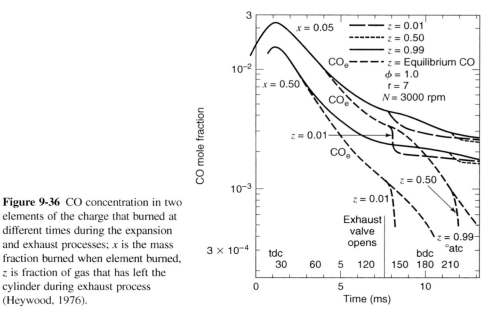

Figure 9-36 CO concentration in two elements of the charge that burned at different times during the expansion and exhaust processes; x is the mass fraction burned when element burned, z is fraction of gas that has left the cylinder during exhaust process (Heywood, 1976).

the exhaust concentrations are so low they are not a practical problem and thus details of these interactions remain largely unexplored.

Thus, the key to minimizing CO emissions is to minimize the times the engine must run rich (such as during start-up). Finally, since diesel engines run lean overall, their emissions of carbon monoxide are low and generally not considered a problem. It does appear that direct-injection diesel engines emit relatively more CO than indirect-injection diesel engines.

9.8 HYDROCARBONS

Spark Ignition Engines

Hydrocarbon emissions result from the presence of unburned fuel in the exhaust of an engine. Hydrocarbon fuels are composed of 10 to 20 major species and some 100 to 200 minor species. Most of these same species are found in the exhaust. However, some of the exhaust hydrocarbons are not found in the parent fuel, but are hydrocarbons derived from the fuel whose structure was altered within the cylinder by chemical reactions that did not go to completion. These are about 50% of the total hydrocarbons emitted. These partial reaction products include acetaldehyde, formaldehyde, 1,3 butadiene, and benzene, which are classified by the U.S. Environmental Protection Agency as toxic emissions. Hydrocarbon emissions from engines have been grouped into a number of classifications for regulatory purposes. Two classifications that are widely used are Total Hydrocarbons (THC) and Non-Methane Hydrocarbons (NMHC).

About 9% of the fuel supplied to an engine is not burned during the normal combustion phase of the expansion stroke. There are additional pathways that consume 7% of the hydrocarbons during the other three strokes of the four-stroke spark ignition engine, so that about 2% will go out with the exhaust (Cheng et al., 1993). As a consequence, hydrocarbon emissions cause a decrease in engine thermal efficiency, as well as being an

Table 9-5 Hydrocarbon Emission Sources (Cheng et al., 1993)

Source	% Fuel escaping normal combustion	% HC emissions
Crevices	5.2	38
Oil layers	1.0	16
Deposits	1.0	16
Liquid fuel	1.2	20
Flame quench	0.5	5
Exhaust valve leakage	0.1	5
Total	9.0	100

air pollutant. Hydrocarbon emissions are greatest during engine start and warm-up, due to decreased fuel vaporization and oxidation.

As listed in Table 9-5, six principal mechanisms are believed to be responsible for the alternative oxidation pathways and the exhaust hydrocarbons appearing: (1) crevices, (2) oil layers, (3) carbon deposits, (4) liquid fuel, (5) cylinder wall flame quenching, and (6) exhaust valve leakage. The crevice mechanism is the most significant, responsible for about 40% of the hydrocarbon emissions.

Crevices are narrow regions in the combustion chamber into which a flame cannot propagate. When a flame tries to propagate into a narrow channel, it may or may not be successful depending upon the size of the channel and a characteristic of fuel-air mixtures called the quenching distance. Crevices have a characteristic size less than the quenching distance. They occur around the piston, head gasket, spark plug, and valve seats, and represent about 1 to 2% of the clearance volume. The largest crevice is the piston ring–liner crevice region. During compression and the early stages of combustion, the cylinder pressure rises, forcing a small fraction of the fuel-air mixture into the crevices. The crevice temperatures are approximately equal to the cooled wall temperatures, so the density of the fuel-air mixture in the crevices is greater than in the cylinder. When the cylinder pressure decreases during the latter portion of the expansion stroke, the unburned crevice gases will flow back into the cylinder.

Wentworth (1971) was one of the first to recognize the importance of the crevice volume around the piston. He designed a special ring package to eliminate it, shown in Figure 9-37. So far it has found application only in research engines where it allows one to make crevice hydrocarbons a negligible source. This allows study of the effects of engine variables on the remaining sources. For example, Wentworth has shown that hydrocarbon from the remaining sources are strongly dependent on the wall temperature.

Oil layers within an engine can also trap some of the fuel and later release it during expansion. Kaiser et al. (1982) added oil to the engine cylinder and found that the exhaust hydrocarbons increased in proportion to the amount of oil added when the engine was fueled on *iso*-octane. They verified that the increased emissions were unburned fuel and fuel oxidation species and not unburned oil and oil oxidation species. They also did experiments in which the engine was fueled with propane and found no increase in the exhaust hydrocarbons when oil was added. Since propane is not soluble in the oil, they concluded that the increase observed is caused by fuel having been absorbed into the oil layer during compression later being released into the cooling burned gas during the expansion stroke. Thus, one can conclude that hydrocarbon emis-

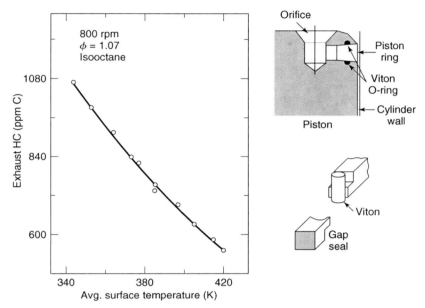

Figure 9-37 Exhaust hydrocarbons as a function of wall temperature: spark timing = 22° btc, intake pressure = 0.60 atm, and compression ratio = 8.51 (Wentworth, 1971). Reprinted with permission © 1971. Society of Automotive Engineers, Inc.

sions from engines will also depend on the amount of oil in the cylinder and the solubility of the fuel in the oil.

With continued use, carbon deposits build up on the valves, cylinder, and piston heads of internal combustion engines. The deposits are porous, and the sizes of the pores in the deposits are smaller than the quenching distance, and as a result the flame cannot burn the fuel-air-residual gas mixture compressed into the pores. This mixture comes out of the pores during expansion and blowdown. Although some of it will burn up when mixed with the hotter gases within the cylinder, eventually cylinder gas temperatures will have dropped to the point where those reactions fail to complete, resulting in hydrocarbons being emitted from the engine.

Fuel injection past an open valve into the cylinder, as can be the case with port fuel injection, allows the fuel to enter the cylinder in the form of liquid droplets. The less volatile fuel constituents may not vaporize, especially during engine start and warm-up, and be adsorbed in the crevices, oil layers, and carbon deposits.

Flame quenching along the surfaces is a relatively minor mechanism. Daniel (1957) showed that as the flame propagates toward the walls in an engine, it is extinguished at a small but finite distance away from the wall. Flame photographs revealed a dark region near the wall of thickness about one half the quench distance. More recent measurements (LoRusso et al., 1981) have shown that these hydrocarbons are subsequently oxidized with a high efficiency as they diffuse into the burned gases during expansion, and thus do not contribute significantly to the engine out hydrocarbon emissions.

Figure 9-38 is a schematic depicting how hydrocarbons are exhausted from the engine. At the end of combustion there are hydrocarbons all along the walls trapped in deposits, oil layers, or the crevice volume. During expansion hydrocarbons leave the crevice volume and are distributed along the cylinder wall. When the exhaust valve opens, the large rush of gas escaping drags with it some of the hydrocarbons released from oil layers

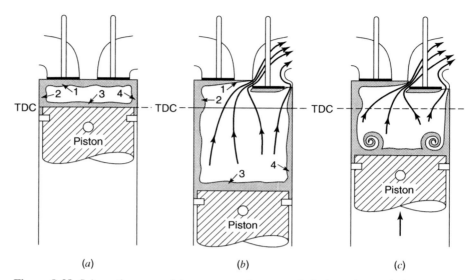

Figure 9-38 Schematic summarizing processes important in hydrocarbon emissions. (Tabaczynski et el., 1972). (Tabaczynski et al., 1972). Reprinted with permission © 1972. Society of Automotive Engineers, Inc.

and deposits. During the exhaust stroke, the piston rolls the crevice volume hydrocarbons that were distributed along the walls into a vortex that ultimately becomes large enough that a portion of it is exhausted.

Tabaczynski et al. (1972) have verified in a water analog experiment that the piston rolls the wall layer into a vortex. They have also measured the hydrocarbon emission mass flow rate as a function of time during the exhaust stroke. Their results, illustrated in Figure 9-39, show that roughly one half the hydrocarbons in their engine are exhausted during blowdown and one half are exhausted during the latter portion of the exhaust stroke. The concentration profile shown with a peak during blowdown and a sudden increase at about 290° (evidently the vortex starts to come out at 290°) is consistent with the process description given.

The hydrocarbon story does not end once the hydrocarbons leave the cylinder. There is considerable burn-up in the exhaust port. Some emission control techniques pump air into the exhaust manifold to further oxidize the hydrocarbons.

Two-stroke engines can produce a significant amount of hydrocarbon emissions. Short circuiting of the fuel-air mixture during the scavenging process is the major source of the hydrocarbon emissions. If crankcase compression is used, the unburnt lubrication oil is a source of hydrocarbons. As discussed in Casarella and Ghandhi (1998), direct injection is increasingly being used in two-stroke engines to eliminate the short circuiting of the fuel.

Compression Ignition Engines

Hydrocarbons from diesel engines come primarily from; (1) fuel trapped in the injector at the end of injection that later diffuses out, (2) fuel mixed into air surrounding the burning spray so lean that it cannot burn, and (3) fuel trapped along the walls by crevices, deposits, or oil due to impingement by the spray (Greeves et al., 1977; and Yu et al., 1980).

The diesel combustion process relies on mixing fuel and air at the time they are intended to burn. As already mentioned, a characteristic time is required for enough

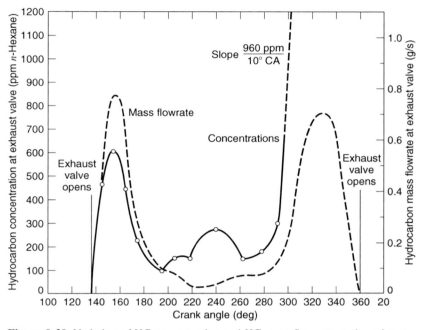

Figure 9-39 Variation of HC concentration and HC mass flow rate at the exhaust valve during the exhaust process (Tabaczynski et al., 1972). Reprinted with permission © 1972. Society of Automotive Engineers, Inc.

precursors to form in order for autoignition to occur. The characteristic time is a strong function of equivalence ratio and is a minimum near stoichiometric proportions of fuel and air. However, since the characteristic time is finite, a fuel-air pocket can be mixed to stoichiometric proportions and then diluted by more air before autoignition occurs in that element. As a result there are contours for lean equivalence ratios and when autoignition occurs there is fuel and air mixed locally to proportions less than the lean flammability limit. Thus this local fuel mixture does not burn and will increase the hydrocarbon emission levels.

There is also fuel mixed too rich to burn at the time of autoignition. However, it will burn later with additional mixing, provided the gases are hot enough. Some hydrocarbons are also produced because some of this fuel does not have a stoichiometric air-fuel ratio to burn until late in the expansion stroke.

Figure 9-40 shows hydrocarbon emissions from both a naturally aspirated direct injection engine and a naturally aspirated indirect injection engine. The results, although unique to the engines in question, illustrate that in general, direct injection engines emit more hydrocarbons than indirect injection engines.

Notice that for the direct injection engine, the hydrocarbons are worst at light load. Thus, hydrocarbon emissions at idle have been a focus of attention. Figure 9-41 shows how they can be influenced considerably by rather small changes in engine or injector geometry. Nozzles (A) and (B) were manufactured by different companies from the same specifications. This illustrates the need for precise control in manufacturing to achieve low emissions. The other variable shown is the piston bowl size, which had a negligible influence on the bsfc, a slight change in NO_x, and a dramatic change in the hydrocarbons.

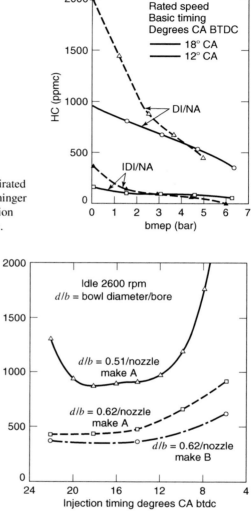

Figure 9-40 HC concentrations, naturally-aspirated direct-injection and prechamber engines (Pischinger and Cartellieri, 1972). Reprinted with permission © 1972. Society of Automotive Engineers, Inc.

Figure 9-41 Effect of combustion bowl diameter and make of nozzle (identical nozzle specification) on high-idle HC emissions (Pischinger and Cartellieri, 1972). Reprinted with permission © 1972. Society of Automotive Engineers, Inc.

9.9   PARTICULATES

A high concentration of particulate matter (PM) is manifested as visible smoke or soot in the exhaust gases. Particulate emissions from engines are regulated since inhalation of small particulate matter can create respiratory problems. Particulates are a major emissions problem for diesel engines, as their performance is smoke limited. With the use of unleaded fuel, particulates are generally not as serious a problem for spark ignition engines.

Particulates are any substance other than water that can be collected by filtering the exhaust. Specifically, the U. S. Environmental Protection Agency defines a particulate as any substance other than water that can be collected by filtering diluted exhaust at or below 325 K. The particulate material collected on a filter is generally classified into two components. One component is a solid carbon material or soot, and the other component is an organic fraction consisting of hydrocarbons and their partial oxidation products that have been condensed onto the filter or adsorbed to the soot. The organic fraction is influenced by the processes that dilute the exhaust with air upon expulsion from the engine. The methods used to measure particulate emissions such as dilution tunnels, light absorption, filter discoloration, and filter paper trapped mass, are discussed in Chapter 5.

Inspection of the soot fraction under an electron microscope reveals it to be agglomerates of spherical soot particles approximately 200 Å in diameter. The agglomerates can resemble a bunch of grapes in a more or less spherical configuration or be branched and chainlike in character. The characteristic dimensions of the agglomerates, on the order of 0.1 μm, pose a health hazard because they are too small to be trapped by the nose and large enough that some deposition in the lungs occurs.

Smoke forms in diesel engines because diesel combustion is heterogeneous. The diesel combustion model presented in Section 9.4 indicated that fuel rich combustion occurs both in the premixed and the mixing controlled combustion phases of combustion. Consider as a simple model the following two stage reaction path for the heterogeneous combustion of a hydrocarbon fuel:

First stage

$$\alpha CO + \frac{\beta}{2} H_2 + (2\chi - \alpha)O_2 \quad \text{clean} \qquad \chi \geq 2\alpha$$

$$C_\alpha H_\beta + \chi O_2 \nearrow \searrow \tag{9.33}$$

$$\frac{\chi}{2} CO + \left(\alpha - \frac{\chi}{2}\right)C(s) + \frac{\beta}{2} H_2 \quad \text{sooting} \quad \chi < 2\alpha$$

Second stage

$$CO + \tfrac{1}{2}O_2 \longrightarrow CO_2$$
$$C(s) + O_2 \longrightarrow CO_2 \tag{9.34}$$
$$H_2 + \tfrac{1}{2}O_2 \longrightarrow H_2O$$

According to this model, combustion takes place in two stages. If in the first stage there is not enough oxygen present to convert all the carbon in the fuel to carbon monoxide, that is, $\chi < 2\alpha$, then soot or solid carbon is produced. This is likely to occur locally within the fuel spray injected into the engine since it takes time for air and its attendant oxygen to be mixed in with the fuel. If there is enough oxygen present, that is, $\chi > 2\alpha$, then the flame is clean since no solid carbon is formed. The second stage burns the soot and other first stage products to completion in a diffusion flame.

The organic fraction results from all the processes that generate hydrocarbons and their partial oxidation products. During the dilution process some of them cool enough to condense or adsorb the soot. In addition, some species originating from the lubricating oil are found in the particulate and may be anywhere from 25% to 75% of the organic fraction (Mayer et al., 1980).

Particulate measurements obtained using a direct-injection diesel with a dilution tunnel are shown in Figure 9-42. The amount of particulates is extremely dependent on the equivalence ratio. As the equivalence ratio is doubled, the particulates increase by an order of magnitude. A mass spectrum of the organic fraction for particulates collected at three different engine speeds, $\phi = 0.5$, and a dilution ratio of 30 is given in Figure 9-43. Notice that the spectrum changes with speed particularly for molecular weights greater than 300. The mean molecular weight of the organic fraction is on the order of 200.

It is frustrating to diesel engine designers that, generally, when a reduction in nitric oxides has been achieved it is at the expense of an increase in smoke. Using the diesel combustion model of Section 9.4, this is due to the fact that a decrease in the temperature

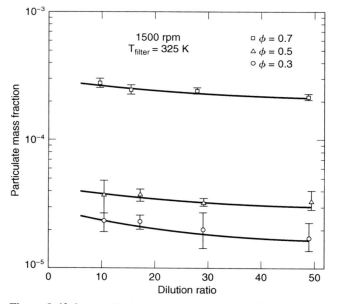

Figure 9-42 Log particulate mass fraction versus dilution ratio, 1500 rpm (Gillette and Ferguson, 1983).

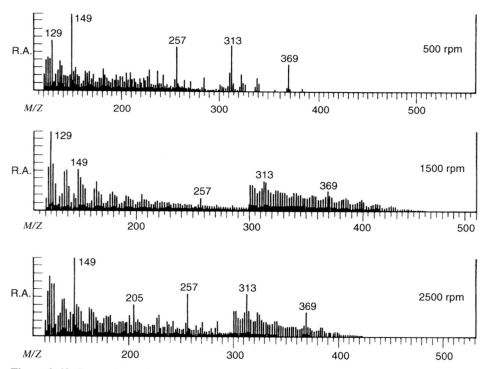

Figure 9-43 Comparison of the chemical ionization mass spectra of diesel particulates collected at three different engine speeds: 500 rpm, 1500 rpm, and 2500 rpm; $\phi = 0.5$. R.A. denotes relative abundance (Wood et al., 1982). Reprinted with permission © 1982. Society of Automotive Engineers, Inc.

of the diffusion flame will decrease the NO_x formation, but also decrease the amount of soot oxidized. Figures 9-44 and 9-45 show that as the timing is retarded, the NO_x decreases, but the particulates increase, creating a tradeoff between NO_x and smoke. One promising technique used to decrease smoke is to increase the in-cylinder turbulence during the late stages of combustion. This increase in turbulence can be accomplished through the use of auxiliary gas injection (Kurtz et al., 2000) or air cells (Mather and Reitz, 1995).

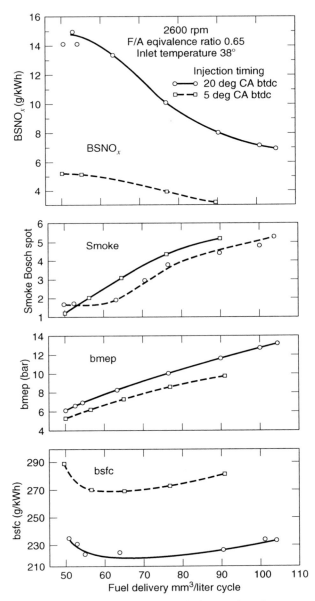

Figure 9-44 Effect of timing on emissions and performance, supercharged research diesel engine, at constant fuel–air ratio and constant air inlet temperature (Pischinger and Cartellieri, 1972). Reprinted with permission © 1972. Society of Automotive Engineers, Inc.

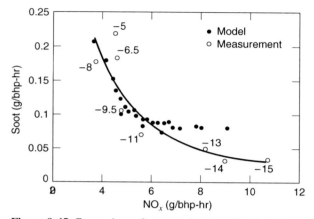

Figure 9-45 Comparison of measured and predicted engine-out NOx and soot as a function of injection timing with injections at 15, −13, −11, −8 and −5 degrees atdc. Solid symbols—measurements, open symbols—predicted (Rutland et al., 1994). Reprinted with permission © 1994. Society of Automotive Engineers, Inc.

9.10   EMISSION CONTROL

Since the mid 1960s, exhaust emissions from engines have been regulated by the U.S. Environmental Protection Agency. Prior to 1966, exhaust emissions from passenger cars were uncontrolled. In 1966, in response to air quality problems, California introduced hydrocarbon and CO emission limits. In 1968, the U. S. adopted nationwide emission regulations. During the intervening years, the requirements have become increasingly rigorous, and engines today are allowed to generate significantly less pollution than their 1968 counterparts. Meeting these emission requirements has been a major challenge and also an opportunity for automotive engineers.

The current and past emission requirements for vehicles are tabulated in Table 9-6, and the emission standards for certification as a low emission vehicle (LEV) or ultra-low emission vehicle (ULEV) are given in Table 9-7.

Note that the emission requirements have units of grams per mile. The current hydrocarbon emission limits have been reduced to 4%, carbon dioxide to 4%, and nitrogen oxides

Table 9-6   U.S. Passenger Car and Light Duty Truck Emission Standards

Year	Emissions (g/mile) HC	NO_x	CO
Pre-control	10.6	4.1	84.0
1968	4.1	–	34.0
1972	3.0	3.1	28.0
1975	1.5	3.1	15.0
1977	1.5	2.0	15.0
1980	0.41	2.0	7.0
1981	0.41	1.0	3.4
1993	0.25	1.0	3.4
1994	0.25	0.4	3.4

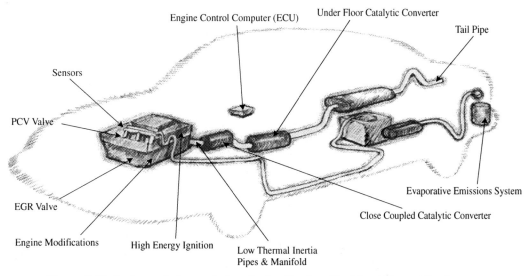

Engine Control Computer (ECU)   Under Floor Catalytic Converter

Tail Pipe

Sensors

PCV Valve

Evaporative Emissions System

EGR Valve

Close Coupled Catalytic Converter

Engine Modifications   High Energy Ignition   Low Thermal Inertia
Pipes & Manifold

Figure 9-46 Engine emission control methods. (Courtesy Englehard Corporation.)

to 10%, respectively, of the uncontrolled pre-1968 values. Internal combustion engines used in applications other than vehicles, for example, engines used in lawn mowers, snow blowers, chainsaws, pumps, and generators, are currently being regulated, since they also have been found to be significant sources of hydrocarbon and carbon monoxide pollution.

As shown in Figure 7-46, there are three basic methods used to control engine emissions:

- engineering of the combustion process,
- optimizing the choice of the operating parameters, and
- using after-treatment devices in the exhaust system.

Application of technological advances in fuel injectors, oxygen sensors, and on-board computers to engines has increased the control and subsequent optimization of the engine combustion process. Two NO_x control measures that have been used in automobile engines since the 1970s are spark retard and exhaust gas recirculation (EGR). The aim of these measures is to reduce the peak combustion temperature and thus the formation of NO_x. A review of the various emission control measures developed and used by automobile manufacturers since the 1960s is given in Mondt (2000).

As shown in Figure 9-28, retarding the spark timing lowers the NO_x since a greater fraction of the combustion occurs in an expanding volume, lowering the peak cylinder pressure and temperature. However, this also decreases the engine thermal efficiency. With the use of exhaust gas recirculation, some fraction of the exhaust gas is routed back into the intake

Table 9-7 Low Emission Vehicle (LEV) and Ultra-Low
Emission Vehicle (ULEV) Standards

| | Emissions (g/mile) | | |
	NMOG[a]	NO_x	CO
LEV	0.075	0.2	3.4
ULEV	0.040	0.2	1.7

[a]Non-methane organic gas.

manifold. The exhaust gas acts as a diluent in the fuel-air mixture, lowering the combustion temperature. The dilution by EGR of the mixture also reduces the combustion rate, so the spark timing is advanced to maintain optimal thermal efficiency. The EGR fraction increases with engine load up to the lean limit, which is about 15% to 20% of the fuel-air flow rate.

Currently, the most important after-treatment device is the three-way catalyst (Kummer, 1981), first installed on the exhaust systems in passenger cars in 1975. It derives its name from the fact that it works on all three of the gaseous pollutants of concern: nitric oxides, carbon monoxide, and hydrocarbons. The operation of the catalytic converter is severely inhibited by lead and sulfur compounds in the exhaust gases, so that vehicular fuels have been reformulated to reduce their lead and sulfur content.

All catalytic converters are built in a honeycomb or pellet geometry (see Figure 9-47) to expose the exhaust gases to a larger surface made of small particles (< 50 nm) of one or more of the noble metals, platinum (Pt), palladium (Pd), and rhodium (Rh). Rhodium

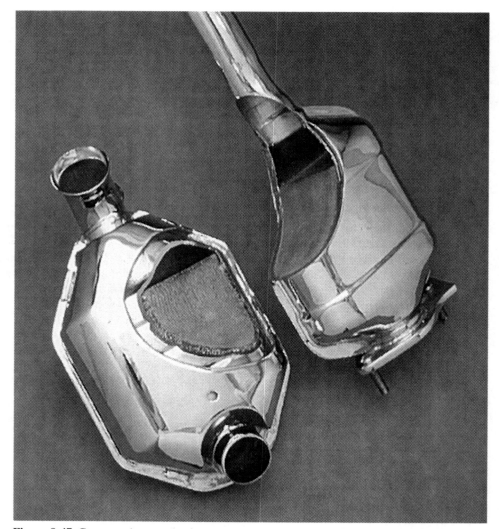

Figure 9-47 Cutaway photograph of a catalytic converter. (Courtesy Englehard Corporation.)

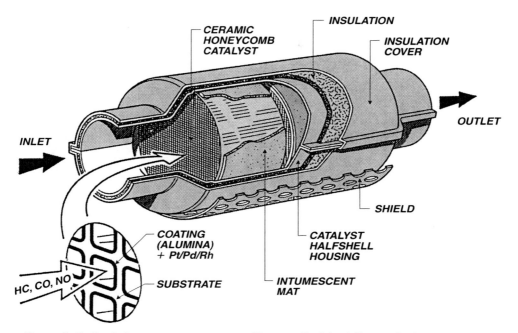

Figure 9-48 Catalytic converter components. (Courtesy Englehard Corporation.)

is the principal metal used to remove NO. Platinum is the principal metal used to remove HC and CO.

Figure 9-48 is a schematic of a three-way honeycomb catalyst. In the converter shown, a thin layer of the noble metals covers a washcoat of inert alumina Al_2O_3 on a cordierite honeycomb foundation. As the exhaust gases flow through the catalyst, the NO reacts with the CO, hydrocarbons, and H_2 via a reduction reaction on the surface of the catalyst. The remaining CO and hydrocarbons are removed through an oxidation reaction forming CO_2 and H_2O products. The oxidation rate of hydrocarbons increases with molecular weight, so that the oxidation of low molecular weight fuels such as methane is very slow in the converter.

A three-way catalyst will function correctly only if the exhaust gas composition corresponds to nearly $(\pm 1\%)$ stoichiometric combustion. If the exhaust is too lean, nitric oxides are not destroyed and if it is too rich, carbon monoxide and hydrocarbons are not destroyed (see Figure 9-49). Herein lies one constraint that emission control imposes upon engine operation; to use a three-way catalyst, the engine must operate in a narrow window about stoichiometric fuel-air ratios. As discussed in Chapter 5, a closed-loop control system with an oxygen sensor is used to determine the actual fuel-air ratio, and adjust the fuel injector so that the engine operates in a narrow range about the stoichiometric set point. Ordinary carburetors are not able to maintain the fuel-air ratio in such a narrow set point range.

Analysis of fuel-air cycles in Chapter 4 showed that lean operation was beneficial to the thermal efficiency of the engine and at first it appears that preclusion of lean operation is a rather severe constraint. However, if one realizes that the excess air in lean combustion is acting as a dilutant, then one can appreciate that exhaust gas recirculation can be used to achieve the same effect. The difference is that the excess air is reactive, and the exhaust is nonreactive. Indeed, the fuel-air cycle computations (see Figure 4-6) showed that efficiency increased with increasing residual fraction.

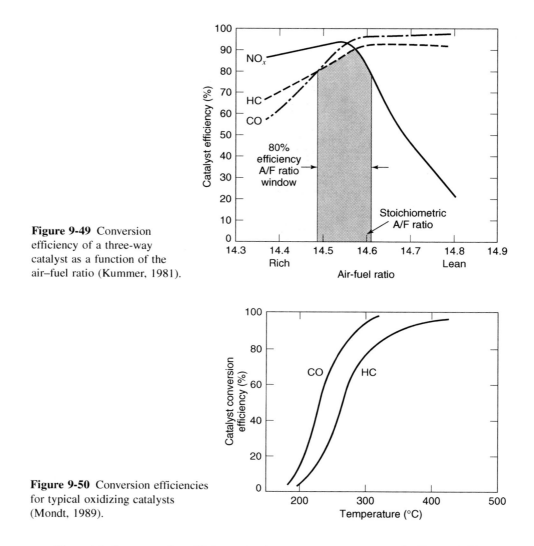

Figure 9-49 Conversion efficiency of a three-way catalyst as a function of the air–fuel ratio (Kummer, 1981).

Figure 9-50 Conversion efficiencies for typical oxidizing catalysts (Mondt, 1989).

The catalytic conversion efficiency is plotted versus temperature in Figure 9-50. The temperature at which a catalytic converter becomes 50% efficient is defined as the light-off temperature. The light-off temperature is about 270°C for the oxidation of HC and 220°C for the oxidation of CO. The conversion efficiency at fully warm conditions is about 98% to 99% for CO and 95% for HC, depending on the HC components. Various measures have been tried to decrease the converter warm up time, including use of an afterburner, locating the converter or an additional startup converter closer to the exhaust manifold, and electric heating, as discussed in Becker and Watson (1998).

With diesel engines, catalytic converters are used to oxidize the HC and CO, but reduction of the exhaust nitric oxides is poor because the engine runs lean. Thus, this pollutant has to be controlled by design of the combustion process and/or the choice of operating conditions. Figure 9-44 shows experimental results for a direct injection diesel engine. Notice that as the injection timing is retarded from 20° to 5° before top dead center, the nitric oxides drop by about a factor of 3, whereas the fuel consumption increases only about 15%. For this reason, diesel engines are usually operated at injection timings slightly retarded from that which produces best fuel economy.

9.11 REFERENCES

ABRAHAM. J., F. WILLIAMS, and F. BRACCO (1985), "A Discussion of Turbulent Flame Structure in Premixed Charges," SAE paper 850345.

AMSDEN, A, P. O'ROURKE, and T. BUTLER (1989), "KIVA-II—A Computer Program for Chemically Reactive Flows with Sprays," Los Alamos National Labs, LA-11560-MS.

BECKER, E. AND R. WATSON (1998), "Future Trends in Automotive Emission Control," SAE paper 980413.

BORMAN, G. and K. RAGLAND (1998), *Combustion Engineering,* McGraw-Hill, New York.

BOWDITCH, F. W. (1961), "A New Tool for Combustion Research: A Quartz Piston Engine," *SAE Trans.,* Vol. 69, p. 17.

CARIS, D., B. MITCHELL, A. McDUFFIE, and F. WYCZALEK (1956), "Mechanical Octanes for Higher Efficiency," *SAE Trans.,* Vol. 64, p. 76–100.

CASARELLA, M. and J. GHANDHI (1998), "Emission Formation Mechanisms in a Two Stroke Direct Injection Engine," SAE paper 982697.

CHENG, W., D. HAMRIN, J. HEYWOOD, S. HOCHGREB, K. MIN, and M. NORRIS (1993), "An Overview of Hydrocarbon Emissions Mechanisms in Spark Ignition Engines," SAE paper 932708.

CHEUNG, H. and J. HEYWOOD (1993), "Evaluation of a One Zone Burn Rate Analysis Procedure Using Production SI Engine Pressure Data," SAE paper 932749.

COOK, H., J. VANDEMAN, and J. LIVENGOOD (1944), "Effect of Several Methods of Increasing Knock-Limited Power on Cylinder Temperatures," NACA ARR E4115 E-36.

DANIEL, W. (1957), "Flame Quenching at the Walls of an Internal Combustion Engine," *Sixth Symposium (International) on Combustion,* p. 886, Reinhold, New York.

DEC, J. (1997), "A Conceptual Model of DI Diesel Combustion Based on Laser-Sheet Imaging," SAE paper 970873.

DEC, J. and C. ESPEY (1995), "Ignition and Early Soot Formation in a DI Diesel Engine Using Multiple 2-D Imaging Diagnostics," SAE paper 950456.

DOUAUD, A. and P. EYZAT (1977), "DIGITAP—An On-Line Acquisition and Processing System for Instantaneous Engine Data—Applications," SAE paper 770218.

DOUAUD, A. and P. EYZAT (1978), "Four-Octane-Number Method for Predicting the Anti-Knock Behavior of Fuels," SAE paper 780080.

ECKBRETH, A. (1988), *Laser Diagnostics for Combustion Temperature and Species,* Gordon and Breach, London, UK.

ESPEY, C. and J. DEC (1993), "Diesel Engine Combustion Studies in a Newly Designed Optical-Access Engine Using High-Speed Visualization and 2-D Laser Imaging," SAE paper 930971.

FLYNN, P., R. DURRETT, G. HUNTER, A. zur LOYE, O. AKINYEMI, J. DEC, and C. WESTBROOK (1999), "Diesel Combustion: An Integrated View Combining Laser Diagnostics, Chemical Kinetics, and Empirical Validation," SAE paper 1999-01-0509.

GERRISH, H. and J. MEEM (1943), "The Measurement of Fuel Air Ratio by Analysis of the Oxidized Exhaust Gas," NACA report 757.

GILLETTE, A. and C. R. FERGUSON (1983), "Measurement and Analysis of the Particulate Emission From a Direct Injection Diesel," *Particulate Sci. and Tech.,* Vol. 1, No. 1, p. 77-90.

GREENHAUGH, D. (1988), "Quantitative CARS Spectroscopy" in *Advances in Nonlinear Spectroscopy,* Vol. 15, p. 193 (Ed. R. Clark and R. Hester), John Wiley & Sons, New York.

GREEVES, G., I. KHAN, C. WANG, and I. FENNE (1977), "Origins of Hydrocarbons Emissions from Diesel Engines," SAE paper 770259.

HESSELBERG, H. and W. LOVELL (1951), "What Fuel Antiknock Quality Means in Engine Performance," *J. SAE, 59,* (April) p. 32.

HEYWOOD, J. (1976), "Pollutant Formation and Control in Spark Ignition Engines," *Prog. Energy Combust. Sci.,* Vol. 1, p. 135–164.

HO, S., D. AMLEE, and R. JOHNS (1996), "A Comprehensive Knock Model for Application in Gas Engines," SAE paper 961938.

HULS, T. and H. NICKOL (1967), "Influence of Engine Variables on Exhaust Oxides of Nitrogen Concentrations from a Multi-Cylinder Engine," SAE paper 670482.

KAISER, E., J. LoRusso, G. LAVOIE, and A. ADAMCZYK (1982), "The Effect of Oil Layers on the Hydrocarbon Emissions from Spark Ignited Engines," *Combust. Sci. and Tech.,* Vol. 28, p. 69–73.

KOMIYAMA, K. and J. HEYWOOD (1973), "Predicting NO, Emissions and Effects of Exhaust Gas Recirculation in Spark Ignition Engine," SAE paper 730475.

KONG, S., Z. HAN, and R. REITZ (1995), "The Development and Application of a Diesel Ignition and Combustion Model for Multidimensional Engine Simulation," SAE paper 950278.

KRIEGER, R. and G. BORMAN (1966), "The Computation of Apparent Heat Release for Internal Combustion Engines," ASME paper 66-WA-DGP-4.

KUMMER, J. (1981), "Catalysts for Automobile Emission Control," *Prog. Energy Combust. Sci.,* Vol. 6, p. 177–199.

KURTZ, E., D. FOSTER, and D. MATHER (2000), "Parameters that Affect the Impact of Auxiliary Gas Injection in a DI Diesel Engine," SAE Paper 2000-01-0233.

KUO, K. and T. PARR (1994), *Non-Intrusive Combustion Diagnostics,* Begell House, New York.

LoRusso, J. (1976), "Combustion and Emissions Characteristics of Methanol, Methanol-Water and Gasoline-Methanol Blends in a Spark Ignition Engine," MS thesis, MIT, Cambridge, Massachusetts.

LoRusso, J., E. KAISER, and G. LAVOIE (1981), "Quench Layer Contributions to Exhaust Hydrocarbons from a Spark Ignition Engine," *Combustion Sci. and Tech.,* 25, p. 121.

MATHER, D. and R. REITZ (1995)," Modeling the Use of Air-Injection for Emissions Reduction in a Direct-Injected Diesel Engine," SAE paper 952359.

MAYER, W., D. LECHMAN, and D. HILDENS (1980), "The Contribution of Engine Oil to Diesel Exhaust Particulate Emissions," SAE paper 800256.

METGHALCHI, M. and J. KECK (1982), "Burning Velocities of Mixtures of Air with Methanol, Isooctane and Indolene at High Pressure and Temperature," *Combustion and Flame,* Vol. 48, No. 2, p. 191–120.

MILLER, J. and C. BOWMAN (1989), "Mechanism and Modeling of Nitrogen Chemistry in Combustion," *Prog. Energy Combust. Sci.,* Vol. 15, p. 287–338.

MIYAMOTO, N., T. CHIKAHISA, T. MURAYAMA, and R. SAWYER (1985), "Description and Analysis of Diesel Engine Rate of Combustion and Performance Using Wiebe's Functions," SAE paper 850107.

MONDT, J. R. (1989), "A Historical Overview of Emission-Control Techniques for Spark Ignition Engines: Part b – Using Catalytic Converters," ASME ICE – Book No. 100294 – 1989.

MONDT, J. R. (2000), *Cleaner Cars: The History and Technology of Emission Control Since the 1960s,* SAE International, Warrendale, Pennsylvania.

NEBEL, G. and N. JACKSON (1958), "Some Factors Affecting the Concentration of Oxides of Nitrogen in Exhaust Gases from Spark Ignition Engines," *J. Am. Pollution Control Assoc.,* Vol. 8, No. 3, p. 213.

PISCHINGER, R. and W. CARTELLIERI (1972), "Combustion System Parameters and Their Effect Upon Diesel Engine Exhaust Emissions," SAE paper 720756.

RASSWEILER, G. and L. WITHROW (1938), "Motion Pictures of Engine Flames Correlated with Pressure Cards," A landmark reprint paper commemorating SAE's 75th Anniversary, SAE paper 800131.

ROBINSON, J. (1970), "Humidity Effects on Engine Nitric Oxide Emissions at Steady-State Conditions," SAE paper 700467.

RUTLAND, C., J. ECKHAUSE, G. HAMPSON, R. HESSEL, S. KONG, M. PATTERSON, D. PIERPONT, P. SWEETLAND, T. TOW, and R. REITZ (1994), "Toward Predictive Modeling of Diesel Engine Intake Flow, Combustion and Emissions," SAE paper 941897.

SIEBERS, D. (1998), "Liquid-Phase Fuel Penetration in Diesel Sprays," SAE paper 980809.

SMITH, J., R. GREEN, C. WESTBROOK, and W. PITZ (1984), "An Experimental and Modeling Study of Engine Knock," Twentieth Symposium (International) on Combustion, Combustion Institute, Pittsburgh, Pennsylvania.

TABACZYNSKI, R., J. HEYWOOD, and J. KECK (1972), "Time-Resolved Measurements of Hydrocarbon Mass Flowrate in the Exhaust of a Spark Ignition Engine," SAE paper 72112.

TABACZYNSKI, R., F. TRINKER, and B. SHANNON (1980), "Further Refinement and Validation of a Turbulent Flame Propagation Model for Spark Ignition Engines," *Combustion and Flame,* Vol. 39, No. 2, p. 111–122.

TAYLOR, C. F. (1985), *The Internal Combustion Engine in Theory and Practice,* Vol. 1, MIT Press, Cambridge, Massachusetts.

TURNS, S. (1996), *An Introduction to Combustion,* McGraw-Hill, New York.

WENTWORTH, J. (1971), "Effect of Combustion Chamber Surface Temperature on Exhaust Hydrocarbon Concentration," SAE paper 710587.

WESTBROOK, C. and F. DRYER (1980), "Prediction of Laminar Flame Properties of Methanol-Air Mixtures," *Combustion and Flame,* Vol. 37, No. 2, p. 171–192.

WESTBROOK, C. and F. DRYER (1984), "Chemical Kinetic Modeling of Hydrocarbon Combustion," *Prog. Energy Combust. Sci,* Vol. 10, p. 1–57.

WITZE, P. and F. VILCHIS (1981), "Stroboscopic Laser Shadowgraph Study of the Effect of Swirl on Homogeneous Combustion in a Spark Ignition Engine," SAE paper 810226.

WOOD, K., J. CIUPEK, R. COOKS, and C. R. FERGUSON (1982), "Characterization of Diesel Particulates by Mass Spectrometry Including MS-MS," SAE paper 821217.

YOUNG, M. (1980), "Cyclic Dispersion—Some Quantitative Cause and Effect Relationships," SAE paper 800459.

YU, R., V. WONG, and S. SHAHED (1980), "Sources of Hydrocarbon Emissions from Direct Injection Diesel Engines," SAE paper 800049.

ZELDOVICH, Y. (1946), "The Oxidation of Nitrogen in Combustion Explosions," *Acta Physicochimica USSR,* Vol. 21, p. 577–628.

9.12   HOMEWORK

9.1   For stoichiometric gasoline-air mixtures it has been found that

$$s_l = 0.2525 P^{-0.13}\left(\frac{T_u}{298}\right)^{2.19}(1 - 2.1f)$$

where s_l is in m/s, P is in bars, T_u is the unburned mixture temperature (K) and f is the residual mass fraction. Assume for a given engine that the pressure and temperature at the time of ignition are given by

$$T_{u,s} = 350\left(\frac{V_{bdc}}{V_s}\right)^{(\gamma-1)/\gamma} \quad \text{where } \gamma = 1.3$$

$$P_s = 0.5\left(\frac{V_{bdc}}{V_s}\right)^{\gamma}$$

Assume further that for the engine design $\Delta\theta_{id} = 25°$ for $r = 8$ and $\theta_s = -25°$. The residual fraction is given by $f = 0.10(8/r)$ and the kinematic viscosity is given by

$$\nu \approx \nu_{air} = \frac{9.47 \times 10^{-10} T_u^{1.7}}{P} \quad (m^2/s)$$

If the bore and stroke of the engine are 10 and 8 cm, respectively, using Equations 1-3 and 1-4 for relating cylinder volume to crank angle, compute the laminar flame speed at ignition as a function of compression ratio and spark timing. Assuming the delay is inversely proportional to the laminar flame speed at the time of ignition. Plot the ignition delay versus θ_s for $-50° < \theta_s < 0°$ and show lines of constant compression ratio for $r = 8$ and $r = 10$.

9.2   A combustion model predicted the following data for an engine operated at wide open throttle on isooctane:

θ (Deg atc)	x	P (bar)	T_u (K)	T_b (K)
-40	0.000	5.6	600	2500
-30	0.001	7.5	650	2572
-20	0.062	12.6	745	2652
-10	0.383	31.6	900	2700
0	0.843	57.8	950	2750
10	0.994	58.6	1000	2800
20	1.000	45.2	975	2747

If the precursor formation rate is (Douaud and Eyzat, 1978)

$$\frac{1}{x_c}\frac{dx_p}{dt} = 50.5 P^{1.7} \exp\left(\frac{-3800}{T}\right)$$

(a) Determine the minimum engine speed for which knock-free operation occurs assuming the table is speed independent. Plot the extent of reaction versus crank angle at that speed. Comment on assumptions implicit or explicit in the analysis.

(b) Assume that throttling the engine reduces all pressures 25% and all temperatures 5%, repeat part (a).

9.3 The existence of a temperature gradient in the burned gas can be explained fairly simply using an ideal gas model in which the fluid is broken into an ensemble of elements. The average pressures and specific volumes of an Otto cycle are represented in Figure 9-a by the diagram 1-2-3-4. All the gas is compressed isentropically from 1 to 2; hence at point 2 the gas is at a uniform temperature T_2.

The first element (infinitesimal) to burn will not influence the cylinder pressure and thus burns at constant pressure to 2'. Thus

$$T_{2'} = T_2 + \frac{q}{c_p}$$

where q is the heat release per unit mass. That gas is then compressed isentropically to

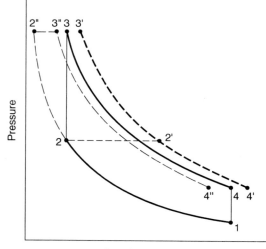

Figure 9-a Illustration for Homework Problem 9.3.

the peak pressure P_3, hence

$$\frac{T_{3'}}{T_{2'}} = \left(\frac{P_3}{P_2}\right)^{(\gamma-1)/\gamma}$$

The last element to burn is compressed isentropically as unburned gas to the peak pressure at $2''$.

$$\frac{T_{2''}}{T_2} = \left(\frac{P_3}{P_2}\right)^{(\gamma-1)/\gamma}$$

The last element then also burns at constant pressure so that

$$T_{3''} = T_{2''} + \frac{q}{c_p}$$

All the elements expand isentropically after the last element burns.

Taking as the average cycle the conditions used in Figure 2-2 of Chapter 2, find
(a) The ratio $T_{3'}/T_{3''}$
(b) The ratio $v_{4'}/v_{4''}$

9.4   Derive Equation 9.18 for the diesel fuel injection rate.

9.5   The rate of change of nitric oxide mass fraction for a fluid element because of chemical reaction is given by Equation 9.25. The mass fraction can also change because of NO convected in and out of the fluid element. Consider the control volume shown in Figure 9-b. Write an expression for the rate of change of nitric oxide mass fraction for this element assuming the fluid entering is devoid of nitric oxides, the fluid leaving has the same properties as fluid in the element, and the generation of NO within the control volume is given by Equation 9.25.

9.6   Emissions data are often presented as the mass flow rate of the pollutant emitted divided by the engine's power. Brake specific nitric oxides for a single cylinder research engine are given in Figure 9-44. These results indicate that to minimize the pollutant emission in grams per hour at a given power level one should operate at as high a bmep as possible. Since the experiments are done at constant fuel–air ratio, they illustrate an advantage of turbocharging as far as nitric oxides are concerned. Present the same data in the different forms found in practice:

- Emission index—Grams of nitric oxides per kilogram of fuel versus bmep.
- Mass fraction—Grams of nitric oxide per kilogram of exhaust versus bmep.
- Concentration—Ppm versus bmep (assume the molecular weight of exhaust is that of air).

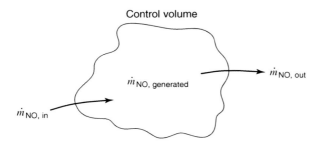

Figure 9-b Illustration for Homework Problem 9.5.

9.7   Use the Equilibrium Combustion Solver Applet to compute the exhaust CO concentration for an engine fueled with C_7H_{17}. Plot CO versus ϕ for two different gas temperatures at the time of exhaust valve opening, $T = 1800$ K and $T = 1500$ K.

9.8   Reaction of hydrocarbons in the exhaust port of an engine is an important process used in determining emissions from either a gasoline or a diesel engine. The rate of change of the mass fraction due to chemical reaction is given by:

$$-\frac{dx_{HC}}{dt} = A x_{HC} x_{O_2} \exp\left(\frac{-E}{RT}\right)$$

Assuming that gases in the port shown in Figure 9-c are well mixed, show that

$$\left(\frac{dx_{HC}}{dt}\right)_{c.v.} = \frac{\dot{m}_{in}}{m}(x_{HC,\,in} - x_{HC}) - A x_{HC} x_{O_2} \exp\left(\frac{-E}{RT}\right)$$

9.9   As an engine warms up, clearance between various parts change because of differing amounts of thermal expansion. Explain how this can affect hydrocarbon emissions from a spark-ignition, homogeneous-charge engine.

9.10   Explain how blowby can affect hydrocarbon exhaust emissions (not crank-case emissions which are no longer a problem). Specifically discuss the influence of engine speed.

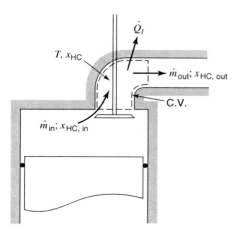

Figure 9-c Illustration for Homework Problem 9.8.

Chapter 10

Fuels and Lubricants

10.1 INTRODUCTION

So far our attention has been on fuels composed of only one chemical species. However, a typical gasoline or diesel fuel may consist of 100 hydrocarbons and another 100 to 200 trace species. This chapter explains why fuels are so complex, how they are manufactured, and how different fuels perform in a vehicular application. This chapter also discusses some of the properties of lubricating oils.

The main source of crude oil is petroleum. The identified worldwide petroleum reserves are estimated by the American Petroleum Institute to be about 1 trillion barrels, with 0.6 trillion barrels remaining to be identified. At present consumption rates it is estimated that petroleum reserves will last for 60 to 95 years. Technological advances in petroleum extraction have created continual increases in the size of the worldwide petroleum reserves. In 1950, the identified worldwide petroleum reserves were estimated to be about 0.09 trillion barrels, so in the last 50 years the identified petroleum reserves have increased tenfold. To put the consumption of petroleum into perspective, about 0.7 trillion barrels of petroleum have been consumed since the advent of the industrial revolution. It is also technically possible, but currently uneconomical, to extract crude oil from coal, shale, and tar sands.

Since petroleum contains carbon, its combustion produces carbon dioxide, a greenhouse gas linked to global warming. There are a number of private and governmental initiatives underway to reduce the amount of greenhouse gas emissions. These initiatives include increased combustion and process efficiency, and increased use of alternative fuels. Further information about hydrocarbon fuels and their use is given in Owen and Coley (1995).

10.2 HYDROCARBON CHEMISTRY

Gasoline and diesel fuels are composed of blends of hydrocarbons, grouped into families of hydrocarbon molecules termed paraffins, olefins, naphthenes, and aromatics. The hydrocarbon families each have characteristic carbon-hydrogen bond structures and chemical formulae.

Paraffins (alkanes) are molecules in which carbon atoms are chained together by single bonds. The remaining bonds are with hydrogen. They are called saturated hydrocarbons because there are no double or triple bonds. The number of carbon atoms is specified by a prefix:

1-meth	4-but	7-hept	10-dec
2-eth	5-pent	8-oct	11-undec
3-prop	6-hex	9-non	12-dodec

Paraffin is designated as an alkane by the suffix *ane*. The general formula for the family is C_nH_{2n+2}. Examples of straight chain paraffins are methane, CH_4, and octane, C_8H_{18},

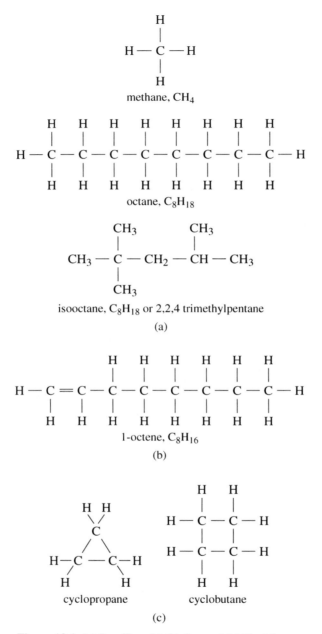

Figure 10-1 (a) Paraffins, (b) Olefins and (c) Naphthenes.

as shown schematically in Figure 10-1. Octane is sometimes called normal octane or
n-octane. Isooctane, as shown in Figure 10-1, is an example of an isomer of octane. That
is, it has the same number of carbon atoms as octane but not in a straight chain. The group
CH_3 attached to the second and fourth carbons from the right is called a methyl radical,
meth because it has one carbon atom and *yl* because it is of the alkyl radical family C_nH_{2n+1}.
Isooctane is more properly called 2, 2, 4 trimethylpentane, *2, 2, 4* because methyl groups
are attached to the second and fourth carbon atoms, *trimethyl* because three methyl radicals are attached, and *pentane* because the straight chain has five carbon atoms.

Olefins (alkenes) are molecules with one or more carbon-carbon double bonds. Monoolefins have one double bond, the general formula C_nH_{2n}, and their names end with *ene*. For example, 1-octene, C_8H_{16} is shown in Figure 10-1. Isomers are possible not only by branching the chain with the addition of a methyl radical but also by shifting the position of the double bond without changing the carbon skeleton. Olefins with more than one carbon-carbon double bond are undesirable components of fuel that lead to storage problems. Consequently, they are refined out and the only olefins of significance in diesel fuel or gasoline fuel are monoolefins.

Naphthenes (cycloalkanes) have the same general formula as olefins, C_nH_{2n}, but there are no double bonds. They are called *cyclo* because the carbon atoms are in a ring structure. Two examples are cyclopropane and cyclobutane, shown in Figure 10-1. Cycloalkane rings having more than six carbon atoms are not as common.

Aromatics are hydrocarbons with carbon-carbon double bonds internal to a ring structure. The most common aromatic is benzene, shown schematically in Figure 10-2. Benzene is a regulated toxic compound, as it is a known carcinogen. Notice that the double bonds alternate in position between the carbon atoms. This makes the molecule hard to break, so that a greater temperature is required to initiate combustion. As a result, aromatics are desirable in gasoline since they increase the octane number. Aromatics are undesirable components of diesel fuels. Some common aromatics (toluene, ethylbenzene, and styrene) have groups such as methyl radicals substituted for hydrogen atoms, and others (biphenyl) have more than one ring. Finally, there are polycyclic aromatic hydrocarbons (PAH) which are aromatics with two carbon atoms shared between more than one ring (naphthalene, anthracene).

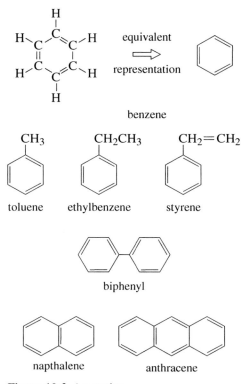

Figure 10-2 Aromatics.

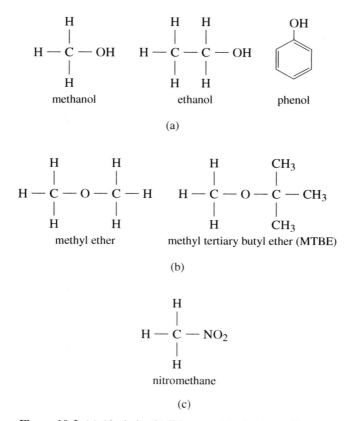

Figure 10-3 (a) Alcohols, (b) Ethers, and (c) Nitroparaffins.

An alcohol is a partially oxidized hydrocarbon, formed by replacing a hydrogen atom with the hydroxyl radical OH. If the hydrogen atom attached to an aromatic ring is replaced by the hydroxyl radical, the molecule is called a phenol. Ethers are isomers of alcohol with the same number of carbon atoms. Some examples, shown in Figure 10-3, are methanol, ethanol, phenol, and methyl ether.

Methyl tertiary-butyl ether (MTBE), shown in Figure 10-3, is an ether with one carbon atom part of a methyl group, CH_3, and the other carbon atom as the central atom of a tertiary butyl group, $C(CH_3)_3$. At ambient temperature, MTBE is a volatile, flammable, and colorless liquid. MTBE is manufactured from the chemical reaction of methanol and isobutylene. MTBE has been used in gasoline since the 1970s. Initially it was added to increase the octane of gasoline, as a lead replacement. Its use has increased with the federal oxygenated fuels and reformulated gasoline programs. MTBE is very soluble in water and adsorbs poorly to soil, so that it migrates through soil at about the same rate as water. Recently, there has been concern about MTBE contamination of drinking water supplies, resulting from leaking underground gasoline storage tanks, resulting in its use being restricted and even prohibited.

Nitromethane (CH_3NO_2) is formed from a paraffinic hydrocarbon by replacing a hydrogen atom with a NO_2 group, as shown in Figure 10-3. It has twice the bound oxygen as monohydric alcohols, and can combust without air. At ambient temperature, it is a liquid, and it is widely used as a drag racing fuel.

Thermophysical Properties of Hydrocarbons

The thermophysical properties of some single hydrocarbons were given in Chapter 3. In general, the equivalent chemical formula of a hydrocarbon of formula $C_\alpha H_\beta$ can be determined from the molecular weight M and the hydrogen to carbon ratio HC, since

$$\alpha = M/(12.01 + 1.008 \cdot HC) \quad \text{and} \quad \beta = HC \cdot \alpha \qquad (10.1)$$

The enthalpy of formation, h_f^o, at 298 K for a hydrocarbon of formula $C_\alpha H_\beta$ can be determined from the heat of combustion q_c. Equation 3.85 can be rewritten as

$$h_f^o = \alpha h_{f,CO_2} + \frac{\beta}{2} (h_{f,H_2O} - (1 - \chi)h_{fg,H_2O}) + q_c \qquad (10.2)$$

where χ is the quality of water in the products. The lower heat of combustion assumes $\chi = 1.0$, whereas the higher heat of combustion assumes $\chi = 0$.

Figures 10-4 and 10-5 show the ideal gas constant pressure specific heat of hydrocarbons (paraffins, monoolefins, aromatics, naphthenes, and alcohols) found in fuels. They show that on a per unit mass basis the specific heat depends on carbon type and is a weak function of carbon number. This is not unexpected since the specific heat of a molecule depends on the number and type of bonds. The results shown are correlated by the following equation:

$$c_{p,i} = a_i + b_i t + c_i t^2 \quad (\text{kJ/kg K}) \qquad (10.3)$$

where $t = T(K)/1000$, and $300 < T < 1500$ K. The specific heat of a motor fuel is then

$$c_p = \sum_i x_i c_{p,i} \qquad (10.4)$$

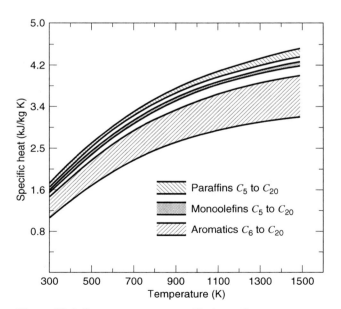

Figure 10-4 Constant pressure specific heat of hydrocarbons found in motor fuels.

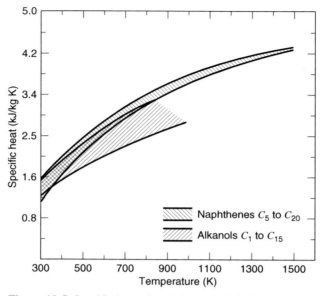

Figure 10-5 Specific heat of naphthenes and alcohols used in motor fuels.

where x_i is the mass fraction of component i. The coefficients of Equation 10.3 are listed in Table 10-1.

The absolute molar entropy of a liquid hydrocarbon fuel of the form $C_\alpha H_\beta O_\gamma N_\delta$ has been correlated by Ikumi and Wen (1981):

$$s = 4.69\alpha + 18.41\beta + 44.55\gamma + 85.97\delta \quad (kJ/kmol\ K) \qquad (10.5)$$

The octane numbers of various single hydrocarbon fuels are tabulated in Table 10-2. In general, it has been found that the octane number is improved by reducing the straight chain length. This can be accomplished by reducing the total number of carbon atoms or by rearranging them into a branch chain structure. These generalizations are illustrated in Figure 10-6 for paraffinic hydrocarbons. The critical compression ratio is determined by increasing the compression ratio of an engine until incipient knock occurs. The correlation with octane number is evident.

Table 10-1   Specific Heat Curve Fit Coefficients

Type	i	a_i	b_i	c_i
Paraffins	1	0.33	5.0	-1.5
Monoolefins	2	0.33	4.6	-1.3
Aromatics ($C_n H_{2n-6}$)	3	0.21	4.2	-1.3
Naphthenes	4	0.04	5.0	-1.4
Alkanols ($T < 1000$ K)	5	0.50	3.3	-0.71

Table 10-2   Knock Characteristics of Single Component Fuels

Formula	Name	Critical Compression Ratio[a]	Octane number[b]	
			Research	Motor
CH_4	Methane	12.6	120	120
C_2H_6	Ethane	12.4	115	99
C_3H_8	Propane	12.2	112	97
C_4H_{10}	Butane	5.5	94	90
C_4H_{10}	Isobutane	8.0	102	98
C_5H_{12}	Pentane	4.0	62	63
C_5H_{12}	Isopentane	5.7	93	90
C_6H_{14}	Hexane	3.3	25	26
C_6H_{14}	Isohexane	9.0	104	94
C_7H_{16}	Heptane	3.0	0	0
C_7H_{16}	Triptane	14.4	112	101
C_8H_{18}	Octane`	2.9	-20	-17
C_8H_{18}	Isooctane	7.3	100	100
$C_{10}H_{12}$	Isodecane	—	113	92
C_4H_8	Methylcyclopropane	—	102	81
C_5H_{10}	Cyclopentane	12.4	101	95
C_6H_{12}	Cyclohexane	4.9	84	78
C_6H_{12}	1,1,2-trimethylcyclopropane	12.2	111	88
C_7H_{14}	Cycloheptane	3.4	39	41
C_8H_{16}	Cyclooctane	—	71	58
C_6H_6	Benzene	—	—	115
C_7H_8	Toluene	15	120	109
C_8H_{10}	Ethyl benzene	13.5	111	98
C_8H_{10}	Xylene-m	15.5	118	115
C_3H_6	Propylene	10.6	102	85
C_4H_8	Butene-l	7.1	99	80
C_5H_{10}	Pentene-l	5.6	91	77
C_6H_{12}	Hexene-l	4.4	76	63
C_5H_8	Isoprene	7.6	99	81
C_6H_{10}	1,5-hexadiene	4.6	71	38
C_5H_8	Cyclopentene	7.2	93	70
CH_4O	Methanol	—	106	92
C_2H_6O	Ethanol	—	107	89

Source: An abridgement of Tables 8-6 and 8-7 from Obert, 1973.

[a]The critical compression ratio is for audible knock at 600 rpm, inlet temperature at 311 K, coolant at 373 K, spark advance and fuel-air ratio set for best power.

[b]Octane ratings above 100 are obtained by matching against leaded isooctane.

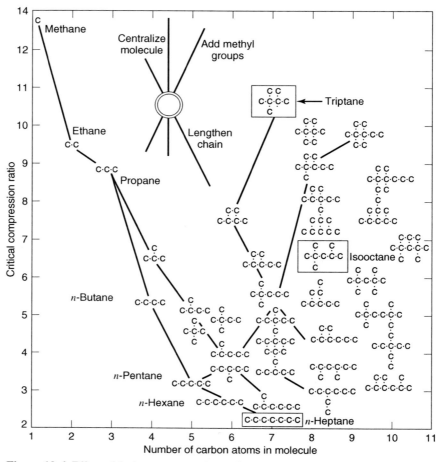

Figure 10-6 Effect of fuel structure on detonation tendency of paraffinic hydrocarbons: CFR engine, 600 rpm, inlet temperature 450 K (Lovell, 1948).

10.3 REFINING

Crude oil contains a large number of various hydrocarbon fractions. For example, 25,000 different compounds have been found in one sample of petroleum-derived crude oil. The compounds range from gases to viscous liquids and waxes. The purpose of a refinery is to physically separate crude oil into various fractions, and then chemically process the fractions into fuels and other products. The fraction separation process is called distillation and the device employed is often called a still.

The generic features of a still are illustrated in Figure 10-7. The sample is heated preferentially boiling off the lighter components. The classification of the various fractions is arbitrary. In the order in which they leave the still, the various fractions are commonly referred to as naphtha, distillate, gas oil, and residual oil. Further subdivision uses the adjectives *light, middle,* or *heavy.* The adjectives *virgin* or *straight run* are often used to signify that no chemical processing has been done to the fraction. For example, since light, virgin naphtha can be used as gasoline, it is often called straight run gasoline. The physical properties of any fraction depend on the distillation temperatures of the products collected in the beaker.

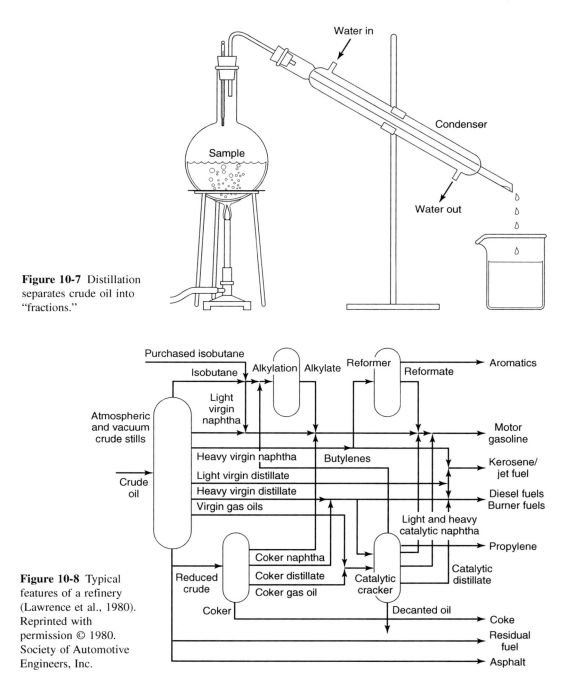

Figure 10-7 Distillation separates crude oil into "fractions."

Figure 10-8 Typical features of a refinery (Lawrence et al., 1980). Reprinted with permission © 1980. Society of Automotive Engineers, Inc.

An example of refinery processing paths is illustrated in Figure 10-8. The chemical processes shown are alkylation, reforming, catalytic cracking, and coking. An actual refinery uses many more processes, but we will limit our study to these most important ones. Likewise, a refinery will produce more products than those shown. The refinery illustrated produces fuels for engines (gasoline, diesel, jet), fuels for heating (burner, coke, kerosene, residual), chemical feedstock (aromatics, propylene), and asphalt. On the average, a refinery will refine about 40% of the input crude oil into gasoline, 20% into diesel and heating fuel, 15% into residual fuel oil, 5% into jet fuel, and the remainder into the other listed hydrocarbons.

A broad cut fraction is collected over a large range of distillation temperatures, a narrow cut over a small range, a light fraction over a low temperature range, and a heavy fraction over a high temperature range. Gasoline fuel is a blend of hydrocarbon distillates with a range of boiling points from about 25°C to 225°C, and diesel fuel is a blend of hydrocarbon distillates with a range of boiling points from about 180°C to 360°C.

Chemical processing is required to convert one fraction into another. For example, a crude might yield, on an energy basis, 25% straight run gasoline but the product demand could be 50%. In this situation the other 25% would be produced by chemical processing of some other fraction into gasoline. Chemical processing is also used to upgrade a given fraction. For example, straight run gasoline might have an octane number of 70, whereas the product demand could be 90. In this case chemical processing would be needed to increase the octane number from 70 to 90.

Alkylation is used to increase the molecular weight and octane number of gasoline by adding alkyl radicals to a gaseous hydrocarbon molecule. Light olefin gases are reacted with isobutane in the presence of a catalyst. Isooctane results from reacting butene with isobutane. This process requires relatively low temperature (275 K) and pressure (300 kPa); and therefore, consumes relatively less energy than other refining processes.

Catalytic cracking breaks molecules in order to convert distillates into naphthas for use as gasoline. The naphtha products of catalytic cracking are high octane gasolines. The reactions are at high temperature (700 to 800 K) and at low to moderate pressure (200 to 800 kPa). Considerable energy is consumed in the process.

Reforming refers to reactions designed to alter molecular structure to yield higher octane gasoline (e.g., conversion of paraffins into aromatic hydrocarbons). This is often done in a hydrogen atmosphere at high temperature (800 K), at high pressure (3000 kPa), and in the presence of a catalyst. Considerable hydrogen is produced as a result of the reaction:

$$C_n H_{2n+2} \longrightarrow C_n H_{2n-6} + 4H_2 \qquad (10.6)$$

Coking is the process used to convert heavy reduced crude fraction to the more usable naphtha and distillate fractions. The reduced crude is heated in an oven. Upon heating, the molecules undergo pyrolytic decomposition and recombination. The average molecular weight of the fraction remains the same, but a greater spectrum of components is produced. The heaviest component, called coke, is a solid material similar to charcoal.

10.4   GASOLINE FUELS

Gasoline has been a dominant vehicular fuel since the early 1900s. It was the fuel used in the first four-stroke engine of Nikolaus Otto in 1876. It has a very high volumetric energy density and a relatively low cost. It is composed of a blend of light distillate hydrocarbons, including paraffins, olefins, naphthenes, and aromatics. It has a hydrogen to carbon ratio varying from 1.6 to 2.4. A typical formula used to characterize gasoline is $C_8 H_{15}$, with a molecular weight of 111. A high hydrogen content gasoline is $C_7 H_{17}$.

Gasoline properties of interest for internal combustion engines are given in Table 10-3. The properties include the octane number, volatility, gum content, viscosity, specific gravity, and sulfur content. The American Society for Testing and Materials (ASTM) has established a set of gasoline specifications for each property, also listed in Table 10-3. The anti-knock index (AKI) is the average of the research (D2699) and motored (D2700) octane numbers and is the number (for example, 85, 87, and 91) displayed on pumps at

Table 10-3   Gasoline Property Specifications

Property	ASTM Method
Distillation, K	D86
Heating value	D240
Specific gravity	D287
Reid vapor pressure, kPa	D323
Gum, mg/ml	D381
Octane, supercharge	D909
Sulfur, wt %	D1266
Hydrocarbons, %	D1319
Octane, research	D2699
Octane, motor	D2700
Benzene, vol %	D3606

service stations. The determination of the octane numbers is discussed in Section 9.3. The octane number for automotive gasoline reached a maximum in the 1960s, with leaded premium gasoline available with AKI ratings of 103 +. The octane number for aviation fuels is based on motored (D2700) and supercharged (D909) test methods.

Knowledge of gasoline volatility is important not only in designing fuel delivery and metering systems, but also in controlling evaporative emissions. The volatility is quantified by three related specifications: (1) the distillation curve (D86), (2) the Reid vapor pressure (D323), and (3) the vapor-liquid ratio (D439). With the D86 distillation method a still is used to evaporate the fuel. The fuel vapor is condensed at atmospheric pressure. The heating rate is adjusted continuously such that the condensation rate is 4 to 5 ml/min. The heating process is stopped when the fuel starts to smoke and decompose, typically around 370°C. The vapor temperature at the top of the distillation flask is measured throughout the test. The volume fraction of condensate is plotted versus temperature to form a distillation curve. The 10% and 90% evaporation temperatures, T_{10} and T_{90}, are used in the volatility specifications. The T_{10} temperature, indicating the start of vaporization, is used to characterize the cold starting behavior, and the T_{90} temperature, indicating the finish of vaporization, is used to characterize the possibility of unburned hydrocarbons. The ASTM driveability index (DI) is also a measure of fuel volatility and is defined as $1.5\,T_{10} + 3\,T_{50} + T_{90}$.

Gum is a product of oxidation reactions with certain molecules often found in fuels. Use of gasoline with a high gum component can lead to sticking of valves and piston rings, carbon deposits, and clogging of fuel metering orifices. Inhibitors are often added to gasoline to reduce the gum formed in such a test under an assumption they will also reduce gum formation in service. The ASTM D381 test method involves evaporating 50 ml of gasoline in a glass dish at approximately 430 K by passing heated air over the sample for a period of about 10 min. The difference in weight of the dish before and after the test is called the existent gum content.

Reformulated Gasoline (RFG)

The U.S. Clean Air Act of 1990 set up two programs, an oxygenated fuels program and a reformulated gasoline program, which resulted in mandated changes in the composition of gasoline. The oxygenated fuels program is a winter program used to reduce carbon monoxide and hydrocarbon levels in major cities that have carbon monoxide levels that exceed federal standards. The oxygenated fuels program requires that gasoline contain at least 2.7% by weight of oxygen. The first cities to use oxygenated gasoline were Denver, CO, and Phoenix, AZ, and it is now required in about 40 cities in the United States.

The reformulated gasoline (RFG) program is a year-round program used to reduce ozone levels. The program requires that gasoline sold year-round in areas that have ozone levels that exceed federal standards have minimum oxygen content of 2% by weight and maximum benzene content of 1%. It is now required in 10 cities in the United States, and an additional 21 areas have voluntarily entered the program. The primary oxygenates used are methyl tertiary-butyl ether (MTBE) and ethanol (EtOH). However, due to MTBE contamination of drinking water supplies, its future use as an oxygenate is questionable. In 1996 California required use of Phase 2 RFG, which has stricter standards than Phase 1 RFG. The properties of various gasolines are compared in Table 10-4. The gasolines listed are:

- Industry average gasoline
- Gasoline oxygenated (2.7 wt % oxygen) with MTBE
- Gasoline oxygenated with ethanol (gasohol)
- Phase 1 reformulated gasoline
- California Phase 2 reformulated gasoline

The volume percentage of olefins and benzene in reformulated gasoline is lower than industry average gasoline. The Reid vapor pressure is reduced in the summer in reformulated gasoline to reduce the emissions due to fuel evaporation. The 90% distillation temperature T_{90} is decreased to increase the vaporization and oxidation of the gasoline, which reduces the hydrocarbon emissions.

The current level of sulfur in gasoline is about 300 ppm. Since sulfur has an adverse impact on the performance of catalytic converters, the California Phase 2 reformulated gasoline specifications reduce the sulfur level to about 30 ppm.

Table 10-5 compares the FTP regulated emissions from industry average gasoline and Phase 2 reformulated gasoline for a group of fleet vehicles. The use of the reformulated gasoline decreased the HC emissions by 26%, NMHC emissions by 27%, CO emissions by 30%, and NO_x emissions by 18%.

Table 10-4   Properties of Gasoline Fuels

	Industry average gasoline	MTBE oxygenated gasoline	Gasohol	Phase 1 RFG	Phase 2 RFG
Aromatics, vol %	28.6	25.8	23.9	23.4	25.4
Olefins, vol %	10.8	8.5	8.7	8.2	4.1
Benzene, vol %	1.60	1.6	1.6	1.3	0.93
Reid vapor pressure, kPa	60-S	60-S	67-S	50-S	46
(S: summer, W: winter)	79-W	79-W	79-W	79-W	
T_{50}, K	370	369	367	367	367
T_{90}, K	440	432	431	431	418
Sulfur, mass ppm	338	313	305	302	31
MTBE, vol %	0.0	15	0	11	11.2
Ethanol, vol %	0	0	10	4	0

Source: Adapted from EPA 420-F-95-007.

Table 10-5   FTP Regulated Emissions (g/mile) from Industry Average and Reformulated Gasoline

	Industry average gasoline	Phase 2 reformulated gasoline
HC	0.226	0.167
NMHC	0.203	0.148
CO	3.22	2.25
NO_x	0.394	0.321

Source: Cadle et al., 1997.

Gasoline Additives

Gasoline additives include octane improvers, anti-icers to prevent fuel line freeze-up, detergents to control deposits on fuel injectors and valves, corrosion inhibitors, and antioxidants to minimize gum formation in stored gasoline. Many compounds have been tested for use as octane improvers in gasoline. Tetraethyl lead was the primary octane improver in general use from 1923 to 1975. Its use in motor vehicles was made illegal in 1995 due to its toxicity and adverse effect on catalytic converters and oxygen sensors. Currently, lead is only used in aviation gas and off road racing gasoline. Alcohols, ethers, and methyl-cyclopentadienyl manganese tricarbonyl (MMT) are now used as octane improvers.

Thomas Midgley of the General Motors Research Laboratory discovered lead additives in 1921. As an aside, Midgley also was the inventor of Freon® (F-12), a refrigerant initially developed for automotive air conditioning systems. Freon® was the most widely used refrigerant in the world until the mid 1990s when it was determined that the decomposition of Freon® in the stratosphere causes depletion of the stratospheric ozone layer. The manufacturing of Freon® in the United States was made illegal in 1998.

10.5   DIESEL FUELS

Diesel fuel consists of a mixture of light distillate hydrocarbons that have boiling points in the range between about 180°C and 360°C, higher than gasoline. It is estimated that there are more than 10,000 isomers in diesel fuel. Like gasoline, diesel fuels are mixtures of paraffinic, olefinic, naphthenic, and aromatic hydrocarbons, but their relative proportions are different. The molecular weight of diesel fuel varies from about 170 to 200. Diesel fuels have about an 8% greater energy density by volume than gasoline, are much less flammable, and are the primary fuel used by heavy duty vehicles.

Diesel fuels are classified both by a numerical scale and by use. The use designations are bus, truck, railroad, marine, and stationary. The American Society for Testing and Materials, ASTM D975, numerical classification scheme for diesel fuels ranges from one to six, with letter subcategories. Diesel fuel number 1D is a cold weather fuel with a flash point of 38°C. Diesel fuel 2D is a diesel fuel of lower volatility with a flash point of 52°C. Diesel 2D is the most common fuel for vehicular applications. Diesel fuel 4D is used for stationary applications where the engine speed is low and more or less constant. The specification chart contained in ASTM D975 is shown here as Table 10-6. The thermodynamic properties of diesel fuel 2D are listed in Table 10-7.

The ignition quality of diesel fuel is given by the cetane number, CN. The higher the cetane number, the easier it is for the fuel to ignite. Current cetane numbers for vehicular diesel fuels range from about 40 to 55. Additives such as nitrate esters can be used to

Table 10-6   ASTM D975 Diesel Fuel Specifications

	ASTM method	No. 1D	No. 2D	No. 4D
Minimum flash point, °C	D93	38	52	55
Cloud point, °C	D2500	local	local	local
Maximum water and sediment, vol%		0.05	0.05	0.05
Maximum carbon residue on 10% res., %	D524	0.15	0.35	—
Maximum ash, wt %	D482	0.01	0.01	0.10
T_{90}, K	D86	561 max	555-611	—
Kinematic viscosity at 40°C (m²/s)	D445	$1.3\text{-}2.4 \times 10^{-6}$	$1.9\text{-}4.1 \times 10^{-6}$	$5.5\text{-}24 \times 10^{-6}$
Maximum sulphur, wt %	D129	0.05	0.05	2.0
Maximum copper strip corrosion		No. 3	No. 3	—
Minimum cetane number	D613	40	40	30

Table 10-7   Thermodynamic Properties of Compression Ignition Fuels

	Diesel	Dimethyl ether (DME)	Rapeseed methyl ester (RME)
Formula	$C_{12}H_{26}$	CH_3OCH_3	
Molecular weight	170—200	46.07	
Liquid density (kg/m³)	820—860	668	882
Kinematic viscosity (m²/s) at 40°C	2.8×10^{-6}	2.2×10^{-7}	4.1×10^{-6}
Lower heating value, mass (MJ/kg$_{fuel}$)	42.5	28.4	37.7
Lower heating value, volume (MJ/liter$_{fuel}$)	34.8—36.5	19.0	
Boiling point at 1 bar (°C)	180—360	−25	
Vapor pressure at 38°C (bar)	0.0069	8	
Cetane number	40—55	55—60	52
Stoichiometric A/F ratio	14.7	9.0	11.2 − 12.5

increase the cetane number. The calculated cetane index (CCI) is an approximation to the cetane number computed from the ASTM D976 empirical correlation for petroleum-based diesel fuels:

$$CCI = -420.34 + 0.016\, G^2 + 0.192\, G \log T_{50} + 65.01(\log T_{50})^2 - 0.0001809\, T_{50}^2 \tag{10.7}$$

where

G = API gravity (ASTM D287)

T_{50} = Midpoint boiling temperature, °F

The cetane index is useful because it is cheaper to obtain than a measurement of the actual cetane number. In addition, it illustrates that not all diesel fuel properties can be specified independently of one another.

The regulated emissions from vehicular diesel combustion include CO, HC, NO_x, and particulate matter (PM). The emissions limits have been tightened in response to concerns about the adverse effect that compression ignition engines have had on ambient air quality, specifically NO_x and PM. Nitrogen oxides are a precursor to ground level ozone formation, and particulate emissions are a respiratory hazard. As discussed in Chapter 9, there is a trade-off between NO_x and particulate matter (PM) emissions from compression ignition engines, as techniques to lower NO_x will generally increase PM, and vice versa.

One component of diesel particulate emissions that has attracted particular regulatory attention is sulfates, due to the adverse impact that sulfates have on air quality. The maximum sulfur content is presently regulated to 0.05%. Reformulated diesel (RFD) has a sulfur content below 0.01%.

10.6   ALTERNATIVE FUELS

Important alternative fuels are methane or compressed natural gas (CNG), propane or liquid petroleum gas (LPG), alcohols, and hydrogen. Alternative fuels are of interest since they can be refined from renewable feedstocks, and their emission levels can be much lower than those of gasoline and diesel fueled engines (Dhaliwal et al., 2000). If there are availability problems with crude oil, due to worldwide geo-political problems, alternative fuels can also be used as replacements. As of the year 2000, the most commonly used alternative fuel for vehicles is propane, followed by natural gas, and methanol.

Alternative fuels are not currently widely used in vehicular applications for both economic and engineering reasons. The cost of alternative fuels per unit of energy delivered can be greater than gasoline or diesel fuel, and the energy density of alternative fuels by volume is less than gasoline or diesel fuel. The smaller volumetric energy density requires larger fuel storage volumes to have the same driving range as gasoline fueled vehicles. This can be a drawback, particularly with dual fuel vehicles, where a significant portion of the trunk space is used by the alternative fuel storage tank. Alternative fuels also lack a wide scale distribution and fueling infrastructure comparable to that of conventional fuels. In recent years, fleet vehicles, such as buses, trucks, and vans have been a growing market for alternative fuels, as they can operate satisfactorily with localized fueling. The market penetration of alternative fuels is currently on the order of 0.5 to 1%. In 1990, there were about 4 million propane fueled vehicles, three million ethanol fueled vehicles, and about one million natural gas fueled vehicles worldwide, compared to about 150 million gasoline fueled vehicles in the United States alone (Webb and Delmas, 1991).

Existing gasoline or diesel engines can be retrofitted fairly easily for operation with alternative fuels. However, various operational considerations need to be taken into account. The different combustion characteristics of alternative fuels require a change in the injection and ignition timing. Also, many alternative fuels, especially those in gaseous form, have very low lubricity, causing increased wear of fuel components such as fuel injectors and valves.

The properties of various alternative fuels are tabulated in Table 10-8, and are compared with the properties of *n*-octane. The first three columns contain gaseous fuels (methane, propane, and hydrogen) and the next three columns are liquid fuels (methanol, ethanol, and *n*-octane). While there is a range of energy densities on a fuel mass (MJ/kg_{fuel}) basis, the energy densities are comparable on a stoichiometric air mass (MJ/kg_{air}) basis. Octane has the greatest energy density by volume (MJ/l). Alternate fuels have higher octane levels than gasoline, so engines fueled with alternative fuels can operate at higher compression levels, and thus at higher efficiency.

Table 10-8   Thermodynamic Properties of Spark Ignition Engine Fuels

	Propane	Natural Gas	Hydrogen	Methanol	Ethanol	Gasoline
Molecular weight	44.10	18.7	2.015	32.04	46.07	~ 110
Vapor pressure (kPa), at 38°C				32	17	62—90
Boiling point (°C), at 1 bar		−160	−253	65	78	30—225
Enthalpy of vaporization, h_{fg} (kJ/kg), at 298 K				1215	850	310
Lower heating value, mass (MJ/kg$_{fuel}$)	46.4	50.0	120	19.9	26.8	44.5
Lower heating value, volume (MJ/l$_{fuel}$)		8.1[a]		15.7	21.1	32.9
Lower heating value, stoichiometric (MJ/kg$_{air}$)	2.98	2.92	3.52	3.09	3.00	2.96
Octane number, research	112	120	106	112	111	90—98
Octane number, motor	97	120		91	92	80—90
Stoichiometric A/F ratio, mass	15.58	17.12	34.13	6.43	8.94	15.04
Vapor flammability limits (% volume)		5.3—15	5—75	5.5—26	3.5—26	0.6—8
Adiabatic flame temperature (K)	2268	2227	2383	2151	2197	2266
Stoichiometric CO$_2$ emissions, g CO$_2$/MJ$_{fuel}$	64.5	54.9	0	69	71.2	71.9

[a]at 15°C, 22 MPa

Source: Adapted from Black, 1991; Unich et al., 1993.

Propane

Propane (C_3H_8) is a saturated paraffinic hydrocarbon. When blended with butane (C_4H_{10}) or ethane (C_2H_6), it is also designated as liquefied petroleum gas (LPG). A common LPG blend is P92, which is 92% propane and 8% butane. In the United States, about one-half of the LPG supply is obtained from the lighter hydrocarbon fractions produced during crude oil refining, and the other half from heavier components of wellhead natural gas.

Propane has been used as a vehicular fuel since the 1930s. In 1993, there were about 4 million LPG vehicles operating worldwide, with the majority in the Netherlands, followed by Italy, the United States, and Canada. There is a relatively extensive refueling network for propane, with over 15,000 refueling stations available in North America. There are a number of original equipment manufacturers that currently sell propane-fueled vehicles, primarily light and medium duty fleet vehicles, such as pick-up trucks and vans. Conversion kits are also available to convert gasoline or diesel fueled engines to dedicated propane or dual fuel use.

In vehicles, propane is stored as a compressed liquid, typically from 0.9 to 1.4 MPa. Its evaporative emissions are essentially zero, since it is used in a sealed system. A pressure regulator controls the supply of propane to the engine, and converts the liquid propane to a gas through a throttling process. Propane gas can be injected into the intake manifold, into the ports, or directly into the cylinder. Propane has an octane number of 112 (RON), so vehicular applications of propane will generally raise the compression ratio.

Table 10-9   LPG Fueled Vehicle Regulated Emissions (g/mile)

| Emission | 3.1 L engine | |
	Propane	Gasoline
HC	0.21	0.37
CO	2.55	5.4
NO_x	0.67	0.42

Source: Bass et al., 1993.

Table 10-10   LPG Fueled Vehicle Toxic Emissions (mg/mile)

Toxic	Propane	Gasoline
Benzene	< 0.1	16.7
1,3 Butadiene	< 0.1	2.5
Formaldehyde	1.2	3.1
Acetaldehyde	0.3	1.5
Total	1.5	23.8

Source: Bass et al., 1993.

As shown in Table 10-8, the CO_2 emissions on an equivalent energy basis are about 90% that of gasoline. Liquid propane has three-fourths of the energy density by volume of gasoline, so that the fuel economy is correspondingly reduced. The volumetric efficiency and the power are also reduced due to the displacement of about 5% to 10% of the intake air by the propane, and the loss of evaporative charge cooling. Propane requires about a 5° spark advance at lower engine speeds due to its relatively low flame speed.

The FTP emissions from an LPG fueled engine are shown in Table 10-9. The engine used was a 3.1 L engine with a LPG conversion system using an intake manifold mixer. The LPG fuel used was HD5 propane (96% propane, 4% ethane). The results indicate that the HC and CO emissions were lower with LPG than gasoline, 43% and 53% respectively, but the NO_x levels were higher. The toxic emissions are shown in Table 10-10. The levels of toxic emissions are typically an order of magnitude less than the baseline gasoline toxic emissions.

Natural Gas

Natural gas is a naturally occurring fuel found in oil fields. It is primarily composed of about 90 to 95% methane (CH_4), with small amounts of additional compounds such as 0—4% nitrogen, 4% ethane, and 1 to 2% propane. Methane is a greenhouse gas, with a global warming potential approximately ten times that of carbon dioxide. As shown in Table 10-8, methane has a lower carbon to hydrogen ratio relative to gasoline, so its CO_2 emissions are about 22 to 25% lower than gasoline.

Natural gas has been used for many years in stationary engines for gas compression and electric power generation. An extensive distribution network of natural gas pipelines exists to meet the need for natural gas for industrial processes and heating applications. Natural gas fueled vehicles (NGV) have been in use since the 1950s, and conversion kits are available for both spark and compression ignition engines. Recent research and development work has included development of bifuel vehicles that can operate either with natural gas and gasoline or diesel fuel. One advantage of a bifuel operation is that the operating range of a vehicle is extended in comparison with a dedicated natural gas vehicle. Currently, original equipment manufacturers are selling production natural gas fueled vehicles, primarily to fleet owners. Natural gas vehicles were the first vehicles to meet the California ULEV emission standards.

Natural gas is stored in a compressed (CNG) state at room temperatures and also in a liquid (LNG) form at $-160°C$. Natural gas has an octane number (RON) of about 127, so that natural gas engines can operate at a compression ratio of 11 : 1, greater than gasoline fueled engines. Natural gas is pressurized to 20 MPa in vehicular storage tanks, so that it has about one-third of the volumetric energy density of gasoline. The storage pressure is about 20 times that of propane. Like propane, natural gas is delivered to the engine through a pressure regulator, either through a mixing valve located in the intake manifold, port fuel injection at about 750 kPa, or direct injection into the cylinder. With intake manifold mixing or port fuel injection, the engine's volumetric efficiency and power is reduced due to the displacement of about 10% of the intake air by the natural gas, and the loss of evaporative charge cooling. Natural gas does not require mixture enrichment for cold starting, reducing the cold start HC and CO emissions.

The combustion of methane is different from that of liquid hydrocarbon combustion since only carbon-hydrogen bonds are involved, and no carbon-carbon bonds, so the combustion process is more likely to be more complete, producing less non-methane hydrocarbons. Optimal thermal efficiency occurs at lean conditions at equivalence ratios of 1.3 to 1.5. The total hydrocarbon emission levels can be higher than gasoline engines due to unburned methane. The combustion process of methane can produce more complex molecules, such as formaldehyde, a pollutant. The particulate emissions of natural gas are very low relative to diesel fuel. Natural gas has a lower adiabatic flame temperature (approx. 2240 K) than gasoline (approx. 2310 K), due to its higher product water content. Operation under lean conditions will also lower the peak combustion temperature. The lower combustion temperatures lower the NO formation rate, and produce less engine-out NO_x.

To meet vehicular emission standards, catalytic converters are used with natural gas fueled engines. Since three way catalytic converters are most effective at stoichiometric conditions, natural gas combustion is maintained at stoichiometric, and exhaust gas recirculation is used to reduce the peak combustion temperatures and thus the nitrogen oxide levels. Table 10-11 gives the exhaust emissions for a 2.2 L bifuel gasoline and CNG engine (Sun et al., 1998), and a 2.2 L dedicated CNG engine (Kato et al., 1999). When the bifuel engine is switched from gasoline to CNG, the non-methane organic gases (NMOG), carbon monoxide (CO), and nitrogen oxide (NO_x) levels were reduced 60%, 34%, and 41%, respectively. The dedicated CNG engine was modified to operate specifically with natural gas, with a higher compression ratio, intake valves with early closed timing, and intake and exhaust valves with increased lift.

Table 10-11   CNG Fueled Vehicles Regulated Emissions (g/mile)

Emission	Toyota 2.2 L engine	GMC 2.2 L bi-fuel engine	
	CNG	CNG	Gasoline
NMOG	0.007	0.027	0.08
CO	0.69	1.01	1.54
NO_x	0.015	0.10	0.17

Source: Sun et al., 1998, Kato et al., 1999.

Table 10-12   CNG Fueled Vehicles Toxic Emissions (mg/mile)

Toxic	CNG	CNG start/gasoline run	Gasoline
Benzene	0.2	14.8	31.2
1, 3 Butadiene	< 0.1	0.1	1.5
Formaldehyde	3.4	4.1	5.9
Acetaldehyde	0.2	0.3	2.0
Total	3.8	19.3	40.6

Source: Springer et al., 1994.

The emissions of FTP toxics from a 1992 0.75 ton light duty truck operated with gasoline and with natural gas are given in Table 10-12. The engine emission control system included a heated oxygen sensor and a standard 3-way catalyst. The same compression ratio of 8.3:1 was used for both fuels. The Table 10-12 indicates that the CNG toxic emissions are much less than the gasoline toxic emissions. The highest mass emissions with gasoline were benzene and formaldehyde, and the highest mass emissions with CNG was formaldehyde, at a level about half of that of gasoline.

Natural gas can replace diesel fuel in heavy duty engines with the addition of a spark ignition system. A number of heavy duty diesel engine manufacturers are also producing dedicated natural gas heavy duty engines. The natural gas fueled engines are operated lean with an equivalence ratio as low as 0.7. The resulting lower in-cylinder temperatures reduce the NO_x levels. Heavy duty natural gas engines are designed to meet LEV emission standards without the use of an exhaust catalyst, and will meet ULEV emission standards with the addition of a catalyst. The emission certification data for three heavy duty natural gas engines is given in Table 10-13.

Natural gas can also be used in compression ignition engines if diesel fuel is used as a pilot fuel, since the autoignition temperature of methane is 540°C, compared to 260°C for diesel fuel. This fueling strategy is attractive for heavy duty diesel applications, such as trucks, buses, locomotives, and ships, compressors, and generators. These engines are also operated with a lean combustion mixture, so that the NO_x emissions are decreased. However, since diesel engines are unthrottled, at low loads, the lean combustion conditions degrade the combustion process, increasing the hydrocarbon and carbon monoxide emissions.

Table 10-13   Heavy Duty Natural Gas Engine Emission Certification Data (g/bhp-hr)

	Hercules GTA 5.6	Cummins L10	Detroit Diesel 50G
Power (hp)	190	240	275
NMHC	0.9	0.2	0.9
CO	2.8	0.2	2.8
NO_x	2.0	1.4	2.6
PM	0.10	0.02	0.06

Source: Owen and Coley, 1995.

Hydrogen

Hydrogen (H_2) can be produced from many different feedstocks, including natural gas, coal, biomass, and water. The production processes include steam reforming of natural gas, presently the most economical method, electrolysis of water, and gasification of coal, which also produces CO_2. Hydrogen is colorless, odorless, and nontoxic, and hydrogen flames are invisible and smokeless. The global warming potential of hydrogen is insignificant in comparison to hydrocarbon based fuels since combustion of hydrogen produces no carbon-based compounds such as HC, CO, and CO_2.

At present the largest user of hydrogen fuel is the aerospace community for rocket fuel. Hydrogen can also be used as a fuel in fuel cells. There have been a number of vehicular demonstration projects, but the relatively high cost of hydrogen fuel has hindered adoption as an alternative fuel. Dual fuel engines have been used with hydrogen, in which hydrogen is used at start up and low load, and gasoline at full load (Fulton et al., 1993) to reduce the cold start emissions levels.

One of the major obstacles related to the use of hydrogen fuel is the lack of any manufacturing, distribution, and storage infrastructure. The most economical method would be to distribute hydrogen through pipelines, similar to natural gas distribution.

The three methods used to store hydrogen are: (1) in a liquid form at $-253°C$ in cryogenic containers; (2) as a metal hydride, such as iron-titanium hydride $FeTiH_2$, or (3) in a pressurized gaseous form at 20 to 70 MPa. The metal hydride releases hydrogen when heated by a heat source, such as a vehicle exhaust system. The most common storage methods are liquid and hydride storage, which have comparable volumetric storage capabilities, both requiring about 10 times the space required by an equivalent 5-gallon gasoline tank, as shown by Table 10-14. At least a 55-gallon tank of compressed hydrogen is needed to store the energy equivalent of 5 gallons of gasoline.

Compressed hydrogen at 70 MPa has one-third the energy density by volume of compressed natural gas, and liquid hydrogen has one-fourth the energy density by volume of gasoline. Use of liquid hydrogen has an additional energy cost, as liquefaction of hydrogen to -20 K requires an expenditure of energy approximately equal to the energy content of the liquid hydrogen. If mixed with air in the intake manifold, the volume of hydrogen is about 30% of the intake mixture volume at stoichiometric, decreasing the volumetric efficiency. The octane rating of hydrogen of 106 RON allows use of an increased compression ratio.

The combustion characteristics of hydrogen are very different from gasoline combustion characteristics, as the laminar flame speed of a hydrogen air mixture is about 3 m/s, about 10 times that of methane and gasoline, and the adiabatic flame temperature is about $100°C$ higher than gasoline and methane. Since it has a wide flammability limit (5 to 75%),

Table 10-14   Comparison of Hydrogen Storage Methods

	Gasoline (5 gallons)	Liquid H_2	Hydride Fe Ti (1.2%)	Compressed H_2 (70 MPa)
Energy (kJ)	6.64×10^5	6.64×10^5	6.64×10^5	6.64×10^5
Fuel mass (kg)	14	5	5	5
Tank mass (kg)	6.5	19	550	85
Total fuel system mass (kg)	20.5	24	555	90
Volume (gal)	5	47	50	60

Source: Kukkonen and Shelef, 1994.

preignition and back firing can be a problem. The flammability limits correspond to equivalence ratios of 0.07 to 9. Water injection into the intake manifold is used to mitigate preignition and provide cooling. Exhaust gas recirculation and lean operation are used to reduce NO_x levels.

Methanol

Methanol (CH_3OH) is an alcohol fuel formed from natural gas, coal, or biomass feed stock. Methanol is also called wood alcohol. It is a liquid at ambient conditions. Its chemical structure is a hydrocarbon molecule with a single hydroxyl (OH) radical. The hydroxyl radical increases the polarity of the hydrocarbon, so that methanol is miscible in water, and has a relatively low vapor pressure. Since oxygen is part of the chemical structure, less air is required for complete combustion. Methanol is toxic, and ingestion can cause blindness and death. Methanol has been used as a vehicular fuel since the early 1900s, and is also used as a fuel for diesel engines and fuel cells.

Pure methanol is labeled M100, and a mix of 85% methanol and 15% gasoline is labeled M85. M85 has an octane rating of 102. Adding gasoline to methanol provides more volatile components that can vaporize more easily at low temperatures. Methanol has been adopted as a racing fuel, both for performance and safety reasons. Since methanol mixes with water, a methanol fire can be extinguished with water, which is not the case with gasoline. The octane rating of methanol of 111 RON allows use of an increased compression ratio. The relatively high enthalpy of evaporation (1215 kJ/kg) of methanol relative to gasoline (310 kJ/kg) produces greater intake air-cooling and a corresponding increase in volumetric efficiency relative to gasoline. The energy density by volume of methanol is about half that of gasoline. However, because of its oxygen content, it has a higher stoichiometric energy density (3.09 MJ/kg air) relative to gasoline (2.96 MJ/kg air). For maximum power, a rich equivalence ratio of 1.6 is used.

Flexible fuel vehicles (FFV) have been developed to use a range of methanol and gasoline blends from regular gasoline to M85. As of 1998, there were about 13,000 flexible fuel vehicles in operation. An optical fuel sensor is used to determine the alcohol content and adjust the fuel injection and spark timing. The engine compression ratio is not increased, to allow for the lower octane level of gasoline. The low vapor pressure of methanol causes cold starting problems. Satisfactory cold starting with M85 requires a rich mixture so that enough volatiles are present to form a combustible mixture. Methanol is corrosive, especially to rubber and plastic, so alcohol tolerant components, such as stainless steel, are required for its storage and transport.

The cetane number of methanol is low at about 5, but it can be used in compression ignition engines with diesel fuel pilot ignition. Methanol burns with a nearly invisible flame, and a relatively high flame speed. Formaldehyde is a significant decomposition product from methanol combustion and is expected to be higher from methanol than other fuels. The formaldehyde emissions are proportional to the equivalence ratio, so rich combustion will produce increased emissions of formaldehyde. Special lubricants also need to be used in methanol fueled engines.

As shown in Table 10-7, the CO_2 emissions of methanol on an equivalent energy basis are about 96% that of gasoline. The change in the average regulated emissions from a fleet of flexible fueled vehicles fueled with M85 and California Phase 2 reformulated gasoline is given in Table 10-15. With a change in the fuel from RFG to M85, the non-methane hydrocarbons (NMHC) and CO emissions decreased by 30% and 17% respectively, and the NO_x emissions remained about the same. The FTP toxic emissions for the methanol- and gasoline-fueled flexible fueled vehicles are given in Table 10-16. In

Table 10-15   Change in FTP
Regulated Emissions with Methanol
Fueled Vehicle Fleet

Emission	California Phase 2 RFG to M85
NMHC	−30%
CO	−17%
NO$_x$	+3%

Source: Cadle et al., 1997.

Table 10-16   FTP Toxic Emissions (mg/mile) from
Methanol Fueled Vehicle Fleet

Toxic	M85	California Phase 2 RFG
Benzene	3.0	6.0
1,3 Butadiene	0.10	0.6
Formaldehyde	17.1	1.6
Acetaldehyde	0.5	0.4
Total	20.7	8.6

Source: Cadle et al., 1997.

comparing the M85 to RFG air toxic emissions there was an 83% reduction in 1, 3-buta-
diene, a 50% reduction in benzene, and a 25% increase in acetaldehyde, and the formaldehyde emissions were almost an order of magnitude higher for M85.

Ethanol

Ethanol (C_2H_5OH) is an alcohol fuel formed from the fermentation of sugar and grain
stocks, primarily sugar cane and corn, which are renewable energy sources. Its properties
and combustion characteristics are very similar to those of methanol. Ethanol is also called
"grain" alcohol. It is a liquid at ambient conditions, and nontoxic at low concentrations.

Gasohol (E10) is a gasoline-ethanol blend with about 10% ethanol by volume. E85
is a blend of 85% ethanol and 15% gasoline. In Brazil, about half of the vehicles use an
ethanol-based fuel "alcool," primarily E93, produced from sugar cane. In the United States,
the primary source of ethanol is currently from starch feedstocks, such as corn, and there
are efforts underway to produce ethanol from cellulose feedstocks such as corn fiber,
forestry waste, poplar and switch grass. The energy density by volume of ethanol is relatively high for an alternative fuel, about two-thirds that of gasoline. The octane rating of
ethanol of 111 (RON) allows use of an increased compression ratio. The cetane number
of ethanol is low, at about 8, but like methanol, it can be used in compression ignition
engines with diesel fuel pilot ignition.

As shown in Table 10-8, the CO_2 emissions on an equivalent energy basis are about
99% that of gasoline. The FTP regulated emissions and toxic emissions are shown
in Tables 10-17 and 10-18 for a fleet of flexible fueled vehicles (Cadle et al., 1997).
Table 10-18 shows the percent change in regulated emissions achieved when switching
fuels. With a switch to E85, the NO$_x$ emissions decreased by 29%, and the CO emissions
increased by 8%. Acetaldehyde is the dominant air toxic.

Table 10-17   Change in FTP Regulated Emissions with
Ethanol Fueled Vehicle Fleet

Emission	California Phase 2 RFG to E85
NMHC	-10%
CO	$+8\%$
NO$_x$	-29%

Source: Cadle et al., 1997.

Table 10-18   FTP Toxic Emissions (mg/mile) from Ethanol Fueled Vehicle Fleet

Toxic	E85	California Phase 2 RFG
Benzene	1.8	5.1
1,3 Butadiene	0.2	0.7
Formaldehyde	4.1	2.1
Acetaldehyde	24.8	0.5
Total	30.9	8.4

Source: Cadle et al., 1997.

Alternative Fuels for Compression Ignition Engines

A number of fuels are being considered as alternatives for diesel fuel. These fuels include dimethyl ether (DME), Fischer-Tropsch (F-T) fuel; and "biodiesel" vegetable oils, such as rapeseed methyl ester (RME), and soybean methyl ester (SME), which are obtained from renewable energy sources. The thermodynamic properties of DME and RME are listed in Table 10.6. DME is an oxygenated fuel produced by dehydration of methanol or from synthesis gas. The volumetric energy density (MJ/l) of DME is about half that of diesel fuel. It burns with a visible blue flame, similar to that of natural gas. It is noncorrosive to metals, but does deteriorate some elastomers. F-T fuel is produced from natural gas using a catalytic reforming process. RME and SME are produced through a catalyzed reaction between a vegetable oil and methanol. The methyl ester is obtained through a process in which the use of methyl alcohol and the presence of a catalyst (such as sodium hydroxide or potassium hydroxide) chemically breaks down the oil molecule into methyl esters of the oil and a glycerin byproduct.

These alternative diesel fuels have a higher cost, and lower volumetric energy density than diesel fuel, but do produce lower emissions. Since they are not formulated from crude oil, they contain essentially no sulfur or aromatics. Fleisch et al. (1995) reported that with the use of a DME fuel in a heavy duty compression ignition engine, all of the regulated emissions were below the 1998 California ULEV standards. Clark et al. (1999) measured the transient emissions of a number of blends of Fischer-Tropsch fuel. All of the regulated emissions were lower in comparison with low sulfur diesel fuel, with 43% lower HC emissions, 39% lower CO, 14% lower PM, and 14% lower NO$_x$ emissions. Krahl et al. (1996) report that RME had about 40% lower HC emissions, 35% lower CO, 35% lower PM, but about 15% greater NO$_x$ emissions.

10.7   ENGINE OILS

Engine oil is used as a lubricant to reduce the friction between the principal moving parts of an engine. In addition to lubricating, engine oil is expected to act as a coolant, to enhance the rings' combustion seal, and to control wear or corrosion. Additives to either

Table 10-19   SAE Specifications for Engine Oils

SAE viscosity grade	Maximum viscosity (cP) at temperature (°C) (SAE J 300)	Maximum borderline pumping temperature (°C) (ASTM D 3829)	Viscosity (cSt) at 100°C (ASTM D445) Minimum	Maximum
0W	3250 at −30	−35	3.8	—
5W	3500 at −25	−30	3.8	—
10W	3500 at −20	−25	4.1	—
15W	3500 at −15	−20	5.6	—
20W	4500 at −10	−15	5.6	—
25W	6000 at −5	−10	9.3	—
20	—	—	5.6	< 9.3
30	—	—	9.3	< 12.5
40	—	—	12.5	< 16.3
50	—	—	16.3	< 21.9

Source: SAE Handbook, 1983.

petroleum or synthetic base stocks include antifoam agents, antirust agents, antiwear agents, corrosion inhibitors, detergents, dispersants, extreme pressure agents, friction reducers, oxidation inhibitors, pour point depressants, and viscosity index improvers. We will restrict our attention to the rheological characteristics of engine oils. The reader interested in the broader picture is referred to the books by Gruse (1967), the SAE Handbook, and the paper by Sieloff and Musser (1982).

The SAE classifies oils by their viscosity. Two series of grades are defined in Table 10-19 from the recommended practice SAE J300. Grades with the letter *W* (Winter) are based on a maximum low-temperature viscosity, a maximum borderline pumping temperature, and a minimum viscosity at 100°C. Grades without the letter *W* are based on a minimum and a maximum viscosity at 100°C. As shown in Figure 10-9, a multiviscosity grade of oil is one that satisfies one of each of the two grades at different temperatures.

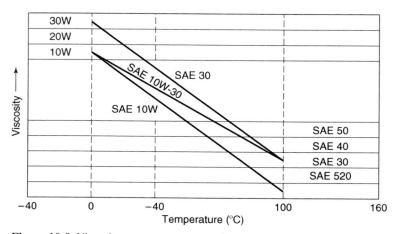

Figure 10-9 Viscosity versus temperature for various SAE oil classifications.

Table 10-20   SAE Engine Oil Carbon Content

	Range	Average
SAE 10	C_{25} to C_{35}	C_{28}
SAE 30	C_{30} to C_{80}	C_{38}
SAE 50	C_{40} to C_{100}	C_{41}

Source: Gruse, 1967

The borderline pumping temperature is measured via a standard test, ASTM D3829, and is a measure of an oil's ability to flow to an engine oil pump inlet and provide adequate oil pressure during warm-up.

The portion of the crude oil refiners use to make lubricants is on the order of 1% and comes from the higher-boiling fraction and undistilled residues which possess the necessary viscosity. Refiners use chemical processing and additives to produce oils with desirable characteristics. Straight-run base stock from petroleum crude oil is referred to as a petroleum oil; whereas those base stocks produced by chemical processing are called synthetic oils. Some synthetic base stocks are compatible with petroleum base stocks and the two types may be blended, in which case the stock is referred to as a blend. Additives range in concentration from several parts per million up to 10%. In terms of carbon content, Gruse (1967) offers the guidelines listed in Table 10-20.

The viscosity of a lubricating oil decreases with increasing temperature and increases with pressure. A Newtonian oil is one in which the viscosity is independent of the shear rate. Shear rates in engines are sometimes high enough that the viscosity decreases, and some oils are deliberately made non-Newtonian via the introduction of wax particles. At some times during hydrodynamic lubrication, the loads increase the oil pressure, which increases the viscosity, increasing the load capacity. It has been suggested that this stabilizing effect is a part of the reason for effects attributed to the "property" oiliness. The kinematic viscosity as a function of temperature and pressure of many oils is correlated by

$$\nu = C_1 \exp\left[C_2/(T - C_3) + P/C_4\right] \tag{10.8}$$

Values of the constants of Equation 10.8 for various SAE grades of engine oils are given in Table 10-21, along with density and specific heat data.

Table 10-21   Engine Oil Properties ($T = 298$ K, $P = 1$ bar)

SAE grade	ρ (kg/m^3)	c_p (kJ/kg K)	ν (m^2/s)	C_1 (m^2/s)	C_2 (K)	C_3 (K)	C_4 (bar)
5W	860	1.99	48.2×10^{-6}	6.44×10^{-4}	900	162	433
10W	877	1.96	86.9×10^{-6}	4.53×10^{-4}	1066	157	296
15W	879	1.95	102×10^{-6}	7.49×10^{-4}	902	173	181
20W	886	1.94	259×10^{-6}	2.63×10^{-4}	1361	150	105
20	880	1.95	129×10^{-6}	5.67×10^{-4}	1028	165	153
30	886	1.94	259×10^{-6}	4.70×10^{-4}	1361	140	105
40	891	1.92	361×10^{-6}	2.17×10^{-4}	1396	151	91.7
50	899	1.91	639×10^{-6}	2.24×10^{-4}	1518	150	75.2

Source: Cameron, 1981.

Finally, it should be mentioned that some two-stroke engines, especially small ones, achieve upper cylinder lubrication by mixing oil with the gasoline. In these cases, additional control of the oil (and the fuel) is required to prevent spark fouling, to assure miscibility with the fuel, and to provide for a hydrodynamic film of the proper viscosity. The fuel-oil mixture that contacts the cold walls separates during compression and combustion, leaving an oil film on the wall.

10.8 REFERENCES

BAILEY, B. and J. RUSSELL (1981), "Emergency Transportation Fuels: Properties and Performance," SAE paper 810444.

BASS, E., B. BAILEY, and S. JAEGER (1993), "LPG Conversion and HC Emissions Speciation of a Light Duty Vehicle," SAE paper 932745.

BLACK, F. (1991), "An Overview of the Technical Implications of Methanol and Ethanol as Highway Motor Vehicle Fuels," SAE paper 912413.

CADLE, S., P. GROBLICKI, R. GORSE, J. HOOD, D. KARDUBA-SAWICKY, and M. SHERMAN (1997), "A Dynamometer Study of Off-Cycle Exhaust Emissions—The Auto/Oil Air Quality Improvement Research Program," SAE paper 971655.

CAMERON, A. (1981), *Basic Lubrication Theory,* Ellis Horwood Ltd., Chichester, England.

CLARK, N., C. ATKINSON, G. THOMPSON, and R. NINE (1999), "Transient Emissions Comparisons of Alternative Compression Ignition Fuels," SAE paper 1999-01-1117.

DHALIWAL, B., N. YI, and D. CHECKEL (2000), "Emissions Effects of Alternative Fuels in Light-Duty and Heavy-Duty Vehicles," SAE paper 2000-01-0692.

DORN, P. and A. M. MOURAO (1984), "The Properties and Performance of Modern Automotive Fuels," SAE paper 841210.

FLEISCH, T., C. MCCARTHY, A. BASU, C. UDOVICH, P. CHARBONNEAU, W. SLODOWSKE, S. MIKKELSEN, and J. MCCANDLESS (1995), "A New Clean Diesel Technology: Demonstration of ULEV Emissions on a Navistar Diesel Engine Fueled with Dimethyl Ether," SAE paper 950061.

FULTON, J., F. LYNCH, R. MARMARO, and B. WILLSON (1993), "Hydrogen for Reducing Emissions from Alternative Fuel Vehicles," SAE paper 931813.

GIBSON, H. J. (1982), "Fuels and Lubricants for Internal Combustion Engines—A Historical Perspective," SAE paper 821570.

GRUSE, W. A. (1967), *Motor Oils: Performance and Evaluation,* Van Nostrand Reinhold, New York.

IKUMI, S. and C. WEN (1981), "Entropies of Coals and Reference States in Coal Gasification Availability Analysis," West Virginia University Internal Report.

KATO, K., K. IGARASHI, M. MASUDA, K. OTSUBO, A. YASUDA, and K. TAKEDA (1999), "Development of Engine for Natural Gas Vehicle," SAE paper 1999-01-0574.

KELLY, K., B. BAILEY, T. COBURN, W. CLARK, L. EUDY, and P. LISSIUK (1996), "FTP Emissions Test Results from Flexible-Fuel Methanol Dodge Spirits and Ford Econoline Vans," SAE paper 961090.

KRAHL, J., A. MUNACK, M. BAHADIR, L. SCHUMACHER, and N. ELSER (1996), "Review: Utilization of Rapeseed Oil, Rapeseed Oil Methyl Ester or Diesel Fuel: Exhaust Gas Emissions and Estimation of Environmental Effects," SAE paper 962096.

KUKKONEN, C. and M. SHELEF (1994), "Hydrogen as an Alternative Fuel," SAE paper 940766.

LAWRENCE, D. K., D. A. PLANTZ, B. D. KELLER, and T. D. WAGNER (1980), "Automotive Fuels—Refinery Energy and Economics," SAE paper 800225.

LOVELL, W. (1948), "Knocking Characteristics of Hydrocarbons," *Ind. Eng. Chem.,* 40, p. 2388–2438.

OBERT, E. F. (1973), *Internal Combustion Engines and Air Pollution,* Harper & Row, New York.

OWEN, K. and T. COLEY (1995), *Automotive Fuels Reference Book,* Society of Automotive Engineers, Warrendale, Pennsylvania.

SAE HANDBOOK (1983), "Fuels and Lubricants," SAE, Warrendale, Pennsylvania.

SIELOFF, F. and J. L. MUSSER (1982), "What Does the Engine Designer Need to Know about Engine Oils?" SAE paper 821571.

SPRINGER, K., L. SMITH, and A. DICKINSON (1994), "Effect of CNG Start-Gasoline Run on Emissions from a 3/4 Ton Pick Up Truck," SAE paper 941916.

SUN, X., A. LUTZ, E. VERMIGLIO, M. AROLD, and T. WIEDMANN (1998), "The Development of the GM 2.2L CNG Bi-Fuel Passenger Cars," SAE paper 982445.

UNICH, A., R. BATA, and D. LYONS (1993), "Natural Gas: A Promising Fuel for I. C. Engines," SAE paper 930929.

WEBB, R. and P. DELMAS (1991), "New Perspectives on Auto Propane as a Mass-Scale Motor Vehicle Fuel," SAE paper 911667.

10.9 HOMEWORK

10.1 What is the chemical structure of **(a)** 3-methyl-3-ethylpentane, and **(b)** 2, 4-diethylpentane?

10.2 If a hydrocarbon fuel is represented by the general formula C_xH_{2x}, what is its stoichiometric mass air-fuel ratio?

10.3 A fuel has the following composition by mass: 10% pentane, 35% heptane, 30% octane, and 25% dodecane. If its general formula is of the form C_xH_y, find x and y.

10.4 If the mass composition of a hydrocarbon fuel mixture is 55% paraffins, 30% aromatics, and 15% monoolefins, what is its specific heat?

10.5 Compute the enthalpy of formation of H_8C_{18}.

10.6 A fuel blend has a density of $700 \, kg/m^3$ and a midpoint boiling index of 90°C. What is its centane index?

10.7 A flexible fuel vehicle operates with a mixture of 35% isooctane and 65% methanol, by volume. If the combustion is to be stoichiometric, what should the mass air-fuel ratio be?

10.8 A four-stroke engine operates on methane with an equivalence ratio of 0.9. The air and fuel enter the engine at 298 K, and the exhaust is at 530°C. The heat rejected to the coolant is 350 MJ/kmol fuel. **(a)** What is the enthalpy of the exhaust combustion products? **(b)** What is the specific work output of the engine? **(c)** What is the first law efficiency of the engine?

10.9 What is the change in volumetric efficiency of an automotive engine when it is retrofitted to operate with propane? Assume standard temperature and pressure inlet conditions, and a representative engine size and geometry.

10.10 Repeat Problem 10.9 for hydrogen and methane.

10.11 Verify the CO_2 concentration values resulting from the combustion of propane, methane, methanol, ethanol, and gasoline given in Table 10-7.

Chapter 11

Overall Engine Performance

11.1 INTRODUCTION

In this chapter we take an integrative view of the performance of the internal combustion engine. We use the information about friction, heat transfer, and combustion presented in the previous chapters to explain and discuss the influence of various factors, such as engine size, compression ratio, speed, part load, and ignition timing. Performance maps for various representative spark ignition and compression ignition engines are introduced. The frictional and aerodynamic drag components of road load are also discussed for application to vehicle performance simulation.

11.2 ENGINE SIZE

The torque an engine will produce, by definition of the mean effective pressure, is

$$
\tau_b = \begin{cases} \dfrac{1}{4\pi}\ \text{bmep}\, V_d & \text{(4 stroke)} \\[2ex] \dfrac{1}{2\pi}\ \text{bmep}\, V_d & \text{(2 stroke)} \end{cases}
\tag{11.1}
$$

The power can also be expressed in terms of the mean effective pressure

$$
\dot{W}_b = \begin{cases} \dfrac{1}{4}\ \text{bmep} \cdot A_p \overline{U}_p & \text{(4 stroke)} \\[2ex] \dfrac{1}{2}\ \text{bmep} \cdot A_p \overline{U}_p & \text{(2 stroke)} \end{cases}
\tag{11.2}
$$

Therefore, for a given stress level (bmep, $\overline{U}_p$), the torque is proportional to the displacement volume V_d and the power is proportional to the piston area A_p

$$
V_d = n_c \frac{\pi}{4}\, b^2 s
\tag{11.3}
$$

$$
A_p = n_c \frac{\pi}{4}\, b^2
$$

Finally, we can also write for four- or two-stroke engines

$$
\dot{m}_f = \begin{cases} \text{bsfc} \cdot \text{bmep} \cdot V_d \cdot N/2 & \text{(4 stroke)} \\ \text{bsfc} \cdot \text{bmep} \cdot V_d \cdot N & \text{(2 stroke)} \end{cases}
\tag{11.4}
$$

$$
\text{bsfc} = \frac{F}{1 + F}\ \frac{e_v\, \rho_i}{\text{bmep}}
\tag{11.5}
$$

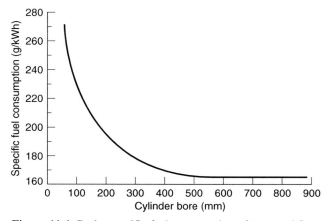

Figure 11-1 Brake specific fuel consumption of two- and four-stroke engines versus cylinder bore (Thomas et al., 1984).

where, as before, F is the fuel-air ratio, e_v is the volumetric efficiency, and ρ_i is the mixture density in the intake manifold. Notice that Equation 11.5 does not explicitly include engine size, so the efficiency of engines is expected to be a weak function of size for a given stress.

The specific fuel consumption versus cylinder bore for representative diesel engines is shown in Figure 11-1. This Figure is based on two- and four-stroke designs with bores from 62 to 900 mm. For bores greater than 500 mm, the thermal efficiency is about 50%. The ratio of the maximum bore to minimum bore is about 15, corresponding to a 3400 to 1 displacement volume ratio; whereas the brake specific fuel consumption varies by only a factor of 1.6. Although it is indeed a weak function with respect to the bore, the change in specific fuel consumption with the bore is significant with respect to fuel economy.

An important factor underlying the trend shown in Figure 11-1 is that the surface to volume ratio of the cylinder is decreasing with increasing bore $[(A/V) \sim b^{-1}]$. This means that there will be less and less heat lost as the bore increases. Another factor working in the favor of large engines is that the rotational speed decreases with the bore size $(N \sim b^{-1})$, so that there is more time near top center for fuel injection and combustion. This means that there will be less of a volume change during combustion and thus the limit of constant volume combustion will be approached.

Von Schnurbein (1981) has reported that the friction mean effective pressure decreases as engine size increases, as shown by Equation 11.6:

$$\text{mmep} = 1.3 \times 10^5 \frac{\overline{U}_p \cdot \mu}{b} + 1.7 \times 10^8 \, (r + 15) \frac{\mu^2}{\rho} \frac{1}{b^2} \tag{11.6}$$

Therefore the friction can be expected to be less in large engines than in small engines.

The brake thermal efficiency of state of the art large ($b > 500$ mm) diesel engines is about 50%. Surely, they are among the most efficient engines in the world. Their low losses due to heat transfer, combustion, and friction have already been mentioned. They also use late-closing intake valves to realize a longer expansion stroke than compression stroke.

The generalizations just drawn ought to hold true for spark ignition engines too, although the point is academic, for they are not practical unless care is taken to stratify the charge. Large homogeneous-charge, spark ignition engines are not practical because their octane requirements are too high. In Chapter 9, it was pointed out that the flame speed is in part controlled by the magnitude of the turbulence; and it was shown in Chapter 7 that the turbulence is proportional to the piston speed. It follows that the combustion duration in crank angle is constant, but in time is inversely proportional to the engine's rotational speed. There is consequently more time for knock precursors to form in relatively low rpm, large engines. For similar reasons, the engineering of small high speed diesel engines is a challenge, as there is little time for autoignition to occur and/or inject the fuel at reasonable pressures.

11.3   IGNITION AND INJECTION TIMING

For spark ignition gasoline engines, the timing parameter is the spark timing, and for diesel engines the timing parameter is the fuel injection timing. A classic plot of the effect of spark timing on the brake mean effective pressure for a number of automotive engines at different chassis dynamometer speeds is given in Figure 11-2. The variations in spark timing have the same percentage effect at all speeds. In fact the data are well correlated by

$$\frac{bmep}{(bmep)_{max}} = 1 - \left(\frac{\Delta\theta}{53}\right)^2 \qquad (11.7)$$

where $\Delta\theta$ is the change in degrees of crank angle from the angle of maximum bmep. Although the data correlated are rather old, they are still representative of today's engines.

Engines today are usually timed to an angle referred to as MBT (minimum advance for best torque). Examine Figure 11-2 and notice how flat the bmep curve is in the vicinity of the maximum. Now, reexamine Figure 9-30 and notice how sensitive the nitric oxide emissions are to variations in spark timing. Clearly, if the timing is slightly retarded, say 5° from that of maximum bmep, then the engine power will hardly suffer; yet under some operating conditions the nitric oxides will be greatly reduced. Retarded timing also somewhat reduces the engine's octane requirement. The term MBT spark timing is widely accepted, yet there is no quantitative definition in terms of how far the spark should be retarded from the point of maximum torque. The fraction of the maximum torque realized for various spark retards, each of which could be a candidate for defining MBT timing, is given in Table 11-1.

We will define MBT timing as a spark retard of 4° from the angle of maximum torque. This definition agrees with values reported in the literature to a tolerance of about ± 2°. Figure 11-3 shows how the MBT timing can be expected to vary with engine speed, equivalence ratio, and residual mass fraction. Because the charge is diluted by either air (in which case it is lean) or exhaust gas, the combustion duration and ignition delay both increase, thereby requiring a greater spark advance to more or less center the combustion

Table 11-1   MBT Timing

$\dfrac{bmep}{bmep_{max}}\%$	$\Delta\theta_{retard}$ (deg)
0.98	7.5
0.99	5.3
0.995	3.8

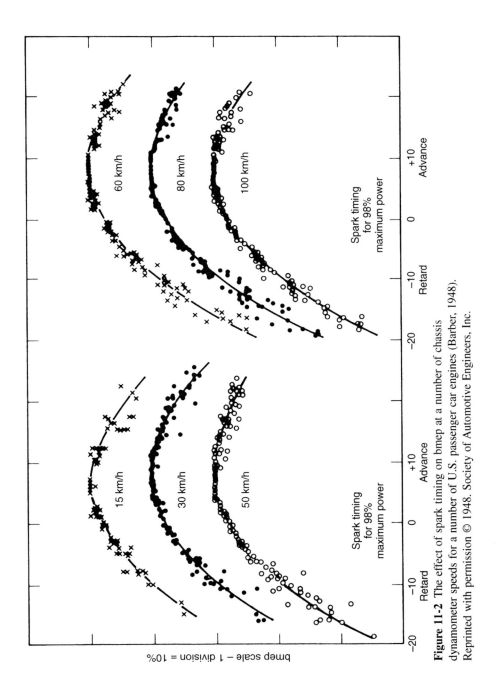

Figure 11-2 The effect of spark timing on bmep at a number of chassis dynamometer speeds for a number of U.S. passenger car engines (Barber, 1948). Reprinted with permission © 1948. Society of Automotive Engineers, Inc.

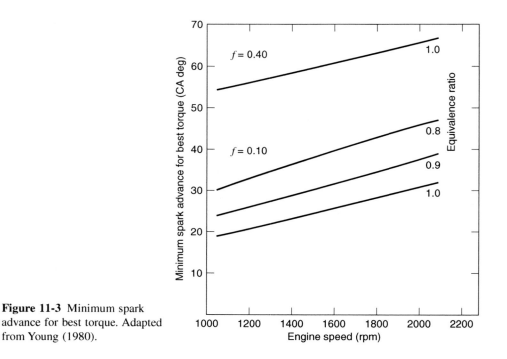

Figure 11-3 Minimum spark advance for best torque. Adapted from Young (1980).

about top center. Likewise, as engine speed increases, ignition delay and the MBT spark advance increase.

One way to illustrate the tradeoffs involved in controlling nitric oxides by retarding the fuel injection timing is to plot the brake specific nitric oxides versus the brake specific fuel consumption at full load as in Figure 11-4. The graph shows the response of five different engines to changes in timing at full load and rated speed (which is the maximum power for

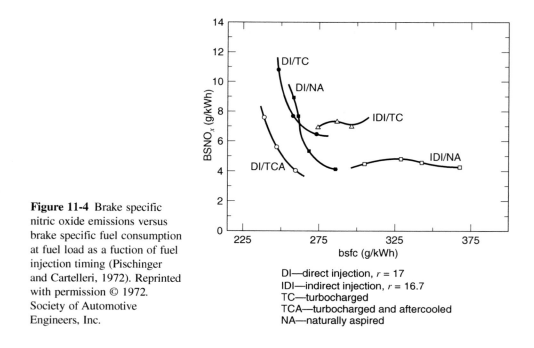

Figure 11-4 Brake specific nitric oxide emissions versus brake specific fuel consumption at fuel load as a fuction of fuel injection timing (Pischinger and Cartelleri, 1972). Reprinted with permission © 1972. Society of Automotive Engineers, Inc.

DI—direct injection, $r = 17$
IDI—indirect injection, $r = 16.7$
TC—turbocharged
TCA—turbocharged and aftercooled
NA—naturally aspirated

continuous operation warranted by the manufacturer). All engines are production, in line six-cylinder, four-stroke diesel engines with V_d = 5.9 L. All tests are at N = 2800 rpm except for IDI/TC which is at N = 3000 rpm. For each of the direct-injection (DI) engines, a significant reduction in the nitric oxide emissions can be realized at the expense of a slight increase in the brake specific fuel consumption. With indirect injection engines (IDI) there is no appreciable change in the nitric oxides; whereas, at part loads these same engines show response curves more like those shown for the direct injection engines.

Retarding the timing is still an effective means of controlling the nitric oxide emissions, but with diesel engines, it is usually at the expense of an increase in the particulate or smoke emissions. Furthermore, retarding the timing does not always reduce the nitric oxides. The results discussed point out that it is more difficult to generalize about diesel engines than about gasoline engines. This is because there are so many more degrees of freedom available in the design of a diesel engine.

11.4   ENGINE AND PISTON SPEED

The effects of engine speed on the power and coolant load of an automotive spark ignition engine at full load, that is, wide open throttle, are shown in Figure 11-5. The graphs

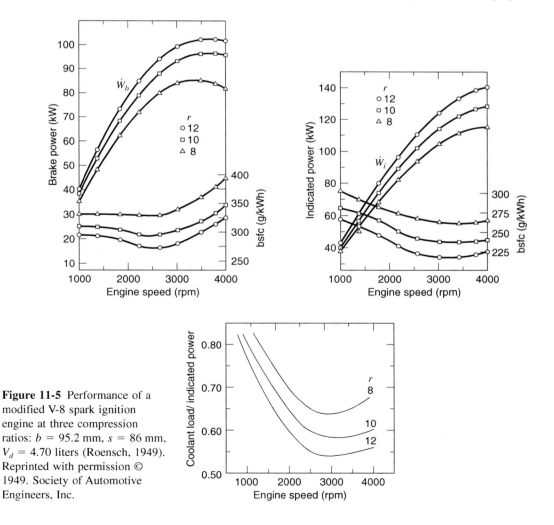

Figure 11-5 Performance of a modified V-8 spark ignition engine at three compression ratios: b = 95.2 mm, s = 86 mm, V_d = 4.70 liters (Roensch, 1949). Reprinted with permission © 1949. Society of Automotive Engineers, Inc.

in Figure 11-5 plot the performance of a V-8 spark ignition engine at three different compression ratios. Note that the indicated specific fuel consumption decreases with increasing engine speed and then levels out. This is mainly because the percentage heat loss to the coolant is decreasing with increasing engine speed. On the other hand, the brake specific fuel consumption is flatter at low speeds and is increasing at the higher speeds. This difference is due to the friction and pumping losses (which though small are still present in an unthrottled engine) increasing with engine speed. The increase in the coolant loads beginning at 3000 rpm can be attributed to the increasing friction.

The power curves in Figure 11-5 are conveniently explained in terms of an expression relating the power of an engine to the volumetric efficiency, the net indicated thermal efficiency, and the mechanical efficiency. The relationship (4 stroke) is as follows.

$$\dot{W}_b = \eta_{mech}\, \dot{W}_i \tag{11.8}$$

$$= \frac{1}{2}\, \eta_{mech}\, \eta_i\, q_c\, \frac{F}{1+F}\, e_v\, \rho_i\, V_d\, N \tag{11.9}$$

Figure 11-5 gives the indicated power as a function of engine speed. Because the engine is unthrottled, we can assume for qualitative purposes that the curve is the net indicated power versus engine speed. All the terms multiplying the mechanical efficiency in Equation 11.9 constitute the net indicated power. If the indicated torque were constant then the indicated power would increase linearly with engine speed.

Torque as a function of engine speed usually reflects the variation in volumetric efficiency with engine speed. A falling off in volumetric efficiency at the higher speeds causes the falling off of the indicated power at high engine speeds. Recall that the speed at which the volumetric efficiency peaks is dependent upon the valve timing. The indicated power is not decreasing as fast as it would if the volumetric efficiency were the only parameter changing with speed. The indicated efficiency is increasing slightly with speed.

The brake power is the product of the net indicated power and the mechanical efficiency. Friction power increases with the square of engine speed, since the friction torque (proportional to bmep) increases linearly with engine speed. The mechanical efficiency therefore decreases linearly with engine speed, causing the brake power to exhibit a maximum even though the indicated power is still increasing.

It was stated earlier that these generalizations are expected to apply to all engines. That statement needs qualification in the case of two-stroke engines, especially carbureted ones. Recall that in two-stroke engines there is a significant difference between the delivered mass and the trapped mass, because of short-circuiting. Any fuel that is short-circuited is wasted and represents a loss not discussed in the context of Figure 11-5, since this effect is usually negligible in four-stroke engines. As the trapping efficiency generally increases with engine speed, the amount of fuel short-circuited will decrease with engine speed.

11.5 COMPRESSION RATIO

As shown in Figure 11-5, increasing the compression ratio decreases both the indicated and the brake specific fuel consumption. The specific results shown are, of course, unique to the particular engine design tested. The compression ratio trends depicted and their underlying causes, however, are typical to all engines, compression or spark ignited, two- or four-stroke.

The indicated specific fuel consumption improves at a faster rate with increasing compression ratio than the brake specific fuel consumption, because both friction and heat losses are increasing with compression ratio. In fact, there is an optimum compression

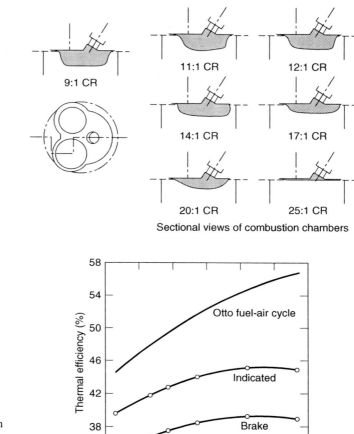

Sectional views of combustion chambers

Figure 11-6 Thermal efficiencies of a gasoline engine at the nominal compression ratios shown (Caris and Nelson, 1959). Reprinted with permission © 1959. Society of Automotive Engineers, Inc.

ratio due to these effects, see Figure 11-6. The spark advance is set for best efficiency as is the fuel-air equivalence ratio at $\phi = 0.91$. Computer simulations of diesel engines show similar results (McAulay et al., 1965). An optimum compression ratio of 12 to 18 is typical and is the underlying reason why direct-injection diesel engines have compression ratios in the same range. The compression ratios of gasoline-fueled, spark ignition engines are less than the optima shown in Figure 11-6 to avoid knock. The compression ratios of indirect-injection diesel engines are greater than optimum to assist in cold starting, which is harder than with direct injection engines because of the high heat loss in the antechamber.

The coolant load relative to the indicated power drops with increasing compression ratio because the increased thermal efficiency results in lower temperatures in the latter part of the expansion stroke thereby reducing the heat loss.

11.6   PART-LOAD PERFORMANCE

The effect of load; i.e., the brake mean effective pressure, on the brake specific fuel consumption is qualitatively the same for both compression ignited and spark ignited engines,

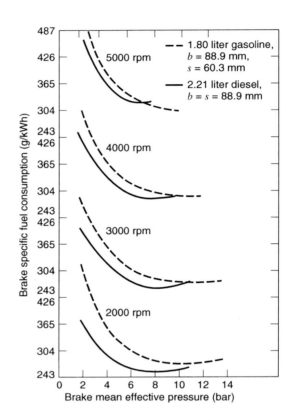

Figure 11-7 Comparison of a SI engine with an IDI-CI engine design to produce equal torque-speed characteristics in an automotive application (Walder, 1965). Reprinted with permission © 1965. Society of Automotive Engineers, Inc.

as shown by representative examples in Figure 11-7, which compares a gasoline and a diesel engine, and Figure 11-8 for a marine diesel with two types of prechambers. In both cases, the bsfc will be infinite at idle since the engine is producing no useful work but is consuming fuel. As the load increases, the brake specific fuel consumption drops, goes through a minimum, and may or may not increase depending on how the load is increased at this point.

In the case of spark ignition engines, opening the throttle and increasing the delivery ratio increases the load. This has little effect on the indicated efficiency, slightly increases the friction, and significantly reduces the pumping losses. Again, the dominant factor is the increase in mechanical efficiency. At constant fuel-air ratio, the brake specific fuel consumption drops with increasing load all the way to the point of maximum load so long as the imep increases faster than the fmep. In engines running at a fuel-air ratio less than that corresponding to maximum power (about $\phi = 1.1$, as we saw in our studies with fuel-air cycles), the load can be increased further by increasing the fuel-air ratio. This causes the brake specific fuel consumption to begin increasing with load once the engine is running rich.

In the case of compression ignition engines, increasing the fuel-air ratio increases the load; although this slightly drops the indicated efficiency and slightly increases the friction, the increase in mechanical efficiency is so great that it improves the specific fuel consumption. Just before the load is about to become smoke limited, the brake specific fuel consumption begins to increase because significant quantities of fuel begin to be only partially oxidized and thus are wasted.

The variable geometry prechamber of Figure 11-8 improved the fuel economy of the marine engine by about 10 g/kWh or about 5%. The 5% improvement is small and does not affect the trend shown for brake specific fuel consumption with load. Nevertheless,

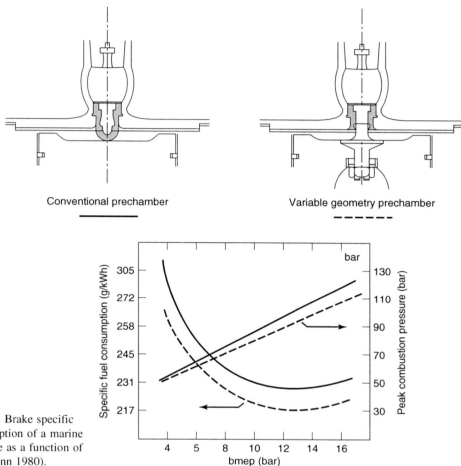

Conventional prechamber

Variable geometry prechamber

Figure 11-8 Brake specific fuel consumption of a marine diesel engine as a function of load (Hermann 1980).

that 5% improvement is very significant since the cost of fuel saved by a ship on a job, although small compared to the total fuel cost, will be comparable to the job's profit.

The effect of the fuel-air ratio on brake specific fuel consumption of a spark ignition engine at constant load is shown in Figure 11-9. The spark ignition engine is most efficient when running stoichiometric or slightly lean. At very lean fuel-air ratios, the engine wastes fuel because of misfire, and at rich fuel-air ratios it wastes fuel since there is not enough oxygen present to liberate all of the fuel's energy. The effect of fuel-air ratio on the brake specific fuel consumption and exhaust emissions of a number of IDI and DI compression ignition engines is plotted in Figure 11-10. The smoke readings are in Bosch Smoke Number (BSN) units, a scale that measures the reflectivity of a piece of filter paper through which some of the exhaust gas is passed.

11.7   ENGINE PERFORMANCE MAPS

A common way of combining the effects of speed and load on engine performance is shown in Figure 11-11. This type of contour plot is called a performance map. Performance maps are generated for both fuel economy and emissions levels. The performance maps of Figures 11-11 to 11-16 show lines of constant fuel consumption in the load-speed plane.

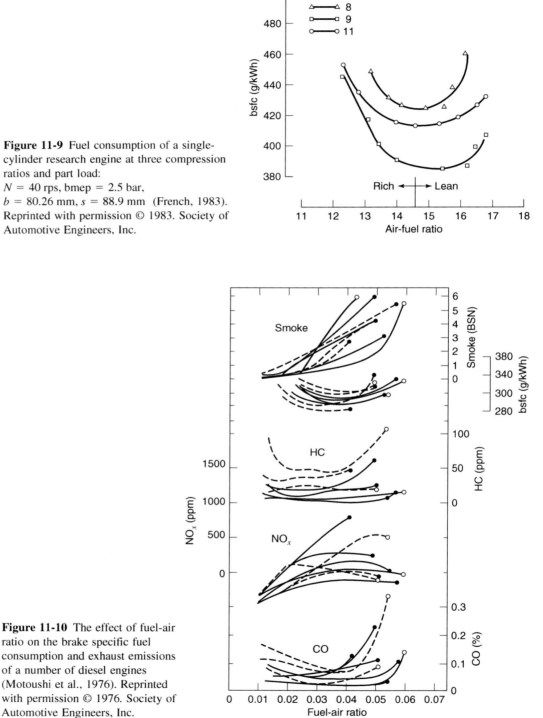

Figure 11-9 Fuel consumption of a single-cylinder research engine at three compression ratios and part load:
$N = 40$ rps, bmep = 2.5 bar,
$b = 80.26$ mm, $s = 88.9$ mm (French, 1983).
Reprinted with permission © 1983. Society of Automotive Engineers, Inc.

Figure 11-10 The effect of fuel-air ratio on the brake specific fuel consumption and exhaust emissions of a number of diesel engines (Motoushi et al., 1976). Reprinted with permission © 1976. Society of Automotive Engineers, Inc.

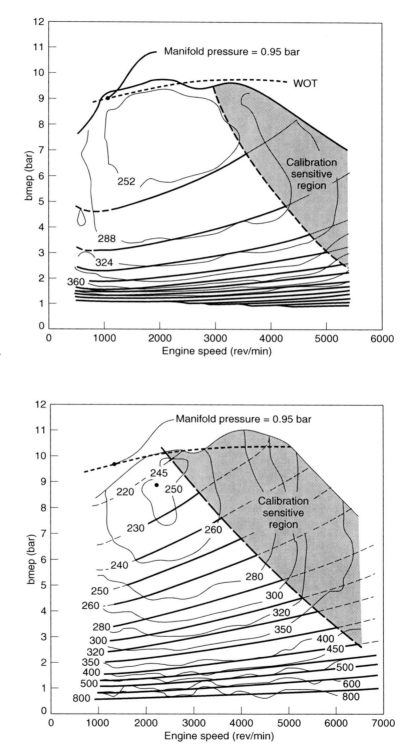

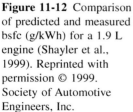

Figure 11-11 Comparison of predicted and measured bsfc for a 1.25 L engine (Shayler et al., 1999). Reprinted with permission © 1999. Society of Automotive Engineers, Inc.

Figure 11-12 Comparison of predicted and measured bsfc (g/kWh) for a 1.9 L engine (Shayler et al., 1999). Reprinted with permission © 1999. Society of Automotive Engineers, Inc.

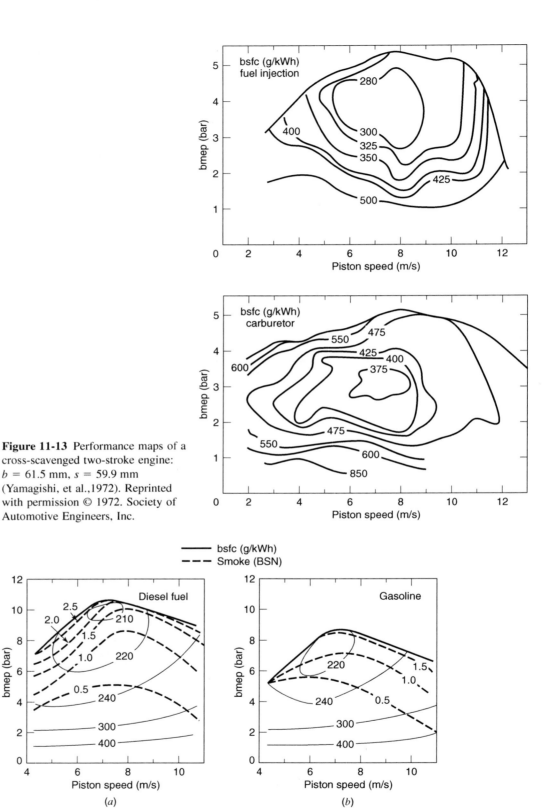

Figure 11-13 Performance maps of a cross-scavenged two-stroke engine: $b = 61.5$ mm, $s = 59.9$ mm (Yamagishi, et al.,1972). Reprinted with permission © 1972. Society of Automotive Engineers, Inc.

Figure 11-14 Performance and smoke number maps of a direct-injection stratified-charge engine, intercooled and turbocharged: $b = 125$ mm, $s = 130$ mm, $r = 20.1$, $N_c = 10$. (a) Diesel, (b) Gasoline (Hardenberg and Buhl, 1982). Reprinted with permission © 1982. Society of Automotive Engineers, Inc.

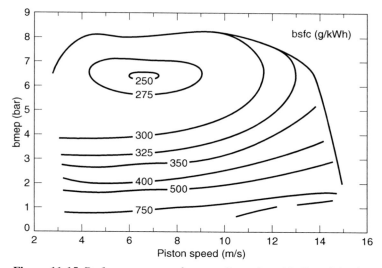

Figure 11-15 Performance map of a naturally aspirated indirect injection diesel engine: $b = 76.5$ mm, $s = 80$ mm, $r = 23$, $n_c = 4$ (Hofbauer and Sator, 1977). Reprinted with permission © 1977. Society of Automotive Engineers, Inc.

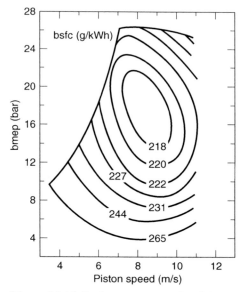

Figure 11-16 Performance map of variable geometry diesel prechamber engine: $b = 200$ mm, $r = 13.7$, intercooled, two-stage turbocharger (Courtesy Société d'Etudes de Machines Thermiques, SEMT).

The upper envelope on the map is the wide open throttle performance curve. Its shape reflects variations in volumetric efficiency with engine speed, although small changes in inlet air density are also involved.

Engine performance maps generally have a single valued minimum bsfc operating point. Starting at the location of minimum bsfc, on the map the fuel consumption increases in all directions. If one increases the engine speed, the fuel consumption increases because of an increase in the friction loss. If one decreases the engine speed, the fuel consumption increases because of an increase in the heat loss. If one increases the load, the fuel consumption increases because the mixture must be enriched beyond stoichiometric. If one decreases the load, the fuel consumption increases because the friction is becoming a larger proportion of the indicated work.

The engine of Figure 11-11 is a 1.25 L DOHC 4-cylinder, 16-valve automotive engine. Since the application is general automotive use, the engine is designed to have the region of minimum bsfc located at relatively low engine speeds (40 to 60% of maximum engine speed) and at relatively high loads (60 to 80% of maximum bmep). Figure 11-12 shows a performance map for a larger 1.9 L SOHC 4-cylinder, 16-valve automobile engine. The minimum fuel consumption of 245 g/kWh occurs at about 90% of the maximum load and 40% of the maximum speed. The general shape of the contours for the engines of Figures 11-11 and 11-12 are remarkably similar.

Maps for a two-stroke gasoline engine in both port fuel injected and carburetted form are given in Figure 11-13. The advantage to fuel injection in this application is clearly evident, yet in each case minimum fuel consumption still occurs at similar speed-load points. The fuel injected version is more efficient than the carbureted version because of a reduction in the short-circuited fuel, and its hydrocarbon emissions are a factor of about ten less.

The performance maps discussed so far have been for homogeneous charge engines. Results for a direct injection, stratified charge engine are given in Figure 11-14. The point of minimum fuel consumption has moved up closer to the maximum load and now occurs near 90% of the maximum bmep and near 60% of the maximum piston speed. The engine produces more power on diesel fuel than on gasoline because the injection system is a jerk pump. Since diesel fuel is denser than gasoline, more diesel fuel is injected at full load.

For a homogeneous charge engine at the point of minimum fuel consumption, increasing simultaneously the fuel-air ratio and the delivery ratio increases the load. For an engine in which the fuel is injected at the time combustion is to occur, solely increasing the fuel-air ratio increases the load; as the engine is unthrottled the delivery ratio remains nearly constant. Since the torque or bmep will depend on the product of the fuel-air ratio and the delivery ratio, when the two variables are increased simultaneously there is a larger change in torque. That is one reason why the point of minimum fuel consumption in Figure 11-14 has moved up relative to the points in Figures 11-11, 11-12, and 11-13.

Another reason relates to the fact that power in heterogeneous charge engines is smoke limited. In a homogeneous charge engine, the minimum fuel consumption occurs at about $\phi = 1.0$, as we have seen. In a diesel engine it occurs at a smaller equivalence ratio because the engine will smoke intolerably at $\phi = 1.0$. If ϕ_m is the equivalence ratio where the fuel consumption is minimum, then at the smoke limit $\phi_s = \phi_m + \Delta\phi$, the manufacturer will set up the fuel injection system so that $\phi < \phi_s$, always. Typically, $\Delta\phi$ is set rather small for emission control and satisfactory engine life, the mixture being enriched by only 5 to 10% beyond ϕ_m for maximum power.

Figures 11-15 and 11-16 show maps for two different diesel engines. They reinforce several of the points made earlier and demonstrate that the relative position of the point of minimum fuel consumption can be moved up or down depending on the degree of mixture enrichment the manufacturer will allow by the choice of $\Delta\phi$.

11.8   VEHICLE PERFORMANCE SIMULATION

Automotive engines are expected to operate well over a wide range of speeds and loads. Figure 5-23 in Chapter 5 shows two driving cycles defined by the U.S. Environmental Protection Agency for regulatory purposes. In each case, vehicle speed as a function of time is specified.

From knowledge of the vehicle's characteristics, such as frontal area, drag coefficient, weight, and gear ratios, the driving cycle can be transformed into a specification of the engine's torque and speed as a function of time. A vehicle simulation can be used to assess the fuel economy performance of various engine and vehicle combinations.

For a vehicle, the power requirements are specified by a road load power equation, Equation 11.10, which includes the effects of aerodynamic drag and rolling resistance.

$$\dot{W}_v = \left(C_r\, m_v\, g + \frac{1}{2} C_d\, \rho_o\, A_v\, U_v^2 \right) U_v \tag{11.10}$$

where

C_r = coefficient of rolling resistance

m_v = mass of vehicle (kg)

g = gravitational constant, 9.81 m/s^2

C_d = drag coefficient

A_v = vehicle front cross sectional area (m^2)

U_v = vehicle speed (m/s)

The total fuel consumed by the vehicle during the driving cycle will be the integrated fuel flow rate

$$m_f = \int_0^t \dot{m}_f(t)\, dt = \frac{A_p}{4} \int_0^t \text{bsfc}\,(t)\, \text{bmep}\,(t)\, \overline{U}_p\,(t)\, dt \tag{11.11}$$

For a two-stroke engine, the factor of four would instead be a factor of two. In order to do the integration on the right of Equation 11.11, one needs bsfc, bmep, and $\overline{U}_p$ as functions of time. The latter two are known since the engine torque and speed is known from the driving cycle requirements and the vehicle characteristics.

The brake specific fuel consumption can be determined for each load and speed point of the cycle from the engine map. If an emissions map is available, a similar computation can be performed to compute the total emissions produced during the driving cycle.

11.9 REFERENCES

AMANN, C. A. (1980), "Control of the Homogeneous-Charge Passenger Car Engine—Defining the Problem," *SAE Trans.*, Vol. 89, paper 801440.

BARBER, E. M. (1948), "Knock Limited Performance of Several Automobile Engines," *SAE Trans.*, Vol. 2, p. 401.

CARIS, D. F. and E. E. NELSON (1959), "A New Look at High Compression Engines," *SAE Trans.*, Vol. 67, p. 112.

FRENCH, C. C. J. (1983), "A Universal Test Engine for Combustion Research," SAE paper 830453.

Hardenberg, H. O. and H. W. Buhl (1982), "The MERCEDES-BENZ OM 403 VA-A Standard Production, Compression-Ignition, Direct-Injection Multifuel Engine," SAE paper 820028.

HERMANN, R. (1980), "PA4-200 Engines with Variable Geometry Precombustion Chamber and Two Stage Turbocharging System," ASME paper 80-DGP-22.

HOFBAUER, P. and SATOR, K. (1977), "Advanced Automotive Power Systems, Part 2: A Diesel for a Subcompact Car," *SAE Trans.*, Vol. 86, paper 770113.

MCAULAY, K., T. WU, S. CHEN, G. BORMAN, P. MYERS, and O. UYEHARA (1965), "Development and Evaluation of the Simulation of the CI Engine," SAE paper 650451.

MILLINGTON, B. W. and E. R. HARTLES (1968), "Frictional Losses in Diesel Engines," *SAE Trans.*, Vol. 77, paper 680590.

MOTOYOSHI, E., T. YAMADA, and M. MORI (1976), "The Combustion and Exhaust Emission Characteristics and Starting Ability of Y.P.C. Combustion System," SAE paper 760215.

PISCHINGER, R. and W. CARTELLIERI (1972), "Combustion System Parameters and Their Effect upon Diesel Engine Exhaust Emissions," SAE paper 720756.

ROENSCH, M. (1949), "Thermal Efficiency and Mechanical Losses of Automotive Engines," *SAE J.*, Vol. 51, p. 17–30.

SHAYLER, P., J. CHICK, and D. EADE (1999), "A Method of Predicting Brake Specific Fuel Consumption Maps," SAE paper 1999-01-0556.

SUTTON, D. L. (1983), "Combustion Chamber Design for Improved Performance and Economy with High Compression Lean Burn Operation," SAE paper 830336.

TAYLOR, C. F. (1985), *The Internal Combustion Engine in Theory and Practice*, Vol. 1, MIT Press, Cambridge, Massachusetts.

THOMAS, F. J., J. S. AHLUWALIA, E. SHAMAH, and G. W. VAN DER HORST (1984), "Medium-Speed Diesel Engines Part 1: Design Trends and the Use of Residual/Blended Fuels," ASME paper 84-DGP-15.

VON SCHNURBEIN, E. and J. BUCHER (1981), "Experience with the Rating and Operation of Medium-Speed, Four-Stroke Engines under Extreme Site Conditions," CINIAC paper D64.

WALDER, C. J. (1965), "Problems in the Design and Development of High Speed Diesel Engines," SAE paper 978A.

YAMAGISHI, G., T. SATO, and H. IWASA (1972), "A Study of Two-Stroke Cycle Fuel Injection Engines for Exhaust Gas Purification," SAE paper 720195.

YOUNG, M. B. (1980), "Cyclic Dispersion—Some Quantitative Cause-and-Effect Relationships," SAE paper 800459.

11.10 HOMEWORK

11.1. A particular car traveling steadily on a level road at 100 km/h requires about 15 kW of power from the engine. For each of the engines in Figures 11-12, 11-15, and 11-16 determine the fuel economy of the vehicle (km/g), the bore required of a four cylinder engine, and the maximum power the engine will produce. Assume in each case that the engine is operating at its best fuel economy point when the vehicle is cruising at 100 km/h, that engine controls limit the piston speed to 10 m/s, and that the maps are size independent.

11.2. The price of large diesel engines is roughly proportional to their rated power. Let

c_1 = engine price per kilowatt per year

c_2 = fuel price per kilogram

At low values of c_1 it pays to buy an engine bigger than required and operate it at its best fuel economy point. For low values of c_2 it pays to buy a smaller engine and run it at its rated power. For the engine characteristics in Fig. 11-18 at what ratio c_1/c_2 will the two different sized engines yield the same total annual cost? Assume the engines are run 20 h/day and their rated power is at $\overline{U}_p$ = 11 m/s, bmep = 25 bars.

11.3. Write an expression resembling Equation on 11.11 for the mass of pollutant species i (given its emission index at any load, speed point) emitted by an engine operated over a duty cycle from $0 < t < t_d$.

11.4 What is the power required to travel up a hill with a 10° slope at 50 mph? Assume a frontal area of 2 m^2, C_d = 0.3, C_r = 0.015, m_v = 1500 kg.

11.5 Derive Equation 11.9.

Appendix A

Physical Properties of Air

Table A-1   Properties of Air at Atmospheric Pressure[a]

T (K)	ρ (kg/m^3)	c_p (kJ/kg · K)	$\mu \cdot 10^7$ (N · s/m^2)	$\nu \cdot 10^6$ (m^2/s)	$k \cdot 10^3$ (W/m · K)	$\alpha \cdot 10^6$ (m^2/s)	Pr
100	3.5562	1.032	71.1	2.00	9.34	2.54	0.786
150	2.3364	1.012	103.4	4.426	13.8	5.84	0.758
200	1.7458	1.007	132.5	7.590	18.1	10.3	0.737
250	1.3947	1.006	159.6	11.44	22.3	15.9	0.720
300	1.1614	1.007	184.6	15.89	26.3	22.5	0.707
350	0.9950	1.009	208.2	20.92	30.0	29.9	0.700
400	0.8711	1.014	230.1	26.41	33.8	38.3	0.690
450	0.7740	1.021	250.7	32.39	37.3	47.2	0.686
500	0.6964	1.030	270.1	38.79	40.7	56.7	0.684
550	0.6329	1.040	288.4	45.57	43.9	66.7	0.683
600	0.5804	1.051	305.8	52.69	46.9	76.9	0.685
650	0.5356	1.063	322.5	60.21	49.7	87.3	0.690
700	0.4975	1.075	338.8	68.10	52.4	98.0	0.695
750	0.4643	1.087	354.6	76.37	54.9	109	0.702
800	0.4354	1.099	369.8	84.93	57.3	120	0.709
850	0.4097	1.110	384.3	93.80	59.6	131	0.716
900	0.3868	1.121	398.1	102.9	62.0	143	0.720
950	0.3666	1.131	411.3	112.2	64.3	155	0.723
1000	0.3482	1.141	424.4	121.9	66.7	168	0.726
1100	0.3166	1.159	449.0	141.8	71.5	195	0.728
1200	0.2902	1.175	473.0	162.9	76.3	224	0.728
1300	0.2679	1.189	496.0	185.1	82	238	0.719
1400	0.2488	1.207	530	213	91	303	0.703
1500	0.2322	1.230	557	240	100	350	0.685
1600	0.2177	1.248	584	268	106	390	0.688
1700	0.2049	1.267	611	298	113	435	0.685
1800	0.1935	1.286	637	329	120	482	0.683
1900	0.1833	1.307	663	362	128	534	0.677
2000	0.1741	1.337	689	396	137	589	0.672
2100	0.1658	1.372	715	431	147	646	0.667
2200	0.1582	1.417	740	468	160	714	0.655
2300	0.1513	1.478	766	506	175	783	0.647
2400	0.1448	1.558	792	547	196	869	0.630
2500	0.1389	1.665	818	589	222	960	0.613
3000	0.1135	2.726	955	841	486	1570	0.536

[a]Adapted from F. Incropera and D. Dewitt (2001), *Fundamentals of Heat and Mass Transfer,* John Wiley, New York.

Table A-2   Curve Fit Equations for Air Physical Properties
(Temperature T in K)

Dynamic viscosity $\mu(\text{kg/m} \cdot \text{s}) = 3.3 \times 10^{-7} \cdot T^{0.7}$

Thermal conductivity $k(\text{W/m} \cdot \text{K}) = a + b \cdot T + c \cdot T^2$

where
$$a = 1.52 \times 10^{-4}$$
$$b = 4.42 \times 10^{-5}$$
$$c = 8.0 \times 10^{-9}$$

Specific heat ratio $\gamma = 1.40 - 7.18 \times 10^{-5} \cdot T$

Table A-3   Additional Physical Properties of Air at Atmospheric
Conditions ($T = 298$ K, $P = 1$ atm $= 1.0133$ bar)

Molecular weight	$M = 28.966$ kg/kmol
Gas constant	$R = 0.28704$ kJ/kg-K
Speed of sound	$c = 345.9$ m/s
Binary diffusion with octane	$D = 5.68 \times 10^{-6}$ m^2/s
Schmidt number	$Sc = \nu/D = 2.77$
Lewis number	$Le = \alpha/D = 3.91$

Appendix **B**

Thermodynamic Property Tables for Various Ideal Gases

Table B-1 Properties of Various Ideal Gases at 300 K (SI Units)

Gas	Chemical Formula	Molecular Mass	R kJ/kg K	c_{po} kJ/kg K	c_{vo} kJ/kg K	γ
Ethane	C_2H_6	30.07	0.27650	1.7662	1.4897	1.186
Ethanol	C_2H_5OH	46.069	0.18048	1.427	1.246	1.145
Ethylene	C_2H_4	28.054	0.29637	1.5482	1.2518	1.237
Helium	He	4.003	2.07703	5.1926	3.1156	1.667
Hydrogen	H_2	2.016	4.12418	14.2091	10.0849	1.409
Methane	CH_4	16.04	0.51835	2.2537	1.7354	1.299
Methanol	CH_3OH	32.042	0.25948	1.4050	1.1455	1.227
Neon	Ne	20.183	0.41195	1.0299	0.6179	1.667
Nitrogen	N_2	28.013	0.29680	1.0416	0.7448	1.400
Nitrous oxide	N_2O	44.013	0.18891	0.8793	0.6904	1.274
n-octane	C_8H_{18}	114.23	0.07279	1.7113	1.6385	1.044
Oxygen	O_2	31.999	0.25983	0.9216	0.6618	1.393
Propane	C_3H_8	44.097	0.18855	1.6794	1.4909	1.126
Steam	H_2O	18.015	0.46152	1.8723	1.4108	1.327
Sulfur dioxide	SO_2	64.059	0.12979	0.6236	0.4938	1.263
Sulfur trioxide	SO_3	80.058	0.10386	0.6346	0.5307	1.196

Source: VanWylen and Sonntag (1996), *Introduction to Thermodynamics: Classical and Statistical,* John Wiley, New York.

Table B-2 Ideal-Gas Properties of N_2 and N (SI Units), Entropies at 0.1-MPa (1-bar) Pressure

	Nitrogen, Diatomic (N_2) $\bar{h}^o_{f,298} = 0$ kJ/kmol $M = 28.013$		Nitrogen, Monatomic (N) $\bar{h}^o_{f,298} = 472\ 680$ kJ/kmol $M = 14.007$	
T K	$(\bar{h} - \bar{h}^o_{298})$ kJ/kmol	$\bar{s}^o$ kJ/kmol K	$(\bar{h} - \bar{h}^o_{298})$ kJ/kmol	$\bar{s}^o$ kJ/kmol K
0	−8670	0	−6197	0
100	−5768	159.812	−4119	130.593
200	−2857	179.985	−2040	145.001
298	0	191.609	0	153.300
300	54	191.789	38	153.429
400	2971	200.181	2117	159.409
500	5911	206.740	4196	164.047
600	8894	212.177	6274	167.837
700	11937	216.865	8353	171.041
800	15046	221.016	10431	173.816
900	18223	224.757	12510	176.265
1000	21463	228.171	14589	178.455
1100	24760	231.314	16667	180.436
1200	28109	234.227	18746	182.244
1300	31503	236.943	20825	183.908
1400	34936	239.487	22903	185.448
1500	38405	241.881	24982	186.883
1600	41904	244.139	27060	188.224
1700	45430	246.276	29139	189.484
1800	48979	248.304	31218	190.672
1900	52549	250.234	33296	191.796
2000	56137	252.075	35375	192.863
2200	63362	255.518	39534	194.845
2400	70640	258.684	43695	196.655
2600	77963	261.615	47860	198.322
2800	85323	264.342	52033	199.868
3000	92715	266.892	56218	201.311
3200	100134	269.286	60420	202.667
3400	107577	271.542	64646	203.948
3600	115042	273.675	68902	205.164
3800	122526	275.698	73194	206.325
4000	130027	277.622	77532	207.437
4400	145078	281.209	86367	209.542
4800	160188	284.495	95457	211.519
5200	175352	287.530	104843	213.397
5600	190572	290.349	114550	215.195
6000	205848	292.984	124590	216.926

Table B-3   Ideal-Gas Properties of O_2 and O (SI Units), Entropies at 0.1-MPa (1-bar) Pressure

T	Oxygen, Diatomic (O_2) $\bar{h}_{f,298}^o = 0$ kJ/kmol $M = 31.999$		Oxygen, Monatomic (O) $\bar{h}_{f,298}^o = 249\ 170$ kJ/kmol $M = 16.00$	
K	$(\bar{h} - \bar{h}_{298}^o)$ kJ/kmol	$\bar{s}^o$ kJ/kmol K	$(\bar{h} - \bar{h}_{298}^o)$ kJ/kmol	$\bar{s}^o$ kJ/kmol K
0	−8683	0	−6725	0
100	−5777	173.308	−4518	135.947
200	−2868	193.483	−2186	152.153
298	0	205.148	0	161.059
300	54	205.329	41	161.194
400	3027	213.873	2207	167.431
500	6086	220.693	4343	172.198
600	9245	226.450	6462	176.060
700	12499	231.465	8570	179.310
800	15836	235.920	10671	182.116
900	19241	239.931	12767	184.585
1000	22703	243.579	14860	186.790
1100	26212	246.923	16950	188.783
1200	29761	250.011	19039	190.600
1300	33345	252.878	21126	192.270
1400	36958	255.556	23212	193.816
1500	40600	258.068	25296	195.254
1600	44267	260.434	27381	196.599
1700	47959	262.673	29464	197.862
1800	51674	264.797	31547	199.053
1900	55414	266.819	33630	200.179
2000	59176	268.748	35713	201.247
2200	66770	272.366	39878	203.232
2400	74453	275.708	44045	205.045
2600	82225	278.818	48216	206.714
2800	90080	281.729	52391	208.262
3000	98013	284.466	56574	209.705
3200	106022	287.050	60767	211.058
3400	114101	289.499	64971	212.332
3600	122245	291.826	69190	213.538
3800	130447	294.043	73424	214.682
4000	138705	296.161	77675	215.773
4400	155374	300.133	86234	217.812
4800	172240	303.801	94873	219.691
5200	189312	307.217	103592	221.435
5600	206618	310.423	112391	223.066
6000	224210	313.457	121264	224.597

Table B-4 Ideal-Gas Properties of CO_2 and CO (SI Units), Entropies at 0.1-MPa (1-bar) Pressure

T K	Carbon Dioxide (CO_2) $\bar{h}^o_{f,298} = -393\ 522$ kJ/kmol $M = 44.01$		Carbon Monoxide (CO) $\bar{h}^o_{f,298} = -110\ 527$ kJ/kmol $M = 28.01$	
	$(\bar{h} - \bar{h}^o_{298})$ kJ/kmol	$\bar{s}^o$ kJ/kmol K	$(\bar{h} - \bar{h}^o_{298})$ kJ/kmol	$\bar{s}^o$ kJ/kmol K
0	−9364	0	−8671	0
100	−6457	179.010	−5772	165.852
200	−3413	199.976	−2860	186.024
298	0	213.794	0	197.651
300	69	214.024	54	197.831
400	4003	225.314	2977	206.240
500	8305	234.902	5932	212.833
600	12906	243.284	8942	218.321
700	17754	250.752	12021	223.067
800	22806	257.496	15174	227.277
900	28030	263.646	18397	231.074
1000	33397	269.299	21686	234.538
1100	38885	274.528	25031	237.726
1200	44473	279.390	28427	240.679
1300	50148	283.931	31867	243.431
1400	55895	288.190	35343	246.006
1500	61705	292.199	38852	248.426
1600	67569	295.984	42388	250.707
1700	73480	299.567	45948	252.866
1800	79432	302.969	49529	254.913
1900	85420	306.207	53128	256.860
2000	91439	309.294	56743	258.716
2200	103562	315.070	64012	262.182
2400	115779	320.384	71326	265.361
2600	128074	325.307	78679	268.302
2800	140435	329.887	86070	271.044
3000	152853	334.170	93504	273.607
3200	165321	338.194	100962	276.012
3400	177836	341.988	108440	278.279
3600	190394	345.576	115938	280.422
3800	202990	348.981	123454	282.454
4000	215624	352.221	130989	284.387
4400	240992	358.266	146108	287.989
4800	266488	363.812	161285	291.290
5200	292112	368.939	176510	294.337
5600	317870	373.711	191782	297.167
6000	343782	378.180	207105	299.809

Table B-5   Ideal-Gas Properties of H_2O and OH (SI Units), Entropies at 0.1-MPa (1-bar) Pressure

T	Water (H_2O) $\bar{h}_{f,298}^o = -241\ 826$ kJ/kmol M = 18.015		Hydroxyl (OH) $\bar{h}_{f,298}^o = 38\ 987$ kJ/kmol M = 17.007	
T K	$(\bar{h} - \bar{h}_{298}^o)$ kJ/kmol	$\bar{s}^o$ kJ/kmol K	$(\bar{h} - \bar{h}_{298}^o)$ kJ/kmol	$\bar{s}^o$ kJ/kmol K
0	−9904	0	−9172	0
100	−6617	152.386	−6140	149.591
200	−3282	175.488	−2975	171.592
298	0	188.835	0	183.709
300	62	189.043	55	183.894
400	3450	198.787	3034	192.466
500	6922	206.532	5991	199.066
600	10499	213.051	8943	204.448
700	14190	218.739	11902	209.008
800	18002	223.826	14881	212.984
900	21937	228.460	17889	216.526
1000	26000	232.739	20935	219.735
1100	30190	236.732	24024	222.680
1200	34506	240.485	27159	225.408
1300	38941	244.035	30340	227.955
1400	43491	247.406	33567	230.347
1500	48149	250.620	36838	232.604
1600	52907	253.690	40151	234.741
1700	57757	256.631	43502	236.772
1800	62693	259.452	46890	238.707
1900	67706	262.162	50311	240.556
2000	72788	264.769	53763	242.328
2200	83153	269.706	60751	245.659
2400	93741	274.312	67840	248.743
2600	104520	278.625	75018	251.614
2800	115463	282.680	82268	254.301
3000	126548	286.504	89585	256.825
3200	137756	290.120	96960	259.205
3400	149073	293.550	104388	261.456
3600	160484	296.812	111864	263.592
3800	171981	299.919	119382	265.625
4000	183552	302.887	126940	267.563
4400	206892	308.448	142165	271.191
4800	230456	313.573	157522	274.531
5200	254216	318.328	173002	277.629
5600	278161	322.764	188598	280.518
6000	302295	326.926	204309	283.227

Table B-6 Ideal-Gas Properties of H_2 and H (SI Units), Entropies at 0.1-MPa (1-bar) Pressure

T	Hydrogen (H_2) $\bar{h}^o_{f, 298} = 0$ kJ/kmol $M = 2.016$		Hydrogen, Monatomic (H) $\bar{h}^o_{f, 298} = 217\,999$ kJ/kmol $M = 1.008$	
T K	$(\bar{h} - \bar{h}^o_{298})$ kJ/kmol	$\bar{s}^o$ kJ/kmol K	$(\bar{h} - \bar{h}^o_{298})$ kJ/kmol	$\bar{s}^o$ kJ/kmol K
0	−8467	0	−6197	0
100	−5467	100.727	−4119	92.009
200	−2774	119.410	−2040	106.417
298	0	130.678	0	114.716
300	53	130.856	38	114.845
400	2961	139.219	2117	120.825
500	5883	145.738	4196	125.463
600	8799	151.078	6274	129.253
700	11730	155.609	8353	132.457
800	14681	159.554	10431	135.233
900	17657	163.060	12510	137.681
1000	20663	166.225	14589	139.871
1100	23704	169.121	16667	141.852
1200	26785	171.798	18746	143.661
1300	29907	174.294	20825	145.324
1400	33073	176.637	22903	146.865
1500	36281	178.849	24982	148.299
1600	39533	180.946	24060	149.640
1700	42826	182.941	29139	150.900
1800	46160	184.846	31218	152.089
1900	49532	186.670	33296	153.212
2000	52942	188.419	35375	154.279
2200	59865	191.719	39532	156.260
2400	66915	194.789	43689	158.069
2600	74082	197.659	47847	159.732
2800	81355	200.355	52004	161.273
3000	88725	202.898	56161	162.707
3200	96187	205.306	60318	164.048
3400	103736	207.593	64475	165.308
3600	111367	209.773	68633	166.497
3800	119077	211.856	72790	167.620
4000	126864	213.851	76947	168.687
4400	142658	217.612	85261	170.668
4800	158730	221.109	93576	172.476
5200	175057	224.379	101890	174.140
5600	191607	227.447	110205	175.681
6000	208332	230.322	118519	177.114

Table B-7   Ideal-Gas Properties of NO and NO$_2$ (SI Units), Entropies at 0.1-MPa (1-bar) Pressure

T K	Nitric Oxide (NO) $\bar{h}^o_{f, 298} = 90\ 291$ kJ/kmol $M = 30.006$		Nitrogen Dioxide (NO$_2$) $\bar{h}^o_{f, 298} = 33\ 100$ kJ/kmol $M = 46.005$	
	$(\bar{h} - \bar{h}^o_{298})$ kJ/kmol	$\bar{s}^o$ kJ/kmol K	$(\bar{h} - \bar{h}^o_{298})$ kJ/kmol	$\bar{s}^o$ kJ/kmol K
0	−9192	0	−10186	0
100	−6073	177.031	−6861	202.563
200	−2951	198.747	−3495	225.852
298	0	210.759	0	240.034
300	55	210.943	68	240.263
400	3040	219.529	3927	251.342
500	6059	226.263	8099	260.638
600	9144	231.886	12555	268.755
700	12308	236.762	17250	275.988
800	15548	241.088	22138	282.513
900	18858	244.985	27180	288.450
1000	22229	248.536	32344	293.889
1100	25653	251.799	37606	298.904
1200	29120	254.816	42946	303.551
1300	32626	257.621	48351	307.876
1400	36164	260.243	53808	311.920
1500	39729	262.703	59309	315.715
1600	43319	265.019	64846	319.289
1700	46929	267.208	70414	322.664
1800	50557	269.282	76008	325.861
1900	54201	271.252	81624	328.898
2000	57859	273.128	87259	331.788
2200	65212	276.632	98578	337.182
2400	72606	279.849	109948	342.128
2600	80034	282.822	121358	346.695
2800	87491	285.585	132800	350.934
3000	94973	288.165	144267	354.890
3200	102477	290.587	155756	358.597
3400	110000	292.867	167262	362.085
3600	117541	295.022	178783	365.378
3800	125099	297.065	190316	368.495
4000	132671	299.007	201860	371.456
4400	147857	302.626	224973	376.963
4800	163094	305.940	248114	381.997
5200	178377	308.998	271276	386.632
5600	193703	311.838	294455	390.926
6000	209070	314.488	317648	394.926

Appendix C

Curve Fit Coefficients for Thermodynamic Properties of Various Ideal Gases and Fuels

Table C-1 Curve Fit Coefficients for Thermodynamic Properties ($300 \geq T \leq 1000$ K)

$$c_p/R = a_1 + a_2 T + a_3 T^2 + a_4 T^3 + a_5 T^4$$

$$h/RT = a_1 + \frac{a_2}{2}T + \frac{a_3}{3}T^2 + \frac{a_4}{4}T^3 + \frac{a_5}{5}T^4 + \frac{a_6}{T}$$

$$s^o/R = a_1 \ln T + a_2 T + \frac{a_3}{2}T^2 + \frac{a_4}{3}T^3 + \frac{a_5}{4}T^4 + a_7$$

i	Species	a_{i1}	a_{i2}	a_{i3}	a_{i4}	a_{i5}	a_{i6}	a_{i7}
1	CO_2	0.24007797E + 01	0.87350957E − 02	−0.66070878E − 05	0.20021861E − 08	0.63274039E − 15	−0.48377527E + 05	0.96951457E + 01
2	H_2O	0.40701275E + 01	−0.11084499E − 02	0.41521180E − 05	−0.29637404E − 08	0.80702103E − 12	−0.30279722E + 05	−0.32270046E + 00
3	N_2	0.36748261E + 01	−0.12081500E − 02	0.23240102E − 05	−0.63217559E − 09	−0.22577253E − 12	−0.10611588E + 04	0.23580424E + 01
4	O_2	0.36255985E + 01	−0.18782184E − 02	0.70554544E − 05	−0.67635137E − 08	0.21555993E − 11	−0.10475226E + 04	0.43052778E + 01
5	CO	0.37100928E + 01	−0.16190964E − 02	0.36923594E − 05	−0.20319674E − 08	0.23953344E − 12	−0.14356310E + 05	0.2955535E + 01
6	H_2	0.30574451E + 01	0.26765200E − 02	−0.58099162E − 05	0.55210391E − 08	−0.18122739E − 11	−0.98890474E + 03	−0.22997056E + 01
7	H	0.25000000E + 01	0	0	0	0	0.25471627E + 05	−0.46011762E + 00
8	O	0.29464287E + 01	−0.16381665E − 02	0.24210316E − 05	−0.16028432E − 08	0.38906964E − 12	0.29147644E + 05	0.29639949E + 01
9	OH	0.38375943E + 01	−0.10778858E − 02	0.96830378E − 06	0.18713972E − 09	−0.22571094E − 12	0.36412823E + 04	0.49370009E + 00
10	NO	0.40459521E + 01	−0.34181783E − 02	0.79819190E − 05	−0.61139316E − 08	0.15919076E − 11	0.97453934E + 04	0.29974988E + 01

Source: Gordon, S. and B. McBride (1971), "Computer Program for Calculation of Complex Chemical Equilibrium Compositions, Rocket Performance, Incident and Reflected Shocks, and Chapman-Jouguet Detonations," NASA SP-273.

Table C-2 Curve Fit Coefficients for Thermodynamic Properties ($1000 < T < 3000$ K)

i	Species	a_{i1}	a_{i2}	a_{i3}	a_{i4}	a_{i5}	a_{i6}	a_{i7}
1	CO_2	0.446080e + 1	0.309817e − 2	−0.123925e − 5	0.227413e − 9	−0.155259e − 13	−0.489614e + 5	−0.986359
2	H_2O	0.271676e + 1	0.294513e − 2	−0.802243e − 6	0.102266e − 9	−0.484721e − 14	−0.299058e + 5	0.663056e + 1
3	N_2	0.289631e + 1	0.151548e − 2	−0.572352e − 6	0.998073e − 10	−0.652235e − 14	−0.905861e + 3	0.616151e + 1
4	O_2	0.362195e + 1	0.736182e − 3	−0.196522e − 6	0.362015e − 10	−0.289456e − 14	−0.120198e + 4	0.361509e + 1
5	CO	0.298406e + 1	0.148913e − 2	−0.578996e − 6	0.103645e − 9	−0.693535e − 14	−0.142452e + 5	0.634791e + 1
6	H_2	0.310019e + 1	0.511194e − 3	0.526442e − 7	−0.349099e − 10	0.369453e − 14	−0.877380e + 3	−0.196294e + 1
7	H	0.25e + 1	0	0	0	0	0.254716e + 5	−0.460117
8	O	0.554205e + 1	−0.275506e − 4	−0.310280e − 8	0.455106e − 11	−0.436805e − 15	0.292308e + 5	0.492030e + 1
9	OH	0.291064e + 1	0.959316e − 3	−0.194417e − 6	0.137566e − 10	0.142245e − 15	0.393538e + 4	0.544234e + 1
10	NO	0.3189e + 1	0.133822e − 2	−0.528993e − 6	0.959193e − 10	−0.648479e − 14	0.982832e + 4	0.674581e + 1

Source: Gordon and McBride, 1971.

Table C-3 Curve Fit Coefficients for Thermodynamic Properties of Selected Fuels ($300 \geq T \leq 1000$ K)

Fuel		a_1	a_2	a_3	a_6	a_7
Methane	CH_4	1.971324	7.871586E − 03	− 1.048592E − 06	− 9.930422E + 03	8.873728
Gasoline	C_7H_{17}[a]	4.0652	6.0977E − 02	− 1.8801E − 05	− 3.5880E + 04	1.545E + 01
Diesel	$C_{14.4}H_{24.9}$	7.9710	1.1954E − 01	− 3.6858E − 05	− 1.9385E + 04	− 1.7879
Methanol	CH_3OH	1.779819	1.262503E − 02	− 3.624890E − 06	− 2.525420E + 04	1.50884E + 01
Nitromethane	CH_3NO_2	1.412633	2.087101E − 02	− 8.142134E − 06	− 1.026351E + 04	1.917126E + 01

[a] High hydrogen content gasoline.

Appendix D

Conversion Factors and Physical Constants

Table D-1   Conversion Factors   *multiply by*

Area	1 m^2	$= 1550.0 \text{ in.}^2$
		$= 10.764 \text{ ft}^2$
Energy	1 J	$= 9.4787 \times 10^{-4} \text{ Btu}$
Force	1 N	$= 0.22481 \text{ lb}_f$
Heat transfer rate	1 W	$= 3.4123 \text{ Btu/h}$
Heat flux	1 W/m^2	$= 0.3171 \text{ Btu/h} \cdot \text{ft}^2$
Heat transfer coefficient	$1 \text{ W/m}^2 \cdot \text{K}$	$= 0.17612 \text{ Btu/h} \cdot \text{ft}^2 \cdot {}^\circ\text{F}$
Kinematic viscosity and diffusivities	$1 \text{ m}^2/\text{s}$	$= 3.875 \times 10^4 \text{ ft}^2/\text{h}$
Length	1 m	$= 39.370 \text{ in.}$
		$= 3.2808 \text{ ft}$
	1 km	$= 0.62137 \text{ mile}$
Mass	1 kg	$= 2.2046 \text{ lb}_m$
Mass density	1 kg/m^3	$= 0.062428 \text{ lb}_m/\text{ft}^3$
Mass flow rate	1 kg/s	$= 7936.6 \text{ lb}_m/\text{h}$
Mass transfer coefficient	1 m/s	$= 1.1811 \times 10^4 \text{ ft/h}$
Power	1 W	$= 1.341 \times 10^{-3} \text{ hp}$
Pressure and stress[1]	1 N/m^2	$= 0.020886 \text{ lb}_f/\text{ft}^2$
		$= 1.4504 \times 10^{-4} \text{ lb}_f/\text{in.}^2$
		$= 4.015 \times 10^{-3} \text{ in. water}$
		$= 2.953 \times 10^{-4} \text{ in. Hg}$
	$1.0133 \times 10^5 \text{ N/m}^2$	$= 1 \text{ standard atmosphere}$
	$1 \times 10^5 \text{ N/m}^2$	$= 1 \text{ bar}$
Rotational Speed	1 RPM	$= 0.10472 \text{ rad/s}$
Specific heat	$1 \text{ J/kg} \cdot \text{K}$	$= 2.3886 \times 10^{-4} \text{ Btu/lb}_m \cdot {}^\circ\text{F}$
Temperature	K	$= (5/9){}^\circ\text{R}$
		$= (5/9)({}^\circ\text{F} + 459.67)$
		$= {}^\circ\text{C} + 273.15$
Temperature difference	1 K	$= 1{}^\circ\text{C}$
		$= (9/5){}^\circ\text{R} = (9/5){}^\circ\text{F}$
Thermal conductivity	$1 \text{ W/m} \cdot \text{K}$	$= 0.57782 \text{ Btu/h} \cdot \text{ft} \cdot {}^\circ\text{F}$
Thermal resistance	1 K/W	$= 0.52750 {}^\circ\text{F/h} \cdot \text{Btu}$
Torque	$1 \text{ N} \cdot \text{m}$	$= 0.73756 \text{ lb}_f \cdot \text{ft}$
Viscosity (dynamic)[2]	$1 \text{ N} \cdot \text{s/m}^2$	$= 2419.1 \text{ lb}_m/\text{ft} \cdot \text{h}$
		$= 5.8016 \times 10^{-6} \text{ lb}_f \cdot \text{h/ft}^2$
Volume	1 m^3	$= 6.1023 \times 10^4 \text{ in.}^3$
		$= 35.314 \text{ ft}^3$
		$= 264.17 \text{ gal}$
Volume flow rate	$1 \text{ m}^3/\text{s}$	$= 1.2713 \times 10^5 \text{ ft}^3/\text{h}$
		$= 2.1189 \times 10^3 \text{ ft}^3/\text{min}$
		$= 1.5850 \times 10^4 \text{ gal/min}$

[1]The SI name for the quantity pressure is pascal (Pa) having units N/m^2 or $\text{kg/m} \cdot \text{s}^2$.

[2]Also expressed in equivalent units of $\text{kg/s} \cdot \text{m}$.

Source: F. Incropera and D. DeWitt (2001), *Fundamentals of Heat and Mass Transfer*, John Wiley, New York.

Table D-2 Physical Constants

Universal Gas Constant:
$$R_u = 8.205 \times 10^{-2} \, m^3 \cdot atm/kmol \cdot K$$
$$= 8.314 \times 10^{-2} \, m^3 \cdot bar/kmol \cdot K$$
$$= 8.315 \, kJ/kmol \cdot K$$
$$= 1545 \, ft \cdot lb_f/lbmole \cdot °R$$
$$= 1.986 \, Btu/lbmole \cdot °R$$

Avogadro's Number:
$$\mathcal{N} = 6.024 \times 10^{23} \, molecules/mol$$

Planck's Constant:
$$h = 6.625 \times 10^{-34} \, J \cdot s/molecule$$

Boltzmann's Constant:
$$k = 1.380 \times 10^{-23} \, J/K \cdot molecule$$

Speed of Light in Vacuum:
$$c_o = 2.998 \times 10^8 \, m/s$$

Stefan-Boltzmann Constant:
$$\sigma = 5.670 \times 10^{-8} \, W/m^2 \cdot K^4$$
$$= 0.1714 \times 10^{-8} \, Btu/h \cdot ft^2 \cdot °R^4$$

Gravitational Acceleration (Sea Level):
$$g = 9.807 \, m/s^2$$

Normal Atmospheric Pressure:
$$P = 101,325 \, N/m^2$$

Index

Printed in the United States
120974LV00001BA/128/A